Practical Forecasting for Managers

Practical Forecasting for Managers

John C. Nash and Mary M. Nash

ARNOLD

A member of the Hodder Headline Group
LONDON
Co-published in the United States of America by
Oxford University Press Inc., New York

First published in Great Britain in 2001 by
Arnold, a member of the Hodder Headline Group,
338 Euston Road, London NW1 3BH
http://www.arnoldpublishers.com

Co-published in the USA by
Oxford University Press Inc.,
198 Madison Avenue, New York, NY 10016

British Library Cataloguing in Publication Data
A catalogue record for this book is available from the British Library

Library of Congress Cataloging-in-Publication Data
A catalog record for this book is available from the Library of Congress

ISBN 0 340 76238 1

1 2 3 4 5 6 7 8 9 10

Commissioning Editor: Liz Gooster
Production Editor: James Rabson
Production Controller: Bryan Eccleshall
Cover design: Terry Griffiths

Printed and bound in Malta by Gutenberg Press Ltd

Layout and typesetting in 10/12 pt Times New Roman by the authors.

What do you think about this book? Or any other Arnold title?
Please send your comments to feedback.arnold@hodder.co.uk

Contents

Chapter 1   Why forecast?   1

Why do we forecast?   1

What can we forecast?   2

Background skills and knowledge   2

How forecasting works   4

    Models and their role   4

Skills and knowledge objectives   5

    Examinable topics   6

    Project, presentation and data analysis skills 7

    Concepts, tools or facts for awareness   7

Examples of forecasting tasks   8

Exercises   9

Chapter 2   Planning the forecasting task   10

Defining the goals of a forecasting activity   10

Types of forecasts   10

Background to the forecast   11

Management of forecasting data   13

Assignment of resources   14

Testing and selecting a model   15

    Basic rules   17

    Trial estimation or modelling period   17

    Example of forecasting with transformed variables 18

    Validation period   18

    Main estimation or modelling period   19

    Forecasts   19

Single and multi-period forecasts   19

A real example   20

What should we report?   20

Good working habits   21

Exercises   22

Chapter 3   Measuring how well forecasting goals are met. Part 1   23

What is a model? Part 1   23

What are we looking for? Pegels' classification   23

Fit of model to data   26

A notation for forecasting   28

Quantification of the size of deviations   29
How good is a model? Part 1.   30
Desiderata and parsimony. Part 1.   32
Exercises   33

Chapter 4   Data search, gathering, documentation and management   35
Data sources. Data collection and validation.   35
Traditional sources of data   35
Library resources on the Internet   38
Other Internet or machine-readable information sources   40
    Examples of Internet searches   42
    Other machine-readable data sources   42
How to reference sources   43
Data edit and imputation   45
    Example   46
Data documentation – metadata   47
Time point conversions   47
Exercises   48

Chapter 5   Qualitative forecasting: long-term   49
Expert opinion   50
Panels of experts   51
Delphi techniques   52
Cross-impact analysis   53
Systems analysis and modelling   55
Simulation and gaming   56
Scenario writing   57
Exercises   58

Chapter 6   Semi-quantitative methods   59
Market versus situational forecasting   59
Technological forecasting   59
Surveys   60
Leading indicators   61
Market penetration models   62
Other extrapolation approaches   64
Media analysis - the Bhopal example   64
Exercises   66

Chapter 7   Forecasting, Risk, and Strategic Management   **68**

The role of forecasting in strategic management   68

Risk anticipation   69

    Example: Damage to automobile   69

Predicting opportunities   71

Deciding a direction for actions   72

Feedback from management to forecasting   74

Exercises   74

Chapter 8   Measuring how well forecasting goals are met. Part 2   **75**

How good is a model? Part 2   75

R_squared – a unitless comparison of model fit   75

Serial correlation   76

Partial autocorrelations   78

Use of autocorrelations   78

Ljung–Box Q statistic and its use   79

    Example of use of the Ljung–Box statistic   81

The Durbin–Watson statistic   84

Desiderata and parsimony. Part 2   85

Exercises   86

Chapter 9   Preliminary data analysis for forecasting   **87**

Graph the data!   87

    Time plot   90

    Histogram or other distributional plot   91

    Boxplots   93

    Scatterplots for possible relationships   93

    Quantile plots   94

Colour, patterns, symbols and shading   97

Adjustment of the data   97

    Special adjustments for unusual events   98

    Imputation for missing or known incorrect data   98

    Scaling of data   98

    Trading day adjustments   98

    First differences   99

    Error checks   99

Transforming data to help us understand it   100

Level and variability   101

Managing the graphs we draw   102

Numerical descriptive statistics   104

Using data subsets   104

Summary and application to a real data set	107
Exercises	108

Chapter 10 The preliminary forecast: concepts and examples — **111**

Ruler forecasts	111
Trend equations	113
Example	114
Data subsetting	115
Example	115
'Simple' seasonal models	115
Trend line calculations for seasonal data	118
Results of 'simple' seasonal techniques	118
Fit	118
Stability of models	119
Deviations and errors	120
Validation	121
Pattern and distribution of residuals	122
Evolution of seasonal factors	124
Housekeeping details	126
Assessment and validation of forecasting models	127
Annual or other long-pattern series	129
Forecasts	130
Exercises	131

Chapter 11 A strategy for performing forecasting data analysis — **132**

Motivations	132
The statistical package approach	133
The spreadsheet approach	134
Special-purpose forecasting software	135
Auxiliary tools	136
Other strategic and tactical issues	139
Making the choice	139
Exercises	140

Chapter 12 Forecasting trend and season I: Multiple regression — **141**

Regression – its purposes	141
Regression jargon	142
Dummy variables	144
A real example	146
Estimating regression models	152
Collinearity	154

Example of collinearity in regression 155

Other uses of regression in forecasting 156

Exercises 157

Chapter 13 Forecasting trend and season II: Smoothing methods 158

Why smooth? 158

Moving averages 158

Properties of the Moving Average MA(n) 160

Trended data 160

Learning about smoothing methods 161

Difficulties with Moving Averages 163

Exponential smoothing 164

Double Exponential Smoothing 166

Brown's 1 parameter ES – a double exponential smoothing 167

Holt's 2 parameters exponential smoothing 168

Reminders 168

Seasonal series – Winters' method 170

Other smoothing methods 174

A real example 175

Alternative seasonal forecasting with smoothing 180

Exercises 182

Chapter 14 Forecasting trend and season III: Time series decomposition 183

How does this differ from Winters' method? 183

Historical notes 183

Classical time series decomposition 184

An example 187

Further analysis of the irregular component 189

A real example 191

Exercises 192

Chapter 15 ARIMA and related models for forecasting 193

What are ARIMA models? 193

Some useful notation 194

Non-seasonal ARIMA models 195

Seasonal models 197

Estimation and use of ARIMA models – the Box-Jenkins methodology 198

A real example 202

Exercises 206

Chapter 16   Using ARIMA models: other issues and examples   **208**

Transforming a series to stationarity   208

Developing candidate ARIMA models   210

Trial estimation of ARIMA models   212

Selecting working models   213

Computing the right measures of fit and error   214

Alternative seasonal ARIMA modelling   215

Exercises   216

Chapter 17 Comparing and combining forecasts   **217**

Descriptive comparison   217

Volume of data and information   217

Type of model   218

Quantitative measures   219

Graphical comparison   220

Reporting honestly   221

Combining forecasts   222

A real example   223

Review of quantitative forecasting methods   231

Exercises   232

Chapter 18 Variations on the theme of seasonal adjustment   **233**

Origins and motivations   233

  Goals of decomposition modelling   234

The underlying model – additive or multiplicative   234

Massaging the data and metadata   235

Missing data in seasonal series   237

Choosing the trend and seasonal filters   237

  Moving medians and SABL – Seasonal Adjustment Bell Labs   237

  Weighted moving averages   238

Some other considerations   238

  Calendar adjustments   238

  Variation in the model components   239

Seasonal adjustment for small to medium organizations   239

Exercises   240

Chapter 19 Mixed and extended models   **242**

Overview   242

Sequential modelling – modelling residuals   242

  Example: predicting population growth of a city   243

Econometric models and their uses   245

    Identification   246

    Estimation   247

    Simulation and forecasting   247

    Revision   247

    When are econometric models useful?   247

    Sources of advice   248

  Dynamic regression models   248

  State space modelling   250

  Exercises   252

Chapter 20 Nonlinear regression modelling   **253**

  Motivations: nonlinear forecasting models   253

  What is a nonlinear model?   253

    An example   254

  Objective functions, constraints and parameters   256

  Methods for nonlinear fitting   257

  Tips for choosing model parameters and bounds   259

  Growth curve examples   262

  Exercises   263

Chapter 21   Artificial neural networks   **264**

  New and improved?   264

  Underlying ideas   264

  Published examples   265

  Assessment of utility of neural networks   266

  Exercises   267

Chapter 22 Building the forecast report   **268**

  Grading our own work   268

  The one-page critique   268

  A marking scheme   268

  Words, numbers and pictures   269

  Generalities and specifics   270

  Exercises   271

Appendix: Tips, tricks and scripts   **272**

1. Importing text data into spreadsheet and statistical software   272

2. Linear interpolation and inverse linear interpolation   272

3. Time point conversions   274

4. Merging seasonal data into a single series   274

5. Separating a time series into seasonal series   276

6. Trading day adjustments 276

7. The ruler forecast 277

8. Long or large (multi-page) time plots 278

9. Multiple plots 279

10. Level adjustment for sudden change 280

11. Compressed scale plots 280

12. Trend line calculations for seasonal data 280

13. Equivalence of additive seasonal models using different base seasons 282

14. Drawing Spread-Level graphs 284

15. Computing measures of fit 284

Bibliography **286**

Index **294**

Preface

This book is about anticipating the future by means of forecasting. It is, in particular, a guide that instructs and provides examples of practical methods that a manager, administrator or owner-operator may use without the assistance of a team of specialists or consultants. To this end, we describe the methods that are applicable to a wide variety of real-world forecasting problems. We support these methods with practical tools – conceptual, graphical, statistical and computational – for the analysis of data and problems and for the application of the methods.

For those techniques we present, we explain the motivation, the background, the applicability, the calculations and the analysis of results so that the reader will know how, why and when to use them. Detailed examples are provided to walk through the methods; outlines of a number of typical forecasting projects are also given. John Nash has taught courses in forecasting almost every year since 1981.

To complement the technical treatment, we show how to report on forecasting exercises and how to prepare objective, short critiques of forecasts made by others. With a clear regard to the difficulties of finding appropriate information upon which to base forecasts, we include a chapter on this subject.

Because forecasting is a broad subject that can draw on ideas from many fields, we provide pointers and introductions to forecasting techniques and approaches which are too sophisticated or specialized to be presented in detail in this book. That is, we explicitly support the manager's active awareness of ideas of potential applicability to forecasting problems, even if such awareness is limited to a rudimentary knowledge of the vocabulary and the most general and simplified perspectives of such approaches.

The book is designed so the reader can learn how to forecast by reading and using the examples. We have sometimes repeated ideas so chapters can be read without extensive reference to previous ones. While designed for independent use, the book is also a textbook for courses about forecasting. Typically such courses last one semester or term, but there is more material in this book than can sensibly be covered in that length of time, since we wish to provide a resource students can continue to use in their working lives.

The major goal of our forecasting courses is to provide students with an appreciation of the dangers of mis-applied methods and of the sensible use of simple techniques. We *teach* some rudimentary forecasting methods. The wider aspects of forecasting are dealt with via presentations, projects and in-class discussions. Most of the projects and exercises will be related to practical questions of general concern.

Examples of calculations will mostly be presented using Minitab statistical software. Sample calculations used in examination questions will also be presented using output from this software. We have also prepared forecasting calculations using other software, for instance, spreadsheets such as Lotus, Quattro or Excel and matrix languages such as MATLAB. If there is demonstrated demand for such tools, we will provide them via additional workbooks, discs, or Internet files.

It has been our practice to provide such aids in our courses using a very active Faculty of Administration Web site. For this book, the publisher is hosting what we will refer to throughout the book as 'the PFM Web site', rather than give a particular address that may change over time. The web address is **http://www.arnoldpublishers.com/support/nash/**.

In courses, we generally use a mix of evaluation techniques to measure how well students have learnt the material. We use an examination for typically 30% of the overall grade to ensure all students meet a basic level of competence in forecasting techniques. The examination emphasizes knowledge of major tools such as exponential smoothing, multiple regression and time-series decomposition, as well as the interpretation of diagnostic computer output. We are not supporters of memorized formulas, so we allow our students to have a sheet of notes for the examination. Our choice is to allow the students to have whatever they want written on their cheat sheets; other professors prefer to provide a standardized sheet.

We expect students to have a calculator and know how to use it. Speed and accuracy are important – the examination usually requires some small calculations to see if students understand and are technically competent in basic forecasting methods. We structure our examinations so that a mistake made in one calculation does not unduly penalize the student in later questions. Such examinations take a lot of care in preparation. We include samples in the instructional materials intended for the professor.

Student grades in our courses typically have the following distribution of the total 100% mark possible:

An Examination	30%
A presentation or essay	10%
(We have, on occasion, had students prepare these in HTML for Web display.)	
Critique of presentations or project (individual or group)	5%
Major Project (group)	30%
Assignments (4 plus 1 critique) (group)	25%

We encourage group work as a way to foster peer instruction, that is, one student helping another to learn. Though there are sometimes cases of easy riders taking advantage of the group situation to avoid work, we have found that the examination and the natural sense of fairness of students generally overcomes major mistakes in assigning final marks. A typical calendar for our course follows, based on a thirteen week term.

Week	Topics
1	Overview of course. Assignment of tasks for the course. Use of computers. Introduction to forecasting, types of forecasts, forecasting models, errors and their measurement. Example use of Minitab. Pegels' Classification. Graphs and their uses. Validation testing of forecasting models.
2	Data sources. Data collection and validation. Data edit and imputation. Introduction to

regression models.

3 Regression models and a brief overview of econometrics. Assignment 1 is due.

4 Discussion of Assignment 1. Exponential smoothing models.

5 Time series decomposition. Assignment 2 is due.

6 Review of quantitative methods of forecasting so far. Introduction to the formal expression of ARIMA models. Discussion of Assignment 2 and project work attempts.

7 Box-Jenkins methods for ARIMA modelling. Assignment 3 is due.

8 Begin discussion of technological forecasting. Assignment 4 due at end of week.

9 Discussion of project progress and difficulties. Discussion of Assignment 4. Review examinable material.

10 Examination. Continue technological forecasting.

11, 12 , 13 Critique of the Assignment 4 of other students is due in week 11. Student presentations and discussion. Project is typically due in the last class for exchange with other students. Critique of other student project due about ten days after last class.

As is evident, we allow a lot of time for actual use of the material learned in the early weeks of the course. Students sometimes find this allocation of time unusual in comparison to other courses where there is a rush to complete material at the end of the course. The early weeks are, we admit, quite hectic, and other instructors may prefer a slower pace with less emphasis on projects.

In concluding this preface, we wish to acknowledge the input of others to this book:

- First, to our students, who have made us learn the material better than we otherwise would have done, and who have pointed out errors and omissions;

- Second, to our colleagues, for their suggestions and ideas, in particular, Bert Waslander, Jérome Doutriaux, John Banasik and Tony Quon;

- Third, to the editorial and production staff at Arnold, namely, Liz Gooster and James Rabson, as well as our initial contact at Arnold, Nicki Dennis.

We welcome feedback to this work and the associated materials to be found on the PFM Web site located at **http://www.arnoldpublishers.com/support/nash/**. We will attempt to put 'the PFM Web site' as a phrase within the site to make it and any mirror sites easy to find by searching. On the PFM Web site the reader will find the latest information on how to contact us to report errors, omissions, improvements or similar matters. We welcome suggestions, but –

given the nature of our lifestyle – cannot undertake to respond quickly. On the other hand, people find us very reliable correspondents. To save explanations later, our policy with regard to queries that require our effort is as follows:

1. We try to fix and report errors as quickly as we can reasonably do so.

2. We will often undertake to actually carry out forecasting or other exercises that are presented as problems if we find them interesting and if there is an occasion to use them in the classroom, or in a technical report or academic article. However, we make no promise as to how quickly we will do this. We treat such tasks as part of ongoing and open scholarly activity in collaboration with colleagues.

3. We carry out actual forecasting assignments presented as problems in other circumstances, e.g., a sales forecast for a company, on the basis of a consulting contract at what we believe are highly competitive rates and with agreed time deadlines.

We trust that you will find this a useful and readable book.

John & Mary Nash
Ottawa, Ontario
August, 2000

1

Why forecast?

Why do we forecast?

In business or administration, the primary reason for forecasting is so we can plan. Planning allows us to profit from advance 'knowledge' of our world or to avoid disasters by virtue of predicting their occurrence. For example, we carry an umbrella when rain is forecast so we do not get wet. An ice-cream vendor, learning of a parade route, places his wagon strategically to take advantage of the crowd. Pilots want to avoid flying into bad weather (see for example, Thomas, 1996).

The relationship between planning and forecasting is central to business and administrative forecasting. We should use the results of forecasting to make decisions that prepare for the predicted outcomes. This is planning under the guidance of the forecasts. Otherwise, why are we spending the time and effort on forecasting? While there is considerable intellectual satisfaction in forecasting correctly, this is not a central business concern. We do not require, however, that forecasts always lead to action, just that they lead to decisions, even if the decision is to do nothing for the moment.

At the beginning of this book, as at the beginning of the many courses we have taught in forecasting, we expect readers will be interested in making predictions that are 'right' in some sense. Certainly we will be using considerable page space to discuss the measurement and assessment of forecast quality. Despite this, our own view of business and administrative forecasting can be summarized as:

Always being not too wrong is better than mostly being right.

By this we mean that the good forecaster never makes absolutely disastrous predictions, even if the predictions made are not exceptionally close to reality much of the time. In business, a cash flow forecast needs only to be precise enough that there is never danger of a shortfall that would precipitate bankruptcy. Similarly, sales forecasts serve as a guidepost. Good managers should be highly suspicious of forecasts that come true to the nearest dollar.

What can we forecast?

The types of situations that we will attempt to forecast are evolutions of existing phenomena or realizations of possible developments. We are not fortune tellers or astrologers, and require that there be some mechanism by which we can advance the videotape of our situation of interest with the clock or calendar. Obviously, our own activities, such as sales or production figures, are a prime candidate. Also of interest are figures relating to our competitors or industry group. Some political and socioeconomic situations are also possible to forecast, though good information may be scarce even if apparently relevant data is plentiful. Changes in technology based on existing flows of research and development will also be of interest.

It is also important to know what not to attempt to forecast. Students considering taking our course in forecasting techniques often give as their reason for registering that 'I want to forecast the stock market and make a lot of money'. Unfortunately, while financial markets offer many opportunities for useful forecasting exercises, the results are usually not so immediately helpful for money-making as the students (and others) would like. This is because it is very difficult to forecast human behaviour with the precision needed for profit. We can usually do quite well in estimating future levels of prices and volumes, and to some extent the variability in those figures. The timing of changes is more difficult to gauge because we need to judge when an individual or relatively small group of people will start doing something to trigger larger events. To underline the difficulty, it can be instructive to treat market data as a source of random numbers and test them for such 'randomness'.

The decisions that initiate market 'movement' lead to a complex sequence of events that ultimately change prices or quantities. Because understanding and modelling complex processes is itself complex, we will have a very large job if we want to forecast such phenomena. Any complex process will be difficult to forecast. Can you predict the delightful and convoluted movements of a kitten?

We are not saying you should simply give up when confronted with the task of forecasting such phenomena. You should, however, recognize what can reasonably be accomplished and focus on obtaining usable and sensible results.

Background skills and knowledge

Forecasting is a very broad subject that draws upon all the knowledge, experience, intelligence and imagination of the forecaster. We clearly cannot include everything you need in this book, even though we will make a serious attempt to provide a comprehensive package of ideas and tools. Some skills and knowledge we will assume you have learned before. It is useful however, especially for those who are learning independently, to provide a list of background items. You do not need to know all these topics at once, nor do you need to know them in great depth, to be able to profit from this book. Many statistics texts cover these topics. Here we provide references to two recent statistics textbooks, namely, Aczel (1996) and Clarke and Cooke (1998).

	Aczel	Clarke and Cooke
Tools for data analysis	Chapter 1	
descriptive statistics (mean, standard deviation, etc.)		Chapters 2 and 4
graphs for exploratory data analysis		Chapter 5
rank-based statistics		Chapters 2 and 4
Probability distributions	Chapter 2	Chapters 7 & 8
binomial distribution	Chapter 2	Sec. 8.6
normal distribution	Chapter 4	Chapter 15
Central Limit Theorem	Chapter 5	Sec. 16.1
t-distribution	Section 6-3	Sec. 17.11
Confidence intervals for mean and proportion	Chapter 6	Chapter 18
Hypothesis tests	Chapter 7 and Chapter 8	Chapters 13, 17 & 18
a mean		
a proportion		
two means (independent, paired)		
two proportions (pooled, non-pooled)		
a variance		
two variances		Sec. 22.4
Linear regression, Simple	Chapter 10	Chapter 23
Multiple	Chapter 11	Chapter 24
Chi-squared test of goodness of fit	Section 14-8	Chapter 19
Chi-squared test of independence	Section 14-9	
Helpful but not essential		
ideas from Statistical Process Control	Chapter 13	n/a
ideas from decision theory (decision trees)	Chapter 15	n/a
ideas from conditional probability, e.g. Bayes' Theorem	Chapter 2	Chapter 6

We assume our students also have some computer skills before taking our forecasting course.
  Computer skills:
    word processing
    plain text editing
    spreadsheet operations
    simple statistical package use
    files, formats, naming, copying, deleting, and conversions
    communications (e-mail, ftp, WWW)
  It would be helpful if students had basic library and research skills before our course

commences. However, we have found it necessary to include some material in this area (see Chapter 4) so that students can use library catalogues, bibliographies, and other reference tools and cite the sources they find.

We will NOT assume that reader has ALL the skills mentioned, but that he/she possesses at least some familiarity with the concepts and a willingness to pursue them further. We may review these topics as we encounter them, but such reviews will necessarily be brief so we can concentrate on their use in forecasting.

How forecasting works

All scientific forecasting methods rely on some principle of continuity. For example, if we roll a ball along the floor, we know that if the ball is a regular solid and the floor is flat and smooth, the ball will continue to roll in a straight line. Moreover, as a first approximation we may assume that it continues to roll at a constant rate, allowing us to forecast not only where it will go, but how fast it will arrive at certain points on the floor.

Our understanding of the rolling motion of the ball and its 'constant' speed allow us to forecast as long as we assume that the motion we observe will continue into the future. A collision with another ball ruins the predictions made based on this simple principle of continuity. However, if we develop a deeper understanding of the situation involving more information about the floor and the other ball, and any other influence on the motion of the ball, then we use the new knowledge to build better forecasts.

Clearly we do not put forecasting in the same basket with fortune-telling, although it is noteworthy that astrology bases its predictions on the unrolling of the pattern of the stars and planets. With astrology, of course, we may challenge the choice of the principle of continuity. This serves as a warning that it is important that the principle of continuity be relevant and applicable to the situation we wish to predict.

Models and their role

We codify our understanding of the forecast situation and the principle of continuity in models. Typically these are equations or graphs or similar constructs that allow us to march time forward more quickly than clocks and calendars to display – or compute – results that forecast what we wish to know. An example:

> Growth of wages is understood to increase at 4% per annum. If our current wage payout is W0, then we expect to pay out $W0*(1.04)^3$ in three years' time.

Much use is made today of the term 'model'. In forecasting, as we shall see, a model is frequently extremely simple. Moreover, it may do nothing more than match the pattern of an observed set of numbers without having any relation to the underlying process that is producing those numbers. Forecasters frequently use such *pattern matching* models. The main justification for their use is that they produce forecasts that are good enough to be of use.

A *causal model* tries to encapsulate what is actually going on inside the processes we wish to forecast. For example, some weather forecasters now try to solve the partial differential equations that describe the earth's atmosphere. This approach, while demanding some of the world's largest supercomputers, is beginning to show promise, although in many cases the 'forecast' is only ready some time after the weather has actually happened. That is, it may take longer to compute the forecast than it does for weather to evolve! Causal models are used in socioeconomic studies to attempt to understand human behaviour, as in the application of econometric models for analysis of government policy proposals.

Sometimes our models will attempt to 'explain' variability or randomness, such as the *random walk* that one may suppose is taken by a very drunk person in the middle of a large, empty parking lot (see, among others, Makridakis et al., 1998, p. 329). While such models may seem at first glance to be unhelpful to forecasting, they do provide limits on the evolution of variables of interest. Thus, forecasters use various models of stochastic or random processes depending on their problem at hand. An example is the attempt to forecast future 'records' in sports or other activity by Tryfos and Blackmore (1985).

Once we have a model, we must be able to extrapolate using it. This means that we need some way to read off the future value of quantities of interest. Some models may be very good at describing what is going on now, but they may not be in a form that allows us to extrapolate. For example, a model may relate many variables, so that the forecast for one of the variables needs future values of the others as inputs. This is clearly unhelpful for forecasting, so we seek models that permit easy extrapolation.

Skills and knowledge objectives

In this section we provide you with a checklist of the skills and knowledge that you should be able to acquire through this book, or a course of studies based on it. Some of these skills are commonly regarded as prerequisites to a course on forecasting, but abbreviated presentations are included here to provide a self-contained guide to forecasting for those who are studying informally. Doing the exercises requires a knowledge of many of these prerequisite concepts.

In our courses we divide the topics into three categories:

- those that can reasonably be tested by an examination,

- those that are demonstrated in more extended works, such as projects, essays or presentations,

- subjects with which a forecaster should have a passing acquaintance, but need not understand in detail.

This last group are 'awareness topics'; they serve to enlarge the possibilities and scope for arriving at better forecasts. All the sets of topics are also intended to be useful outside forecasting.

Examinable topics

Subjects usually taken as prior knowledge
    Distributional properties of errors
- use of histogram, Stem and Leaf and normal quantile plot to check normality
- boxplots, including stacked boxplots

    Hypothesis tests
- general structure
- format of a test (statement of hypotheses, assumptions, test statistic, comparison values or prob. value, decision)
- tests for regression coefficients (t-test)
- test of utility of regression model (F-test for at least 1 non-zero slope)

Forecasting topics
    Principle of continuity
    Types of forecasts
    Forecast horizon
    Errors and their measurement
    Norms of errors
    Graphs and their uses
    Validation testing of forecasting models
    Understanding of the output of forecasting or statistical software.
    In courses, we present examples using just one or two software packages (e.g. Minitab) to provide a predictable and familiar format for examinations. We do NOT examine students on the active use of software commands, but do expect them to be able to understand what the commands do when reading output. Asking students to write the commands that will carry out particular operation upsets them and results in papers that are difficult to grade.
    Multiple regression models
    Dummy variables for additive seasonal models
    Serial correlation
        autocorrelation coefficients (ACF and PACF diagrams)
        Durbin–Watson test
        Ljung–Box test, i.e. Modified Box–Pierce test (using chi-squared distribution)
    Moving averages

Exponential smoothing models
    single or simple
    double exponential smoothing (Holt or Brown)
    Winters' seasonal method
Time series decomposition
Formal expression of ARIMA models
    non-seasonal models
    seasonal models
    Differencing and logs to render series stationary
    Use of exp() to reverse log() transformation
Estimation and forecasting using ARIMA models (Minitab output)

Project, presentation and data analysis skills

Getting data into computers
Moving data between applications
Documenting work
Putting reports together
Critique of forecasts and forecasting reports
Finding information in libraries, on the Internet, and elsewhere

Concepts, tools or facts for awareness

Data sources
Data collection and validation
Data edit and imputation
Technological forecasting
Overview of econometrics
Internet information search
Commercial forecasting services
Data movement tools
Market forecasting
Developing a forecasting strategy
Various statistical and forecasting software

Throughout our treatment, and especially in regard to software, we urge the use of simple tools and features as much as possible. We avoid any tool that requires a special machine configuration or operating system, since others will not be able to replicate our work or methods.

Examples of forecasting tasks

As examples of the types of tasks that we believe are part of managerial forecasting, the following are some of the projects we have assigned our students over the last 20 years.

Provide a forecast of AIDS cases and deaths for the next (number, e.g. 5) years for some jurisdiction.

Forecast smoking habits of some population by year until (a given year). Consider age/gender/location variations in such forecasts.

Forecast the distribution of types of living accommodation and form of tenure (i.e. rent, own, cooperative) for a geographic region and/or parts thereof for the next decade.

Forecast the student population in (a named jurisdiction's) post-secondary institutions to (a given year).

Forecast for the next 5 years, if possible by season, the number and percentage of homeless persons in (named region) and/or the number of places for emergency shelter.

Forecast some weather variables for (named place) for the next 25 years, e.g. mean temperatures, precipitation, number of storms etc.

Forecast the population of the city of (name) for the next 10 years.

Forecast the gambling industry in (name) and some of its impacts on society.

Forecast some measures of price and performance for personal computers for a time 5 years hence.

These topics are deliberately open-ended. They are intended to force the students to gather data and integrate it into a coherent picture of the situation under study. Some of the topics pose a severe shortage of data of any sort, while others give difficulties because there is such a lot of data, but possibly little real information. We have also chosen to restrict attention to certain jurisdictions, since businesses and governments work in such domains and clearly want forecasts that apply to them. On the other hand, for a population forecast, we have used the city of Calgary because it is not bordered by other cities to confuse the forecasting question. Despite recent amalgamations of municipalities such as Toronto, many urban areas are a confluence of several municipalities.

Most of these projects do not lend themselves very well to any of the 'standard' approaches to forecasting. It is our contention that it is important that students recognize that the real world is like this, requiring ingenuity and inventiveness in adapting methods so that a forecast can be achieved.

We also provide students with exercises that are related to the standard approaches. Throughout this book we will use a data series derived from an early term project. This is a Canadian traffic fatalities series. We have actually aggregated a monthly series into a quarterly one to render it more suitable for expository purposes. Nevertheless, it is a real series. Many other possibilities exist, in the literature, on the Internet, and those generated artificially. In the course of the chapters that follow, some of these series will be introduced or referenced.

Finally, we believe that it is important to be able to learn new techniques quickly. We often ask students to prepare essays or presentations on general or methodological topics related to forecasting. Some of these overview or perspective topics:

Mathematical filtering methods applied to forecasting

Forecasts and forecasting services available commercially

Technological forecasting – forecasting future ways of life and doing things

Econometric forecasting – its role in business and government

Forecasting software (a review of one or several packages)

Tools and data for market forecasts

Exercises

E1.1. Use the list of topics under 'Background skills and knowledge' as a checklist for your own preparedness. If appropriate, check the topics against textbooks you have used.

E1.2. Write down three or four topics for forecasting projects. Expand these topics with brief abstracts of the forecasting objectives and, in particular, the managerial importance of these objectives.

2

Planning the forecasting task

Defining the goals of a forecasting activity

Before starting any project, and particularly forecasting activities, it is essential to define the goals. We need to decide what we want to forecast, whether we need to know the level (quantity or value) or variability (spread, volatility, uncertainty) or both for this forecast, and the time horizon during which the forecast will apply.

Refining our goals, we will eventually need to decide how our forecasts will be measured or tested and usually will want to develop the subsidiary objectives of our forecast effort relating to the ease of presenting the method and the results to potential clients or users of the forecasts. Accuracy is a quality most people desire in a forecast, but as we shall see, there are many ways to measure how close a forecast is to the corresponding reality. Especially important is the possibility that the forecast is intended to predict change points, that is, the time point(s) where the behaviour of a situation will change in some clear manner.

Types of forecasts

We make forecasts within a framework or environment such as a business or government department. The results of the forecasting exercise are not simply abstractions – we will use them to plan for real-world situations. So that we can describe what we are doing or want to do, we characterize the forecasts we want. There are several possible classifications. One considers the nature of the forecasts we make and refers to them as:

- Qualitative, e.g. it looks like rain,

- Quantitative, e.g. we will get 10–12 mm of rain, or

- Mixed, e.g. there will be 12–15 cm of snow with the possibility of freezing rain

Another important categorization of forecasts concerns the forecast horizon, that is, the time ahead for which we want the forecast results. In business and administrative forecasts, typical choices are:

- Short term, meaning less than one year ahead,

- Medium term, meaning from one to five years ahead, and

- Long term, meaning greater than 5 years ahead.

The dividing lines between short, medium and long term are human inventions. The choices suggested here are in line with those of several authors and practitioners, but readers should carefully note the usage in a particular context. Bowerman and O'Connell (1979, p. 22) choose 3 months and 2 years as the break points between the categories, for example. We attended one seminar on forecasting ocean currents where 'medium term' was 200 years.

A third categorization we have already met. That is the use of structural (or causal) versus descriptive (or pattern-matching) approaches. Clearly, this classification refers more to the methods than to the forecasts themselves.

Background to the forecast

Before we actually start forecasting, we should consider the context in which our forecasting situation arises or exists. One of our complaints about some books and software for forecasting is that they presume there is a set of data and that by putting it into one end of a pipe and pushing some buttons we will get a forecast out of the other end of the pipe. Perhaps some authors have seen *The Wizard of Oz* too many times. We strongly believe that knowing the background to our task will help us to do a better job. We will want to know about:

- Data and its quality;

- Situations that give rise to the data; and

- People and their agendas.

In forecasting, much of the data is in the form of time series. The most obvious and simple form of this is a single (or univariate) series of numbers that correspond to equispaced time points. Quite often we index the time points from 1 to n, where n is the number of points in the series. An example of such a series is the Hobbs weed infestation data (Nash, 1979, p. 121), which gives the density of weeds per square metre in each of 12 successive years (Table 2.1).

While such time series are commonly associated with forecasting, so that colleagues in mathematics and statistics departments will frequently confuse 'forecasting' with 'time series analysis', we caution that data for forecasting has many forms. First, even time series data may be recorded at irregular time points. This is sometimes treated as if there are missing values, as many of the methods for forecasting are conceived on the assumption of regularly spaced time points. We believe this to be a particularly grave oversight on the part of developers of forecasting methods, but we appreciate the difficulties of developing more flexible methods. The data set in

Table 2.1. Hobbs' Weed infestation data

Year	Weeds / m^2	Year	Weeds / m^2
1	5.308	7	31.443
2	7.24	8	38.558
3	9.638	9	50.156
4	12.866	10	62.948
5	17.069	11	75.995
6	23.192	12	91.972

Table 2.2 represents the overall percentage of the adult population in Canada that were considered to smoke tobacco products in the years indicated, which are the years of surveys available to students who attempted to forecast the proportion of smokers in the population (see Wright, 1985 and 1986). What the table does *not* show is that the surveys were not all conducted at the same time of year, and that there appear to have been different criteria used for developing the samples.

Table 2.2. Percentage of smokers in Canada data

Year	Smokers (%)	Year	Smokers (%)	Year	Smokers (%)	Year	Smokers (%)
1965	43.8	1970	40.6	1977	35.9	1983	31.1
1966	42.8	1972	39.8	1981	32.7	1986	28.2

For many forecasting situations, we may need to use cross-sectional information. For example, one of the projects we recommend our students undertake is that of predicting the proportion of the population that smokes. In such exercises, many aspects of the population(s) of interest may affect the propensity to smoke. For example, the Canadian situation has a marked variability with geography and education, although there is some evidence that variables used to represent these qualities of members of the population are correlated.

As already suggested, we may expect that we may be unable to find data for every time point, or data for every variable at every time point. Missing values are always a concern in statistical analysis. Dealing with them requires particular attention to the reasons why the data are missing. Are we simply asking for data before there is a reasonable hope of collecting it? Do people or agencies have reasons to suppress information?

A more dangerous case is incorrect or imperfect data, since when data are missing we are at least aware that something is lacking. When there are numbers available, we generally presume they have some relationship to reality. We can be misled and produce false forecasts. This is where it is important to collect and keep track of qualitative or anecdotal information that may inform us of the relative reliability of data elements. In this category of data we include the situational and anecdotal information that enrich our capability to make informed forecasts. In addition, we should note here the known or suspected motivations and agendas of the major personalities who may drive the evolution of the processes under study.

We are acutely aware of the growth of 'data mining' (for example, *Communications of the Association for Computing Machinery*, vol. 39, no. 11, November 1996 issue, or Penner and Watts, 'Mining Information', *The American Statistician*, vol. 45, no. 1, pp. 4–9, 1991). Uncritical use of data mining may promote poor quality data to the unwarranted status of valuable information. The function of data mining is essentially forecasting, usually of the behaviour of consumers of products or services. In particular, financial institutions use data mining to rate credit risks, which are a forecast of the probability a loan will be repaid.

Management of forecasting data

Most students in our courses, and as far as we can tell most people in the business and administrative world, have rather poor skills in managing the data they must use for forecasting or other analyses. This is likely a consequence of the general lack of training in data handling. (Jane Gentleman, formerly of Statistics Canada, has on several occasions delivered a delightful seminar 'Data's perilous journey' guaranteed to make a sceptic of any listener.) Data handling is, unfortunately, almost always tedious and demanding of discipline. To save yourself much grief, 'Keep it simple'.

- Document your data. Some packages make this relatively easy, but it still requires you to actually put in the comments that tell you where you got the data, when it was published, when you entered it, what changes you may have made to it, such as changing the units from miles to kilometres, etc.

- At the risk of annoying our readers, we'll ask you to note the first point again.

- Keep data, if at all possible, in *plain text* form, sometimes called ASCII (American Standard Code for Information Interchange). Actually different operating systems have slightly different ways for recording text, but the differences are largely in the ways lines are ended. This involves concepts such as the 'carriage return' or CR character, the 'line feed' or LF character, and their use individually or in combination to end lines. For example, Microsoft MS-DOS uses the pair CR LF to end lines. The end of such a file in MS-DOS is marked with the character Control-Z (character number 26 decimal in the ASCII sequence). If this sounds horribly technical, remember that this is the simplest format for files. It is also one of the more compact formats. A student once sent us a file of the (monthly) Canadian traffic fatality data for a single year (an update). The file had been entered in Microsoft Word, which we did not have installed on our computer. It was about 16,000 bytes long. When we asked for it in a form we could read, we got a Microsoft Excel file of 10,000 bytes. Excel wasn't installed either (we do have over 100 statistical and mathematical packages, however). Then we got an electronic mail with the data in plain text – 250 bytes!

• Back up your data religiously. This means arranging that your important files are physically kept in separate locations. One team of students some years ago had their main and backup disks in the same box, which they set beside the monitor of the workstation where they planned to work. When the disks were put in the drive to read, there was no readable data on the disks!

The quality of the data files that forecasters prepare is highly variable, reflecting the character and temperament of the individuals themselves. We once asked students to enter data series from a printed collection, intending to use the resulting files for use in classroom examples. However, we discovered that students generally had a very poor idea of how to enter data and commentary so that the resulting files could be usable by Minitab, the package we were using at the time. The exercise was not wasted, however, as presenting the different files and pointing out the 'bugs' allowed everyone, including the professor, to learn the sources of problems. These were primarily:

• Lack of knowledge of acceptable syntax for data and comments;

• Use of tab characters rather than spaces;

• Inclusion of punctuation marks where not permitted by the statistical package;

• Lines that were too long to fit on a screen or printed page;

• Omission of comments and suitable identification.

Assignment of resources

Whenever people are assigned work tasks, they must manage their resources to complete these tasks along with all other responsibilities. We are acutely aware that students never take only one course. Similarly, we know of no real-world business or organization where administrative staff are given just one thing to do. Forecasts, in particular, are but one part of the background to managerial decision-making. A rational assignment of our resources to the diverse tasks we must accomplish is therefore part of our 'job'. Failure to allocate our resources properly is equivalent to choosing poor forecasting methods or making mistakes in analysis.

By extension, we all know that occasions arise where we must limit the resources that can be allocated to a given task. This will restrict the methods, data, and detail of our forecasting efforts. Throughout this book, we will attempt to structure our discussion of methods so that users can make sensible choices between them when faced with resource constraints. We also suggest that the reader always keeps in mind the necessity of balancing goals with resources. Some particular computing strategies are presented in Chapter 11.

There is also an issue of the mix of resources we allocate to forecasting. The resources managers control include time, people (that is, the choice of staff to assign to tasks), and money. They sometimes have assets, such as data, computers, communication technology and software that may be applied to the tasks, but only if the staff know how to employ them profitably. In our experience, forecasting is generally best done by small teams of well-trained people using relatively straightforward approaches. We believe that forecasting tasks can rarely be accomplished well in less than a few days, and this after data has been gathered. Our basis for this statement is that while the actual calculations and analysis take only a few hours, most forecasts should be revisited after a day or so of other activities to provide a perspective and opportunity for judgement.

Testing and selecting a model

We have already pointed out that forecasting, at least as discussed here, depends on some principle of continuity that we model and then extrapolate in order to forecast. To choose a model from a set of potential candidates, we will, if possible, set aside some of our data in order to test how well different methods or models work. That is, we divide our complete data set into two parts, which we will label the test estimation and validation periods. Forecasters sometimes call this 'ex-ante forecasting' and 'ex-post testing', but we feel it makes more sense to use 'validation testing'. In general, this strategy is called *cross-validation* in the statistical literature. Note that we estimate with data that is 'in sample' and validate with data that is 'out of sample'.

Although the idea of dividing the data in order to validate a method is simple, its application can lead to confusion if you do not clearly document what you do. We want to test how well our forecasting methods work. We usually set aside the few latest observations, and use the rest of the data to 'forecast' these. We can then compare our forecasts with what actually was observed. That is, if we have data for 1949 to 1997, we could use the data from 1949 to 1990 to predict 1991 to 1997 and compare with what actually happened. To do our 'real' forecasts, we choose a method that worked well in the trials, then use *all* our data (1949 to 1997) to predict for 1998 onwards. We will term the estimation of the model using all the data as the *main estimation*. The predictions for 1998 onwards are the *forecasts*. This is illustrated in Figure 2.1.

How much data should be set aside? This is a perennial question in the classroom and often difficult to answer clearly. If you have plenty of *relevant* data, then the data set aside should correspond to the forecasting horizon, that is, how far ahead you want to actually forecast. But, the environment in which the phenomena studied exists may have changed, so we may not have data that is 'relevant' for a long enough period. Many situations of interest just do not have much good data. Therefore you may not always be able to do such testing for as long a period as is

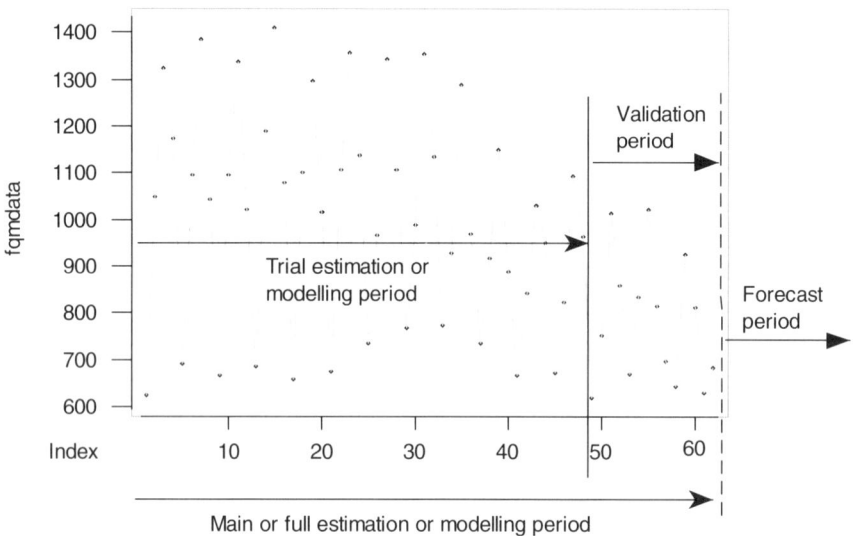

Figure 2.1. Illustration of the division of data for testing, validation, main modelling and forecasting.

desired, or may not be able to do it at all. Indeed, there are some studies that suggest that good results in trial estimation and validation may not necessarily predict how well the forecasts work.

The next question that arises is 'How do we compare methods in testing?' We can use all the measures that exist for assessing the size of errors or deviations. These are presented in Chapter 3. We will also want to display the deviations between different quantities of interest in graphs over time and sometimes other dimensions. The use of colour or different symbols allows for quick comparisons if done sensibly.

Once we have chosen one or more 'best' methods, we then try to estimate the models (or run the methods) using all the data, then compute the final forecasts. When using data for the main estimation period, it may turn out that one or more of the candidate methods does not work properly and must be discarded. The final *forecasts* should, of course, be checked for consistency with our understanding of the situation, and they should be clearly reported. On more than one occasion, students have submitted a highly competent analysis of a situation they were assigned to forecast, but the forecasts themselves were nowhere to be found in their reports!

We want to emphasize the importance of taking a wide perspective, and pose the question:

> *Should the 'best' candidate method as indicated by*
> *validation testing be used for forecasting?*

Our answer is: '*Not necessarily.*' We will have measures of fit and graphical evidence of the suitability of the forecasting model over the period in which the validation models are estimated (test estimation period), the validation test period (the balance of the data) and the period used to estimate the final forecasting model (final or full estimation period). The last period is most often the entire set of data, but does not always have to be so if there are reasons to believe that early data is no longer relevant.

The measures of fit (test estimation, validation, main estimation) can and should be tabulated for comparison. If there are large differences between the measures of fit for the two estimation periods, then something has changed and we should be suspicious of any decisions made based on the validation tests. This is generally *not* the case, but we should check, since several measures of fit are generated automatically by most forecasting methods. We can also include commentary on the graphical evidence of suitability in our table to provide a summary of our analysis and modelling.

Anticipating the techniques of analysis that will come later, we will provide a 'recipe' here. You may find it helpful to use this as a checklist for your own and other forecasting reports. Note that we will be presenting most of the actual material in later chapters.

Basic rules

- Document clearly what you do!

- If possible, use a validation forecasting period that corresponds to the forecast horizon for which a forecast is desired. This may not always be possible owing to lack of data or lack of data that is appropriate. For example, the system governing the phenomenon of interest may have changed at some point. If you cannot use the validation test period that is ideal, then justify the choice(s) you make.

- Watch how you define quantities. For example, both

    residual = actual - model   and   alternative_residual = model - actual

  are used by different (or even the same) authors. Remember to include notes that tell which form is used. Similarly, watch out if you transform data (e.g. take logarithms) so you know which particular transformation (e.g. log vs ln) and which columns are affected. You will have to back-transform to compute error measures for the original data in order to compare models properly.

Trial estimation or modelling period

- Our first concern is to see if the model(s) *fit* the data using (see Chapters 3 and 8)
    Mean square error
    R_squared
    maximum absolute error

> percentage errors
>
> graph(s) showing fit and pattern and distributional shape

- Note that you may have to compute residuals explicitly, that is, (actual - model) to allow measures of fit to be computed if the modelling has been done with transformed variables. That is, remember to back-transform if needed, and to compute measures of fit in the *original* scale. An example is provided below.

- Similarly, you may also need to compute the residuals (and back-transform) to draw a distributional plot of the residuals, e.g. histogram or stem and leaf. These graphs show us where our 'errors' lie relative to zero. A boxplot could also be used, but it provides less information.

- The distributional graphs also let us see if the 'errors' are consistent with sampling from a Gaussian distribution. We do this because, if such an assumption can be made, there are statistical tests we can use to examine model parameters (regression coefficients, ARIMA parameters) or model qualities (Ljung–Box statistic, F–test), introduced later.

- We will usually want to make the tests just mentioned, at least informally, and report the outcome if it is important to our forecast results.

Example of forecasting with transformed variables

The following edited output was obtained starting with some data that grows exponentially. It is in the column 'expeg' below. By using linear regression (Chapter 12) to fit a straight line model to the natural logarithm of this data, labelled 'ln expeg' below, we get fitted values in 'log mod'. If we subtract 'log mod' from 'ln expeg' we get the residual 'resinlog'. This has a very small sum of squares, 0.141339. We must 'back transform' the log-form residuals by taking the exponential function of them to get 'explnmod' and the true residuals 'trueres' in the original scale. The true residuals are 'expeg' minus 'explnmod'. The sum of squared residuals is now 116596604. This may seem unfortunate, but the total sum of squares (the sum of squared deviations from the mean of the original 'expeg' data) is 61083376410, so that our R_squared (see Chapter 8) is 0.998091.

Row	expeg	explnmod	trueres	ln expeg	log mod	resinlog
1	11	15	-3.61	2.4365	2.7110	-0.274508
2	22	25	-2.73	3.1021	3.2180	-0.115945
3	49	41	7.77	3.8969	3.7250	0.171824
4	75	69	5.98	4.3153	4.2320	0.083294
...						
19	133601	138311	-4709.94	11.8026	11.8373	-0.034647
20	220259	229641	-9382.15	12.3026	12.3443	-0.041714

Validation period

- We compare forecast and actual values for the validation test period.

graph of errors

mean squared error

maximum absolute error or % error

any trends in errors and also whether forecast is always or usually high/low. Descriptive or summary statistics of errors can quickly tell us if our forecasts are high or low.

- On the basis of good fit and other properties in estimation period, and good fit in the test period, we choose a candidate model or models.

Main estimation or modelling period

- All the same considerations regarding fit and quality apply as in the validation estimation period.

- In addition, we should compare the model parameters for validation and full estimation periods. If the phenomenon we are forecasting is more or less stable, we expect the parameters to have changed little. If they have changed a lot, our forecasting exercise is put in doubt and some discussion of approach is needed. In some forecasting methods it is possible that the parameters themselves (seasonal factors and trend lines can be treated as parameters) may not be available. However, we can then generate the values predicted by the models and compare these.

Forecasts

- Don't forget the final forecasts that are our ultimate goal. Graphs are helpful, and it is worthwhile commenting on the forecasts if they are unusual in any way.

- Some methods allow confidence bands to be placed on forecasts. These are often very poor estimates, but they may provide a scale to our lack of knowledge about the future values of some variable.

Single and multi-period forecasts

Many methods for forecasting are intended to be used to forecast *one period ahead*. The naive forecast – use this month's sales as a forecast for next month – is a simple example. Methods that require us to have data at time $(t-1)$ to forecast for time t may give us difficulties in our validation. The temptation is to 'run' the methods and get the forecasts for the validation period, then provide the measures of quality of these forecasts. What we may overlook is that the method (and the computer programs that implement it) have quietly used data *in the validation period* to compute forecasts.

This is clearly *not* our intention. We want to prepare forecasts for the validation period based only on data up to the last observation in the test estimation period. We then make *multi-period forecasts* for the entire validation period. Otherwise, we are really getting residuals, as in the test estimation period.

A real example

To provide a continuing example that we can use throughout the text, we now present a file of data that we call the 'Quarterly Traffic Fatalities in Canada' data, Table 2.3. File FQ98BYQ.MTB is a Minitab executable script (see the Appendix) that loads a set of data for road fatalities in Canada by calendar quarter when executed. The data set has been created from monthly data. Our experience with the monthly series was that the level of detail made it more difficult for students to see what was going on in our forecasting efforts. We simply aggregated the appropriate observations to get a quarterly series, although the word 'simply' hides the care that is needed to ensure a correct output series.

The script mentioned above for loading the Quarterly Traffic Fatalities in Canada data puts the observations for each quarter into a separate column (or variable) in the Minitab work-sheet. To work with the time series, we will generally want to have the data in a single series in time period order. In the Appendix, we suggest a script that will generally be used as our starting point for most of the analyses in this book.

What should we report?

Styles for reports vary. We have never been rigid in requiring students to follow a particular format. We do, however, believe there are some elements that are generally found in good reports.

- A report summary, often called an executive summary or abstract, is helpful to readers in quickly deciding if the report is relevant to their needs.

- Since forecasting, at least in the context of this book, relies on models and methods, the report should clearly indicate which methods and models have been tried. Brief justifications for their potential utility should be mentioned.

- For the same reason, include a brief mention of methods or models that might reasonably be considered but which have been discarded in the present report. Again, we should explain, in a few words, why these approaches are not appropriate.

Table 2.3. Quarterly Traffic Fatalities in Canada.

Year	Q 1	Q 2	Q 3	Q 4	Year	Q 1	Q 2	Q 3	Q 4
1975	1043	1457	1999	1564	1987	674	1107	1357	1138
1976	840	1377	1679	1411	1988	734	967	1345	1106
1977	786	1340	1735	1392	1989	768	988	1355	1135
1978	821	1353	1822	1433	1990	774	929	1288	970
1979	985	1460	1836	1582	1991	734	916	1147	888
1980	969	1371	1849	1272	1992	666	841	1030	949
1981	961	1369	1732	1321	1993	670	823	1093	964
1982	624	1049	1324	1172	1994	618	752	1013	858
1983	691	1095	1386	1044	1995	668	834	1021	815
1984	666	1095	1338	1021	1996	697	643	926	813
1985	686	1190	1411	1078	1997	630	683	*	*
1986	657	1102	1296	1016					

- The development and testing of candidate models should be reported in graphical or numeric form or both. Key measures of the quality of the approaches should be given in a comparable form. We want to convince the reader that the work has been well done.

- The full estimation of selected models and their forecasts should be reported in graphical or numeric form or both. This is the goal of our work, and should be presented prominently, with a précis in the abstract.

- A final selection of the 'best' forecast in our opinion, with justification, or else some combination of the candidate forecasts should be given. It is appropriate and worthwhile here to attempt to provide an estimate of the reliability of our forecasts if possible.

- We should properly reference materials or sources used (see Chapter 4).

Good working habits

Since computers are used as data handling devices as much as anything else, we think it is very important to develop good habits and standards so effort and time are not wasted. We suggest, in partial repetition of some ideas already mentioned:

- Data are best kept as plain text (ASCII) files, since these can be easily printed, communicated by electronic mail, and imported into almost all software.

- Although it may complicate the importing of data into software, text files can easily be annotated with NOTEs or similar constructs that document the origin, changes or additions, time stamp, and other information (often called meta-data) that can clearly influence our use of the data. For example, a particularly low sales figure may correspond to severe weather forcing closure of business.

- Back up your files in a regular and disciplined manner, making sure that the original and duplicate files are kept physically separate.

- To simplify the organization of your work, do devise a file naming convention. Files named 'RESULTS', 'TEMP', 'DATA' are not helpful. We need to differentiate the different files we create and use as we advance our forecasting exercise.

- In a similar fashion, it helps to keep a log-book or diary of your work and files. This is a tough job. We admit that we do not always manage to follow our own advice in this regard.

- If possible, run calculations from scripts (macros, programs, etc). This lets you easily make small modifications to data or controls and re-run the forecast. Inevitably such near-repetition is needed, either to confirm analyses, when new data arrives to permit our forecast to be updated, or when we must prepare presentation material and want to save output. The script also serves, if properly commented, as one of the better forms of documentation of our work.

Exercises

E2.1. Read the Quarterly Traffic Fatalities data into a spreadsheet package (Excel, Quattro, 1-2-3) and try to create a single data series from the data for the individual quarters. Create series giving the year and the quarter number aligned with your new time series.

E2.2. Find a data series based on your own organization or work and repeat E2.1.

3

Measuring how well forecasting goals are met. Part 1

What is a model? Part 1

We mentioned in Chapter 1 that our forecasts will be based on extrapolating models. Almost all scientific work is built on *models*. This includes statistical research and analysis, of which forecasting is a part. We often find that our students have an unclear understanding of what models are.

A model is any mechanism that allows us to predict the behaviour of the phenomenon of interest. This can be as simple as a narrative description. It can be a physical model; Renaissance astronomers sometimes built an orrery, a mechanical reproduction of the solar system that shows the relative positions and movements of the planets and their moons. Models can be expressed as graphs, diagrams or tables. They can be coded as computer simulations. Most often they are represented as mathematical expressions, including integral and differential equations.

In this book, we will almost always see models as functional expressions. That is, we will write an expression of the form

$$y(t) = f(t, X(t), \boldsymbol{b})$$

where $y(t)$ is the value of a variable we wish to forecast, t is the time index, $f()$ is some mathematical function, $X()$ is an array of data relevant to the forecast (which may include previous time values of $y()$), and $\boldsymbol{b}$ is a vector of parameters. The parameters $\boldsymbol{b}$ are numbers that we can adjust to alter the shape of the function $f()$. We use them to make our model more appropriate in some way for forecasting purposes.

What are we looking for? Pegels' classification

It is only rarely that we come to a data set with no forecasting agenda. Because many forecasting tasks involve similar types of data – sales, production, traffic flows, etc. – we would like to get to the meat of the matter quickly. One shortcut that has been found useful is Pegels' classification.

Pegels (1969) described a useful classification of time series and models for time series. This is based on the recognition that a lot of business and administrative data reflect the calendar by which we mark time. The first breakdown that Pegels' classification uses is therefore based on noting whether or not data has seasonality. This does not merely include the usual Spring, Summer, Fall (or Autumn) and Winter of our traditional calendar year, but also cycles in daily data recorded hourly, weekly sales variation with the days of the week, and similar regular patterns that can be easily linked to the scheme by which time is measured. The other major subdivision of series is decided on the basis of their stationarity, or in more positive terms, whether the series has trend. Stationarity is the absence of long term movement in the mean of a series. Trend, conversely, is the long term change in the mean.

Table 3.1. The matrix of combinations of trend and seasonality in Pegels' classification

	Seasonality		
Trend	1) None	2) Additive	3) Multiplicative
A) None (Stationary)	Mean	Mean + seasonal_term	Mean * seasonal_factor
B) Additive (Linear trend)	$(a + b*\text{time})$	$(a + b*\text{time}) +$ seasonal_term	$(a + b*\text{time}) *$ seasonal_factor
C) Multiplicative (Exponential trend)	$(a * \text{rate}^{\text{time}})$	$(a * \text{rate}^{\text{time}}) +$ seasonal_term	$(a * \text{rate}^{\text{time}}) *$ seasonal_factor

Pegels' classification breaks down both trend and seasonality one further stage. Regular movement in data can be in absolute or relative, that is, percentage, fashion. A regular absolute movement gives rise to an additive change in the data. If we regularly change data by a fixed percentage, we get a multiplicative change in the data. The changes can be either in the trend or in the season.

There are a number of details about trend–season models that we can usefully discuss here. First, although we use the term 'additive', the amounts added can actually be negative, so that an additive trend can actually decrease the numbers in a series by having the slope, b, negative. Similarly, the additive seasonal_term can also be negative. Some authors use the description 'linear' to describe additive trend, while using 'exponential' to describe a multiplicative one. The models show why this terminology is used. The model

$a + b*\text{time}$

is a straight line. The exponential trend model is

$a * (\text{rate}^{\text{time}}) = a * \exp(\text{time} * \ln(\text{rate}))$

where $\ln()$ is the natural or Napierian logarithm function to base e (an irrational number approximately equal to 2.7183). Since the exponential function (e to the power of the argument) appears, we have an exponential model. Students seem to find it helpful to apply an indexing scheme to the classification, and we have labelled the trend possibilities as A, B and C while the seasonality is indexed by 1, 2 or 3. The categories are none, additive and multiplicative, respectively, for both dimensions.

Another detail worth mentioning is that we distinguish the seasonal_term, which is additive, from the multiplicative seasonal_factor or seasonal index. For the seasonal part of the additive seasonal models, we assume that the mean of the seasonal_terms is zero. By contrast, for the seasonal parts of the multiplicative seasonal models, we want the 'mean' of the seasonal_factor to be 1. Strictly speaking, we may want to use the geometric mean here. The geometric mean of n numbers is the nth root of their product. Thus, the effective or 'average' interest earned on an investment that doubles in four years is

$$100 * \{\ 2^{0.25} - 1\ \} \quad \text{or approximately } 18.9207\%$$

rather than the 25% that is obtained by reasoning that we have gained 100% in 4 years.

Some practitioners use a seasonal index scaled so that the base level is 100, in which case we must remember to divide by 100 when applying such indices to a trend model. A lot of government statistics are presented in the form of such indexes. They can be troublesome if the 'base period' is changed in the middle of the time period for which we are developing a forecast. Our advice when indexes are encountered is to be cautious about their definition and to convert them to factors, that is, adjustments that simply multiply our model to make it work better.

In teaching forecasting, we have found Pegels' classification to be a concept that students understand quickly and easily in its generality. The details sometimes prove troublesome, in particular the use of logarithm and exponential functions, and dealing with indexes where the base period has been altered. However, we caution that Pegels' classification is simply a way to provide a preliminary categorization of the data or situation at hand. It does not, and should not be expected to, provide a forecast. Nor does it account for special events or for other layers of data structure such as long term business cycles. It does help us to eliminate from consideration those methods and models that will not suit our situation. Thus, it helps us to avoid trying 'every trick in the book' and to focus on those techniques likely to be of use.

Both in teaching and in practising forecasting, our opinion is that it is very difficult to be sure whether trend and season are additive and multiplicative if there is any 'error' or noise in the data. Note that such 'error' may itself be specified as additive or multiplicative. To illustrate the Pegels' classification in Figures 3.1 and 3.2, we created data that perfectly matches the different categories in the classification, then added additive error or multiplied by appropriate multiplicative disturbances. (See exercise E3.4 at the end of the chapter.)

Some authors have come up with variations on Pegels' theme. In particular, Gardner (1985) adds another row to the Pegels matrix to allow for *damped trend*. This is a trend that curves upward or downward towards an asymptote. In our experience, such trends are relatively rare, although we do consider nonlinear models in Chapter 20 that have similar characteristics.

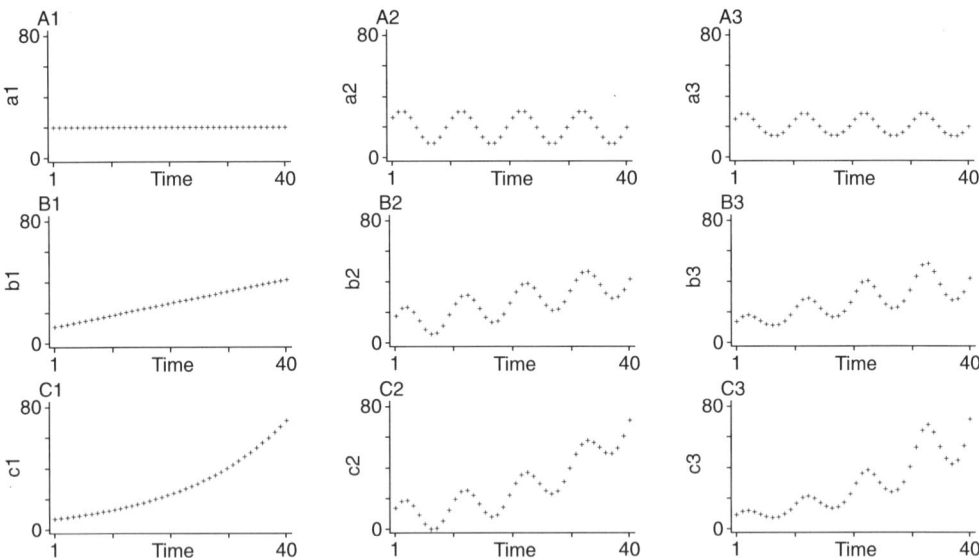

Figure 3.1. Pegels' classification illustrated. No error. The functions used to generate these particular patterns are described in an exercise for this chapter.

Fit of model to data

The usual adjustments of the parameters are intended to improve the 'fit' of the model to the data. As in tailoring made-to-measure clothing, we look at the deviations between the data (the body to be clothed) and the model (the garment under construction). *Deviations* are, in fact, key to almost everything we do to measure the fit of models.

In the literature of forecasting, and of statistics generally, deviations are usually called either *errors* or *residuals*, by some authors interchangeably. We will distinguish the terms here. *Errors* will be the deviations between quantities of interest and their true, but possibly unknown values. *Residuals* will be the deviations between observed values and model (or forecast) values. In fact, we will usually adopt the convention that the ith residual is

$$r(i) = y(i) - \text{model}(i, X, b)$$

that is,

   residual = data - model

This is the usual convention employed by practitioners. However, early statisticians, notably Gauss and Legendre, preferred a 'model - data' choice for the sign of the residuals. We have used this convention ourselves for nonlinear modelling, where it helps to avoid sign errors when taking the derivatives of the residual in Gauss–Newton methods for nonlinear least squares (Nash and Walker-Smith, 1987, and references therein).

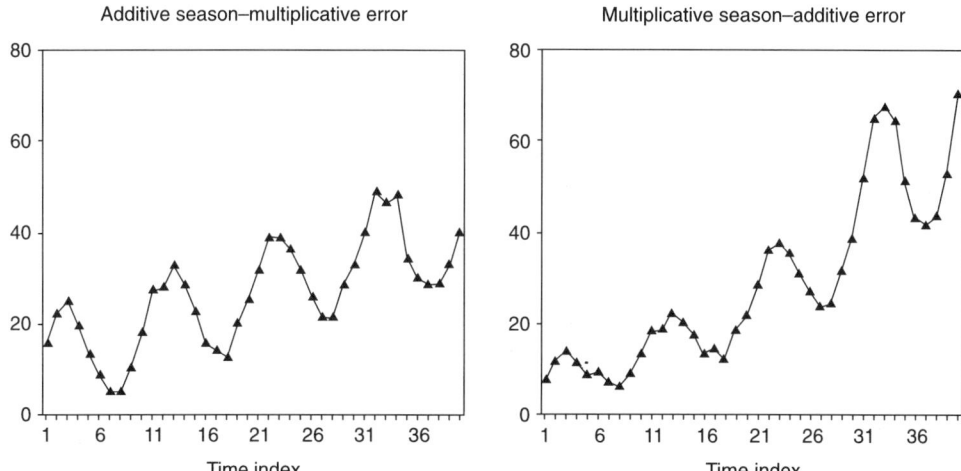

Figure 3.2. Pegels' classification confusion possibilities. The data from Figure 3.1 have been modified by adding or multiplying by 'errors' or perturbations. The left-hand plot shows data with an additive trend and additive seasonality perturbed by a multiplicative disturbance. The right-hand plot shows data with a multiplicative trend and multiplicative seasonality perturbed by an additive error. While the original classifications are still perceptible here, the different categories can easily be confused when the disturbances are only slightly larger.

An issue that may arise in some forecasting situations to complicate matters is that of inaccuracies in measurement or observation. This is, of course, the basis for much of statistics, but in a forecasting context our audience may presume that 'the numbers' are truthful to an extent that cannot be supported by reason. Furthermore, much data that is used in forecasting is provided in preliminary form(s) for which updates are provided when and if they become available. An example is import/export figures for commodities. These are frequently refined as government agencies check the assignment of products to categories and as various ports of entry submit their figures.

Whatever the source of data inaccuracy, we will want to be careful to note whether our forecasts are attempting to predict the actual phenomenon, true but possibly unknown, or else the measured and reported values. It may not be in our interest to attempt to forecast the actual quantities if our managers treat the observed as true. If our sales personnel are rewarded by reported sales, then they are likely to be interested in forecasts for those quantities, not for 'actual' sales adjusted for returns, refunds etc.

Deviations exist then for:

- Forecast vs. actual (usually called errors)

- Forecast vs. observed or measured (usually called residuals), and

- Observed vs. actual, i.e. inaccuracies in measurement (almost always called errors).

A notation for forecasting

In the nearly three decades that we have been associated with forecasting, the notations used by most authors in their presentation of forecasts have almost always left us dissatisfied. Recall our emphasis on documenting one's work. This implies that the practice of labelling the forecast for time point t as $F(t)$, no matter what model or method is used, will provide a source of confusion for both the developers and users of the forecast. We will almost always want to use more than one method for forecasting. Moreover, we will usually develop a sequence of forecasts over time as new data and information become available. Think of yearly sales forecasts. In 1990, we may have projected our sales forward for 5 years. In 1991 we will do so again. There are clearly four overlapping years of forecasts, and it would be very surprising if the same numbers were suggested in the two forecasting exercises.

To overcome such difficulties, we recommend a notation that explicitly includes the following elements:

*   The time point for which the forecast is intended, t ;

*   The time point, T, at which the forecast could have been prepared, that is, the time point at which the most recent data used in the forecast became available;

*   A label giving the particular method used to make the forecast. This should be unique to the method and should include modifications to flag any special conditions that were applied, for example, the deletion of data deemed to be unreliable.

    Example forms of this notation would be

    F_{LABEL}(timepoint of forecast, timepoint when forecast made)

or

    F(timepoint of forecast, timepoint when forecast made, LABEL & documentation)

where the choice of brackets and commas or subscripts/superscripts is largely a matter of taste. We happen to prefer to use brackets, e.g., $F(Label, t, T)$, since such forms can be easily included in plain text files suitable for the simplest of editing programs on computers and also for electronic mailing programs. Caution is needed to ensure that the correct meaning of all the labels and notations is understood. However, we believe such a notation is far more suitable for permitting the easy comparison of forecasts. Note that the LABEL can also be used to identify the author or analyst of a particular forecast. This may be essential if we wish to compare the results of different forecasting groups.

Quantification of the size of deviations

To measure 'fit' between a model and the data it is supposed to approximate, we want ways to summarize how 'large' is the deviation between them. Since we have several, say k, data points in most cases, we need measures that apply to a set, or vector, of observations. Typically, we will have a vector, e, of residuals where

$$e = (e_1, e_2, ..., e_k)^T$$

Note that we use e rather than r to avoid later confusion with the symbol used for autocorrelations. The superscript T (^{T}) denotes matrix transposition, so the vector is a 'tall' one rather than a 'fat' one.

The sum of the deviations e_i is rarely appropriate as a measure of fit, since we must worry about the sign of the deviations. The sum of the deviations divided by their count, k, is the average deviation size. If we overestimate our sales by a million dollars one month and underestimate it by the same amount the next, our average forecast is perfect, but we are unlikely to be considered good forecasters unless we work for a company with hundreds of millions in sales.

Nevertheless, the sum of the deviations or their average may be useful to show how much our forecasts are 'over' or 'under' on average. With the convention that residual = data - model, a positive sum or average implies that we have under-forecast, while a negative measure implies we have overshot the mark.

As in statistics generally, one way to deal with the sign of the deviations is to square them. The sum of squared deviations, or more simply the sum of squares, is perhaps the most common measure of fit used in estimating model parameters. The mean value, often called the Mean Square Error or MSE, allows us to compare forecasts that apply to different numbers of periods. To get this to the same scale as the original data – something we recommend – we can take the square root and get the RMSE, which corresponds to the standard deviation of a data series and the standard error of estimate in linear regression. For reasons we do not fully understand, forecasters often quote the MSE rather than the RMSE.

Squared deviations have the unfortunate property that large deviations contribute disproportionately to the MSE. For instance, if one deviation is twice the size of another, it will contribute four times as much to the MSE. Many approaches have been suggested to overcome this difficulty.

- The sum of absolute deviations retains the original data scale. Similarly, we can compute the mean absolute deviation, or MAD of the vector of deviations.

- A *winsorized* sum or average of squared deviations is computed by dropping one or more of the largest deviations in magnitude from the set of values before the calculation is made. A serious issue in this approach is to decide how many deviations we will set aside. Unless we have supportable reasons for discarding data, we should not do so. A similar calculation to compute a mean results in a *trimmed mean*.

•   Various techniques from robust statistics (Hoaglin *et al.*, 1983) could be considered that down-weight large deviations.

Those with a mathematical background may recognize the vector norms that carry the names L1, L2 and L-infinity. L1 is a vector norm based on the sum of absolute deviations. The L2 vector norm is essentially the square root of the sum of squared elements. L-infinity is simply the maximum absolute deviation.

Rather than treat the deviations equally, we may decide that our forecast should be better in the short rather than long term (or conversely). To suitably measure our forecasts to match such objectives we could compute *weighted averages* or else averages of deviations selected by time period. With any such specialized measure, we should be careful to document what we do, and why. We are trying, like W.S. Gilbert in his libretto for The Mikado, to 'make the punishment fit the crime'. One of our forecasting trainee groups was employed to forecast how much telecommunications infrastructure would be needed in new real-estate developments. Those preparing the forecasts had particular concerns about underestimating the need, since they would lose their jobs in that case. However, they did not want to overestimate, since they would then lose bonuses or possibly suffer a pay reduction. We can easily develop measures of fit that satisfy the interests of these forecasters. For example, we can weight underestimates three times greater than overestimates by forming the MSE of some 'new' deviations:

$$
\begin{aligned}
new\_e(i) \quad &= \quad 3 * e(i) \qquad & \text{if} \quad e(i) > 0, \\
&= \quad e(i) \qquad & \text{if} \quad e(i) <= 0
\end{aligned}
$$

In many real-world situations, we may be more interested in relative or percentage deviations. All the above ideas apply, but now we use

$$p(i) = 100 * e(i) / e_{base}$$

where e_{base} is the magnitude (that is, absolute value) of some reference or base value of the data or forecast. That is, the base value can be an observed or predicted value, or an average of some kind, e.g. we can measure percentage deviation of monthly sales from the average for the year.

How good is a model? Part 1

'Fit', as we have seen, may be defined and quantified in a number of ways, and for different sets of time periods – trial, validation and full. Each measure allows us one comparison between different forecast methods. However, we do not only want good 'fit', although it is generally the most important criterion. We will also be interested in *how* our forecasting models deviate from the actual or observed data. That is, we wish to have a model that is free of any time, or temporal, pattern about the data. If there should be such a pattern remaining, we may have an opportunity to remove it by 'adding' extra features to the model. For example, if our model is always 10% too high, we can simply adjust by that amount and get a better model.

If there is no time pattern, we will then want to look at the distribution of the errors or residuals themselves. The most common pattern we hope to find is the familiar bell curve of the Gaussian (normal) distribution. If indeed we can make an assumption of Gaussian errors about the model, we can generally carry out a number of statistical 'quality checks' of our method, such as the significance of the parameters, no time relationship between the errors, or similar statistical measures that lend credibility to our model and hence to the forecasts generated.

We find that students often write 'the residuals look normal' (i.e. Gaussian) even when they do not! The proper approach is to look for (a) symmetry of the distribution and (b) a mound shape or, ideally, a bell shape using a histogram or Stem and Leaf plot. This is rare in practice, but one often finds data that are 'not extremely skewed' and 'not too dispersed' (not too much of the data in the tails of the distribution). In such cases, you can state that the distribution of residuals over the estimation period *does not rule out* an underlying Gaussian distribution of errors. With the growing availability of capable software, it is easy to prepare the normal probability plot or an equivalent graph (normal quantile plot or normal scores plot), and possibly conduct a statistical test.

If the distribution *is* skewed or otherwise unsatisfactory, then you have to say so. In such cases, the statistical measures that we use to judge the usefulness or suitability of our models must be treated as suspect. We don't throw the model or method away, but are careful not to treat the values computed for statistical tests or confidence intervals, or the conclusions based upon them, as anything like 'truth'.

Examples of tests we will use once a decision that the underlying errors are Gaussian has been made are:

* t-ratios for regression or ARIMA parameters. Typically, an absolute value less than 2 suggests that the true parameter could really be zero, although for small data sets we need to use a more precise measure of the critical value of Student's t. In some cases we can re-try the model without this parameter.

* F–test that at least 1 regression coefficient is non-zero.

* Ljung–Box test that all autocorrelation coefficients up to a given lag are really 0.

In practice, patterns can occur in many ways. We may have a pattern over time when we look at a graph of the data versus time, or a pattern relative to the predicted quantity when we plot data versus this quantity. Some patterns may be complex. In looking at patterns over time, we are concerned that there should be no *serial correlation* or *autocorrelation* – relationships between data at different points in time. This is considered later in Chapter 8.

Having made such a fuss about the importance of the distributional properties of the residuals, we must also admit that students sometimes get concerned that they do not have 'nice' residuals when they have found a model that fits the data extremely well. In such cases, the residuals are very small relative to the size of the data. Indeed, for data that is not near zero in mean value, the percentage error will be small. The computation of the residuals, that is, the subtraction of model from data, is almost certain to be cancelling digits. For example, if the data value is 234.5 and the model is 232.6, the residual is just 1.9. Note that the first two digits of the data have been cancelled. Our residuals will often have only one or two digits of precision, so they are rather like integers. And integers cannot be distributed like a *continuous* Gaussian random variable. We need not worry too much about this – our model fits well. The purpose of the statistical properties and tests is to help us to deal with the variation that is left when the fit is not very good.

Desiderata and parsimony. Part 1

Our list of desiderata for forecasting models is getting longer.

1. Small residuals over the estimation period, whether for the trial or main model.

2. Zero mean residuals over the estimation period.

3. No 'pattern' left in the residuals over the estimation period (more details about this will emerge later).

4. Small errors for validation forecasts.

5. Zero, or at least small, mean for validation forecast errors.

    On top of all these, we would like:

6. As simple a model as possible, either in terms of as few parameters as possible, or in the sense that we can use and explain the forecast method easily. Models with few parameters are said to be 'parsimonious'. We know of no technical term for models that are easily explained and used, but note that time series decomposition models developed by the ratio-to-moving-averages or Winters' techniques (Chapters 10 and 11) are *not* parsimonious but are quite easy to use. Simplicity is in the eye of the beholder.

It is worth mentioning, while we have our objectives before us, that modern software tools often allow us to try to achieve these objectives directly via optimization. The Excel and Quattro spreadsheet packages we have used include optimization, but other spreadsheets may not. Such tools provide one way to fit models, even including constraints, although the underlying optimization codes have some weaknesses. We provide some details in Chapter 20.

Exercises

E3.1. The following data give a quantity for 25 consecutive periods.

```
46.1   48.4   50.9   53.6   56.5   59.6   62.9   66.4   70.1   74.0   78.1   82.4   86.9   91.6
96.5   101.6   106.9   112.4   118.1   124.0   130.1   136.4   142.9   149.6   156.5
```

A model for this data is   `data = 32.3 + 4.60 t`, where t takes values 1, 2, ..., 25. Compute residuals for this model and discuss whether the model is satisfactory or not.

E3.2. On the PFM Web site we provide data (E3-2dta.txt) for a forecasting problem with two sets of forecasts for this exercise. Which of the sets of forecasts do you prefer and why?

E3.3. For discussion: when should you prefer percentage error measures to absolute ones?

E3.4. Using ideas from Table 3.1 and any software with which you are familiar (we suggest a spreadsheet), supply some trial values for parameters in the various trend and season models and attempt to draw the graphs of the Pegels' classification. In this book:

0. We used $n = 40$ observations.
1. A set of Gaussian random numbers with zero mean and unit variance was generated.
2. We set the additive disturbance 'aerr' to be 1.0 times the Gaussian random numbers.
3. The 'time' variable was set to be a count from 1 to n.
4. The stationary series 'staser' has all elements equal to 20.
5. A linear trend series was created as addtr=10+0.8*time.
6. A multiplicative trend series was created as multr=7.5*(4^((time-1)/24)) The symbol '^' means 'to the power'.
7. A seasonal pattern was created as ssn=sin(8*atan(1)*time/10)
      Atan(1)= $\pi/4$ radians or 45 degrees, that is, an angle whose tangent is 1.
8. A multiplicative error was created as   merr=1.05^aerr.
9. An additive seasonality pattern was created as asn = 11*ssn.
10. A multiplicative seasonality was created as msn=1.45^ssn.
11. The stationary (no trend) Pegels series were created as follows:
      a1 = staser;   a2 = staser+asn;   a3=staser*msn
12. The linear (additive) trend Pegels series were created as follows:
      b1=addtr;   b2=addtr+asn;   b3=addtr*msn
13. The multiplicative trend Pegels series were created as follows:
      c1=multr;   c2=multr+asn;   c3=multr*msn

Each series was plotted against time, and the nine graphs put together in an array. **Warning**: if you have a lot of data, your word processor may happily import the graphs and refuse to print them! We spent over two days' worth of work with this 'little' problem.

14. For graphs with additive error, simply add 'aerr' to each of the nine series and replot.

15. For the multiplicative disturbance, multiply each of the nine series by 'merr' and replot.

4

Data search, gathering, documentation and management

Data sources. Data collection and validation

Very little of what is discussed in this chapter is traditionally found in books on forecasting. Unfortunately, it is also rarely found elsewhere. The knowledge and skills related to data finding are not usually included in the program of study for most degree programmes, the primary exception being library science, where forecasting is not on the menu. In our experience, however, data finding and gathering is one of the most difficult aspects of forecasting. We have always included at least one formal lecture and several discussion sessions on finding data in our courses, perhaps because one of us (MN) has formal training in library science.

Traditional sources of data

Public, university and government libraries are perhaps the most common traditional places to seek forecasting data. While some of these libraries have reasonable subject catalogues, we generally have not found these catalogues to be our approach of choice to finding forecasting data. Instead, we are more likely to focus on government or institutional document collections where data pertinent to situations we are investigating are likely to be found. This reflects our desire to find *numerical data* rather than to learn about a subject in general.

To get *descriptive* information about a subject, it is clear that a subject catalogue may be helpful, but rather than using the general catalogue of a library, we may instead look to see if there are existing bibliographies related to the subject we are interested in. In the past, these were typically produced on paper or microfiche (e.g. Nash, 1976b). More recently, the growth in online bibliographic services has seen a decline in published bibliographies, although occasionally we have seen specialized reference lists done by libraries and posted on the Internet. In using such bibliographies or online bibliographic services, our data search is a two-stage process:

- Find the references to the sources of data;

- Find those sources of data, if they are available to us.

For both numerical and descriptive data, we will be interested in recent as well as historical information. The latest material is likely to be found in periodicals and journals, most of which will not have subject catalogue entries that will lead us to them. Once again, we need to find the pointers and then the data sources themselves.

There are a number of ways to find references to periodicals. Some guides that have long been available in paper form are:

- The Reader's Guide to Periodical Literature, published in the USA, covers mostly periodicals from that country. You may need to consult more local aids; in our case we use the Canadian Periodical Index.

- Engineering Index contains references to scientific and engineering articles from around the world.

- Social Science or Science Citation Indexes are useful in tracking down material related to items we have already found.

Most of these are now available for searching in electronic form. If you are affiliated with a university or college, free access to a wide selection of online databases may now be available on CD-ROM or via the Internet. There are also several online fee-based bibliographic information services offering access to a wide selection of databases, and most libraries subscribe to one or more of these. There is, as already stated, usually a fee for use and it can be relatively expensive to do so, especially if one does not know how to interrogate the database efficiently. Some of the databases we have used over the years are:

- Magazine Index (Information Access Company) this includes full-text of approximately 100 titles. (We used the paper version of some years ago.) There is a wide range of publications.

- Compendex is now called Ei Compendex*Plus and provides abstracts and full bibliographic citations for worldwide engineering and technical literature and encompasses all engineering disciplines, as well as related fields in science and management. The records in the database are drawn from over 2600 published journals, conference proceedings and individual conference papers, technical reports, monographs, and other materials.

- Medline, which is produced by the U.S. National Library of Medicine. The MEDLINE database is widely recognized as the premier source for bibliographic and abstract coverage of biomedical literature. MEDLINE encompasses information from Index Medicus, Index to Dental Literature, and International Nursing, as well as other sources of coverage in the areas of allied health, biological and physical sciences, humanities and information science as they relate to medicine and health care, communication disorders, population biology, and reproductive biology. More than 9.5 million records from more than 3900 journals are indexed, plus selected monographs of congresses and symposia (1976–1981). Abstracts are included for about 67% of the records.

- SciSearch. Among other citation indexes, the Institute for Scientific Information now publishes the Arts & Humanities Citation Index, which provides access to current bibliographic information and cited references. The Arts & Humanities Citation Index covers more than 1140 of the world's leading arts and humanities journals in a broad range of disciplines. This coverage includes articles, letters, editorials, notes, meeting abstracts, discussions, errata, poems, short stories, plays, musical scores, excerpts from books, chronologies, bibliographies, and filmographies; as well as reviews of books, films, records, art exhibits, television and radio programs, and dance, musical and theatrical performances.

The second stage to finding data is tracking down the references gathered in our search. We can start with the catalogue of the library we are using, and our references will probably have author and title information to speed this search. Often, of course, our library will not have the item we are looking for. This is increasingly common as libraries shed less-used periodicals that come with high subscription prices or cannot afford to buy all the latest books. This is not necessarily the end of the road, however, as libraries have cooperative arrangements to assist their readers to obtain material through Inter-Library Loan (ILL). Among the most helpful tools to determine where material is, are *union lists*, which are listings of the aggregated document holdings of groups of libraries. National libraries throughout the world created and maintain such lists or now act as gateways to the catalogues of other libraries on the Internet. We will discuss various national gateways further in the section on Internet data sources.

If we have a known source of data, we may look for articles or documents that have made reference to it. This is the use of a *Citation Index*. The process now involves three stages based on our known 'good' item. First, we use the citation index to get a list of articles that have referenced the known item. Then we use the *source index* to check the title and possibly other information about these articles. For example, we may be able to tell from the title that some articles are of no interest to us. Others may be in obscure publications that we will not be able to access, or they may be by authors whose attitudes we do not share. We may even find some articles we have already seen and used and can therefore ignore these as well. Finally, we must find the articles that have cited our known item. If you are familiar with your library's collection, you can save time by first locating items in books and periodicals available there and then trying to find other items through ILL.

In the same spirit, we find it quite useful to browse quickly journals or books to which searches have pointed us. Often there will be related articles, commentaries, or otherwise useful information. Some of this will lead in unhelpful directions, but we may still pick up some useful ideas or terminology that will aid in future searching.

Library resources on the Internet

The Internet gives access to data sources in ways that were not possible before. We examine some of these below, but first will look at the use of the Internet as a tool to access traditional libraries and library materials, possibly in non-traditional ways. An electronic or e-journal is a journal that exists only in electronic form. Electronic journals can include articles, letters and news sections. Many traditional scientific journals exist in parallel electronic form, for example, those of the Society for Industrial and Applied Mathematics (SIAM). In our own case, we have benefited from the electronic access provided by the University of Ottawa Library to some forecasting and related journals through its Polaris and Orbis systems.

Entire national collections are available to be perused online. For the forecaster, it is important to be able to access national information, since the context of most business forecasting is, despite globalization, still regional. For example, in the humanities and social sciences area, the National Library of Canada (www.nlc-bnc.ca/)gives free access to over 2 million brief records on items in its collection. For a modest monthly fee, the library's AMICUS system provides access to 20 million bibliographic records and 562,000 authority records. It also gives information on 35 million holdings in more than 500 Canadian libraries. On the natural sciences and engineering side, the Canada Institute for Scientific and Technical Information (CISTI) provides a wealth of information through its catalogue and associated services. CISTI has a collection of over 50,000 serial titles and more than 600,000 books, reports and conference proceedings in science, technology, engineering and medicine. The CISTI Catalogue also includes the records for the Canadian Agriculture Library (CAL), Main Library. CAL's collection of over 30,000 serial titles and 60,000 books, reports and conference proceedings provides access to worldwide agricultural information. You can find all this information at www.nrc.ca/cisti/.

Many of the resources of the Library of Congress in Washington, DC are also available on the Internet at www.loc.gov. The Library of Congress Online Catalog (catalog.loc.gov) is a database of approximately 12 million records representing books, serials, computer files, manuscripts, cartographic materials, music, sound recordings, and visual materials in the Library's collections. The Online Catalog also provides references, notes, circulation status, and information about materials still in the acquisitions stage. In addition, this website gives access to the catalogues of many other libraries in the US and worldwide.

For United Kingdom materials, the British Library (BL) is the prime resource. The Library's services are based on its outstanding collections, developed over 250 years, of over one hundred and fifty million items representing every age of written civilization, every written language and every aspect of human thought. At present, individual collections have their own separate catalogues, often built up around specific subject areas. Many of the Library's plans for its collections, and for meeting its users' needs, require the development of a single catalogue database. This is being pursued in the Library's Corporate Bibliographic Program which seeks to address this issue.

At present, however, OPAC 97 (Online Public Access Catalogue 97) can provide information about the contents of a number of the Library's major collections. When researching a particular topic it is advisable to consult with the collection areas involved to ensure that you have discovered all possible resources since some may only appear in printed catalogues or indexes. Many areas have their pages on Portico or can be contacted by e-mail, fax or telephone. The BL website is at www.bl.uk.

Blaise is The British Library's Automated Information Service, providing access to 21 databases containing over 18.5 million bibliogfaphic records. Blaise is an online information retrieval service that includes access via a new user-friendly graphical interface on the World Wide Web. A direct online link to the British Library Document Supply Centre, the world's leading supplier of documents, allows you to transmit requests for individual items quickly and easily.

Also from the UK, *Current Serials File* is a service that allows you to search the DSC's Current Serials Received file. This file contains the titles of over 60,000 serials currently received by the British Library Document Supply Centre (BLDSC) and Science Technology & Business (STB) and believed to be current in 1999.

Gabriel is the World Wide Web server for Europe's National Libraries represented in the Conference of European National Librarians (CENL). Gabriel aims to help bring national libraries in Europe closer together by providing a single point of access for the retrieval of information about their functions, services and collections.

Inside offers a fully integrated current awareness and document ordering service, via the Web, which allows you to search, order and receive documents held at the British Library. You can search 20,000 of the world's most valued research journals and over 70,000 conference proceedings at paper title level. *Inside* is a massive database and expands by over two million articles every year.

The National Library of Australia, accessible at www.nla.gov.au, is responsible under the National Library Act 1960 for preserving a comprehensive collection of documentary materials relating to Australia and Australians. Their Australiana collections have developed into the nation's single most important resource of materials, recording the Australian cultural heritage. They also have considerable collections of general overseas and rare book materials, and world-class Asian collections, which augment the Australiana collections. Besides being able to search the catalogue, the library also makes available Australian electronic journals, free of charge.

A union list of sorts for Australian material is provided by a service called Kinetica, which provides access to the national database of material held in Australian libraries. You can search for any item and locate which library in Australia holds it. Gateways to other major library databases are provided. It is available from 0600 to 2400, 7 days a week.

The following databases can be searched:

the National Bibliographic Database (NBD),

- the Australian National Chinese, Japanese, and Korean Service, and

- RLIN: the Research Libraries Information Network, based in the United States, which includes access to the Library of Congresss Name Authority (LCNA) and Library of Congress Subject Headings (LCSH).

While there may be similar facilities for non-English language countries, we have yet to use them, so will not discuss them here.

Other Internet or machine-readable information sources

Beyond the confines of libraries, there are many information sources available on the Internet that may be useful to forecasters. Sadly, there are many distractions and false leads, along with false information, when we use search engines such as Alta Vista, Excite, Lycos, Google, or others. There are several reasons for this:

- The 'It's free!' mentality of the Internet has encouraged an advertiser-pays model for revenue generation. Therefore, site creators embed popular search keywords in their content. On the Alta Vista Search Engine on 20 April 2000, 'Future' had 13,851,479 hits, while 'forecast' had 934,339. Even the phrase 'exponential smoothing' found 2,622. Requiring that 'exponential smoothing' be NEAR 'software' reduced this to 52, and specifying that we only wanted items in English after 1 January 1999 left us with 27, most of which related to university courses in Management Science.

- Even with search terms that are relatively focused, we may end up with items well outside our zone of interest. Both psychology and mathematics have an interest in 'group theory', but the groups are very different!

- Automatic searching is indiscriminate, even with the best of software. Again, we waste a lot of time. This is where a reputable, fee-for-service index service would be valuable, but the 'free' nature of many so-called services on the Internet has not encouraged the use of indexing services.

- At the other extreme we may find highly relevant information, but the purpose of the site is to sell it. For example, *The Forecasting Report: A Comparative Survey of Commercial Forecasting Systems,* Pages: 646, Published: September 1999, Price: Euro 1450.00 , US$ 1595.00 (plus Tax). Actually you can get it cheaper by the dozen! And a document with a similar description and publication date is available from the magazine *OR/MS Today* and is on the publisher's web site (www.lionhrtpub.com).

- The 'search engines' available on the Internet (at least the freely available ones) make their money from advertising, so their goal is to keep the advertisements in front of your eyes as long as possible. As a searcher, you want to minimize the time searching. Is it surprising that the typical Internet search is slow?

Thus, we can only reluctantly recommend a 'search from scratch'. More useful, we feel, are those Internet sites that have collected pointers to sites that relate to a topic. There are several of these for forecasting, many for statistics and various aspects of computation and software. Similarly, in subject areas, there will be sites that contain data useful to our tasks. Because the Internet is dynamic, we will put our own collection of links on the PFM Web site.

Internet information may be obtained online through the World Wide Web, through Newsgroups and list servers, through a terminal connection (Telnet), or as files that can be downloaded using the File Transfer Protocol (FTP). Most users now use a single browser such as Netscape or Internet Explorer as a single tool for access, although specialized tools are often more efficient. Despite some annoyances, the availability of data, especially in a form that is machine-readable, is attractive. Our opinions on Internet data sources are:

- The Internet often has useful information or pointers leading to such information, but finding it requires time, care and effort.

- There are unlikely to be good indexes or search aids for our particular forecasting purposes.

- People putting information on the Internet want you to see their site, whether or not it is relevant.

- We recommend downloading and saving relevant files and archiving them with a forecasting project as we encounter them. This is quite a lot of work – we need a catalogue or index of our downloads, even handwritten! Some tools exist to download whole sites, but they tend to grab too much material.

- References to Internet sources are troublesome, as the source can disappear, either temporarily or permanently.

- Time series are rare – especially for *your* topic!

In teaching, we are willing to give bonus marks for well-documented uses of Internet facilities to find forecasting data, methods or forecasts, but have not yet awarded any bonus marks. A number of interesting 'finds' have been made, but more by accident than design. Much material has been downloaded from the Internet, but a high proportion has been found as a result of non-Internet search. Electronic mail exchanges with colleagues have often played a role, as have some specialized list-server broadcasts. Clearly these latter 'tools' are a continuation of personal contacts and communications that are the essence of collaborative research.

Examples of Internet searches

To illustrate the type of information that can be gleaned from the Internet, we undertook a search (November 1998) using the Alta Vista search engine (Compaq Computer, formerly Digital Equipment Corporation). Alta Vista allows a phrase to be entered. We used this facility in Table 4.1. Note that we limited the search to our own geographical area in order to illustrate typical searches needed for real business and administrative forecasting. Our message is this: finding information *appropriate* for a given forecasting task is almost always difficult.

Table 4.1. Example results of Internet searching

Topic	Search Terms	What we found
AIDS Cases and Deaths	*aids & deaths*	AIDS deaths in different countries from various sites
Smoking habits (Canada)	*smoking & Canada*, with some refinements	1994 Health Canada survey, but VERY difficult to find time series
Ownership of housing (Canada)	*housing in Canada*	housing starts 1998 / history of Cooperative housing
Student population in Canada	*student population in Canada*	no relevant information found
Homeless people in Canada	*homeless persons in Canada*	National Welfare Council Study, homeless in Calgary
Weather variables for Ottawa	*climatology and Ottawa*	no relevant information found
Population of Calgary, Alberta	*demographics Calgary*	Calgary Economic Dev. Authority
Gambling industry in Canada	*gambling industry in Canada*	National Welfare Council report, 1996

Other machine-readable data sources

While the Internet is now the most popular information source for our students, other forms of machine readable data exist. Traditionally, large-scale statistical information was distributed on magnetic tapes, now superceded by CD ROM. The existence of inexpensive writeable CD or similar high-capacity media, however, makes it possible for data to be archived and distributed in this way. Some commercial and government CD data products applicable to forecasting exist, but given the rapidly changing nature of the offerings, we will not specify them here.

For teaching and learning purposes, of course, the 'CD in the back of the book' has replaced the diskette (first 5.25", then 3.5") as a vehicle for data sets and software. Now web sites are more common, as they allow material to be updated and the production costs related to the CD and the carrier pouch are avoided.

Because of the possibility that web sites 'move' (that is the URL changes), change in content, or disappear, we feel a secondary use of high-capacity media is to capture material as it is found. We have already noted the difficulty of documenting and cataloguing such captured information. Now we turn to the issue of citing our sources in reports.

How to reference sources

A skill that seems to be left out of many educational programmes is that of correctly citing references to sources you have quoted or consulted. Given this lack, we will briefly make some suggestions about how references should be presented. For those interested, there are style guides such as the Canadian Style, the Chicago Manual of Style and the MLA Handbook for writers of research papers. This last one is particularly geared to university students. Here we present the style and content that we like to see in reports submitted in our courses.

*   This book (Nash and Nash, 2000) was written by John and Mary Nash (2000). The previous sentence shows two ways to reference it. More text follows, etc.

    For three or more authors, use: Firstauthor, I. N. *et al.*, al though some writers may prefer to list all the authors in the bibliography, while the citation is kept short.

*   How should a referenced work appear in the bibliography?

    First, items should be in the proper alphabetical order, by author, and have the name, date, title, publication, volume and number, and pages. For example,

    Koplan, J. F. (1986) Assessment of new vaccines in immunization programs. *Israel Journal of Medical Sciences*, vol. 22, no. 3/4, pp. 272–276.

    If an author has more than one article in one year cited, we can use the form:

    -------------- (1986a) etc.
    -------------- (1986b) etc.

    Rather than the ellipsis '---------', you may simply prefer to repeat the name. The repetition makes life easier for the writer when using word processing and moving items around, and for the reader when the list crosses a page break.

    If there no author, fill the author field with 'Anonymous' (if a journal or periodical article) or repeat the publisher or publishing organization (if a book), for example,

- Microsoft Corporation (1983). *Microsoft BASIC reference manual*. Bellevue, WA: Microsoft Corporation.

- In our courses we tell students 'DO NOT USE footnotes, ibid's, op cit's, etc.' Few of us really know the Latin well enough to use them correctly.

When preparing a bibliography, there are always situations that give trouble. The main issue is that the bibliography should be as *consistent*, *complete* and *accurate* as possible. Here are several ways to deal with 'problems':

- No date of publication. Example:

    Nicholas *et al.* Is Poliomyelitis a serious problem in developing countries? *BMJ*, April ???, pp. 1005–1014.

- If you do not know the standard abbreviation for a journal title, write it out completely!

- Wrong or ambiguous journal title. For example, 'Spectrum' and 'Liaison' have been the names of many publications. (One of us has been Editor and Managing Editor of one such publication, and dearly wished it had a different name!) If possible, provide the name of the publishing organization, such as

    Nash, J.C. (1992) Editorial: Guns and Numbers / Des Chiffres et des Armes, (Statistical Society of Canada) *Liaison*, vol. 7, no. 2, pp. 3–4, October.

    Do disposable lenses solve the problems of extended wear?, *Spectrum*, April, 1989.

- Singular and plural nouns can cause confusion. Examples:

    Various speakers (1982). Cost benefit analysis of vaccines (a discussion). *ReviewS of Infectious DiseaseS*, vol. 4, no. 5, Sept–Oct. pp. 978–983.

- Incomplete page lists. Example:

    Martin, J.F. (1984). Consequences of the introduction of the new inactivated polio virus vaccine into the expanded program in immunization. *Reviews of Infectious Diseases*, vol. 6, suppl. 2, May–June. pp. S481, S482.

- Electronic references: Internet links are a very useful tool while online, but frequently useless in a bibliography. This is because many source documents are moved or deleted before the report you are preparing is finished. Worse, a trend to the use of on-the-fly creation of 'documents' means that the reference is a one-time wonder. Any reference that includes the string 'cgi-bin' or '.asp' (for Active Server Pages) is unlikely to be permanent, since it refers to programs that run on an Internet server. If possible, try to find references to the paper document and make reference to it, though you may wish to add 'found at http://...' where you consider it likely the link will remain helpful to readers. We also note that the form of links that first appeared were FTP (File Transfer Protocol) addresses. Later the gopher protocol emerged, followed by the URL (Universal Resource Locator). However, the pattern of change we have seen is apparently continuing. At the time of writing, DOIs (Digital Object Identifiers) are a fashionable topic. These are identifiers that allow the author some control on the manner of use; for example, to demand payment or to allow only viewing, but not extraction or quotation. As our only references to this topic are electronic, we will not labour you with them.

We strongly recommend against the use of 'Private communications'. There is simply no value for the readers in seeing that there may be a source of information that is denied to them. Sometimes we are forced in forecasting to use data from sources that cannot be verified. In such cases, where possible, it is helpful to readers to include the actual data in your report, which if published, becomes the source for others, despite potential weaknesses. If we are barred by conditions of use from including data or a usable reference, then the source and use of data are best described in the text of your report, without attaching a reference.

Students often include many appendices to their reports, such as photocopied source material. For the professor, this is generally a minor nuisance. We don't read the appendices. Unless the material is unusual or hard-to-find, we are unlikely to be interested.

Finally, we sometimes see both a bibliography and list of references in a report. They should be the same, and only one should be needed, unless one of the lists (the bibliography) is to material not mentioned in the text but included for further information.

Data edit and imputation

There are occasions when data is missing or 'obviously' wrong. 'Obvious' is a slippery concept. Make sure you rigorously justify any changes to published data. Moreover, ensure that your changes are well-documented so that the process and outcome of any editing is available to both yourself and others. We find that this demands a great deal of discipline, and is especially onerous when there are deadlines to meet, either for students or workers. And after the deadline, it is rare that anyone carries out the documentation, so altered data remains in files to trap the unwary, including the person who originally did the changes, in later forecasting exercises.

Nevertheless, we will sometimes need to massage our data so that it is suitable for forecasting. Let us consider some of the situations, then examine a few possibilities for actually carrying this out.

• Missing data are a nuisance when our methods require regularly spaced observations. Most forecasting methods require such equi-spaced data, including ARIMA modelling, time series decomposition and the smoothing methods, although there are some variants of these methods that may be suitable on occasions for irregularly spaced data. If we regard a missing point as providing a double-length time gap, we can treat the situation with a method for irregular time spacing. Usually, however, we will want to provide or *impute* a value for the missing observation. That is, we devise a mechanism to compute a value for the period where data are missing. Clearly we should be able to justify and defend the chosen mechanism.

- Outliers, especially those due to known, documented events, will distort models and forecasts. We will want to *edit* the values. This is essentially removing the observation and providing an appropriate (imputed?) value. The example below suggests how this may be done.

- In many forecasting situations, we can find data for time periods or jurisdictions that do not match those for which we wish to forecast. The boundaries of time or space we want to consider are different from those for which data are available. This requires us to *aggregate, disaggregate* or *interpolate* to provide suitable observations for forecasting.

- There are some situations where the data we want just is not available – the number of homeless people in a city – and we are forced to seek *proxies* or surrogates for the 'real' data to allow forecasting to be attempted.

Methods for edit and imputation have generally emerged in the context of large national surveys, typically in government statistical agencies (Sande, 1982), and in scale and perspective are beyond the scope this book. We can, however, suggest that our readers use simple but justifiable approaches. Of these, simple interpolations are perhaps the most useful. This is a very old technique for filling in blanks in tabular data. See the Appendix for some details of interpolation calculations.

Example

Suppose that an observation in an otherwise complete time series is missing or 'known' to be unreliable; for instance, sales data for areas in Canada affected by the January 1998 ice-storm. To provide a value for the January 1998 period, we could first graph our available data and if there is a relatively smooth pattern month by month, we could average the values for December 1997 and February 1998 and put the result in the January 1998 position. If the data is seasonal, then averaging January 1997 and January 1999 values may be a suitable imputation. Whatever we decide to do, we should be sure to include documentation of our actions in our data set.

Data documentation – metadata

We will, throughout this book, preach many sermons about the necessity of documenting data and what we do with it. Good forecasts need good understanding of where data comes from and what has happened to it on the way to the forecast report. The 'data about the data' are called *metadata*, and pose one of the major challenges to all data analysts. Particularly when comparing or combining data from different jurisdictions, we need to know the context and the transformations (Fairgrieve and Brannen, 1998).

We find it best to keep the metadata attached to the numerical data in some way. When working with Minitab, we prefer to have our data as executable scripts. These have the file extension MTB. They are plain text files. Since there is a NOTE command, we can include the metadata via comments in such note lines. We urge that the name and contact information of anyone altering the data set be included along with a description and rationalization for the change. In spreadsheets, we can easily include metadata information by placing the commentary in cells near the data.

We note that many of the standard files for Minitab and similar packages frequently lose commentary. Recent versions of Minitab do have provision for commentary in 'Project' files, but one needs to use Minitab itself to see them, while the plain script MTB files can be viewed, edited, and imported into other software.

The trouble with metadata is that it is a big nuisance to include it and update it. We have encountered descriptions of systems that will automate part of the work, and believe that large statistical agencies have systems designed to keep track of changes and updates to data. Such tools are unlikely to be generally available to our readers in the foreseeable future. The only alternative is planning and discipline, which take effort. Worse, they require *continuing* effort, since modern computing facilities make it very easy to highlight and copy data, then paste it elsewhere. What data have we then copied and used for our analysis? Was it the most recent data? Or last year's figures? We know of a case where a government publication was printed and distributed with the wrong year's figures. Even though there was a 'recall', we can anticipate that this mistake had continuing consequences, as it is almost certain that not all the faulty publications were recovered.

Time point conversions

An issue that is treated as 'trivial' in some forecasting texts is the appropriate adjustment of the time index when we are working with seasonal time-series data. In our experience, this gives those learning forecasting a great many headaches. The issue of converting time indexes is an occasion of many mistakes that wreck otherwise satisfactory forecasting efforts. (We use 'indexes' here, but others prefer 'indices'.) The central manipulations are not difficult, but they are tedious and require attention to detail. We have not seen this topic covered in other books.

Consider a set of time series data organized as a single column. Generally we *index* the series from the first to the last observations, with the first labelled as 1 and the last as n. We often call this index 'time', but it clearly is not expressed as a usual time or date. For example,

```
Time           Data
 1            12.34
 2            16.51
   .  .  .
 25            45.32
```

In our work, we may need to consider a given observation, say the kth and decide which year, y, and which season, s, it belongs to. We have seen many, many mistakes made where an observation was put in the 'wrong box', that is, assigned to the wrong year or season. Moreover, people do not remember major events as occurring in 'period 23'. They recall the 'Summer of 1945' etc. We will, as indicated in the next paragraph, use a number to index the season so that it is easier to compute. Once we have the season number, we can, of course, look up or print out the name of that season.

Before we can do the manipulations, we need to know the *periodicity* or number of seasons in a year for our data. We will label this L. We will label our seasons 1, 2, ..., L. Thus quarterly data has $L=4$. We also need a convention to number the seasons. For many situations we will use season 1 for January to March, season 2 for April to June, season 3 for July to September, and season 4 for October to December. This is, however, merely a labelling convention we have applied. *Be very careful with series where the 'seasons' have special limits.* Examples are fiscal years that start at non-calendar year boundaries. As far as we can determine, the University of Ottawa uses a fiscal year 1 May to 30 April but an academic year 1 July to 30 June. There are plenty of traps here for the unwary. The Appendix gives more details on time point conversions.

Exercises

E4.1. Re-run the searches of Table 4.1 for your own geographic region. (The authors would be interested to hear of any particularly interesting or difficult searches and their results.)

E4.2. Draw up a search plan for one or more forecasting projects, such as those prepared in Exercise 1.2.

5

Qualitative forecasting: long term

Long term forecasting of qualitative variables or scenarios poses a different challenge to the forecaster. Much of the published material in this area has been the domain of those who call themselves *futurists*, that is, people who aim to describe one or more possible futures for an industry, country or society. Some of the forecasting has serious underpinnings, but there is plenty of fantasy and wild conjecture, and a good deal that is a mixture of both types of effort. In our own opinion, there is worthwhile material to be found in *The Futurist*, which is the house journal of the World Future Society, www.wfs.org. However, the reader must carefully question the methods and credentials of the authors of articles, that are clearly written for a general audience. Examples include such topics as:

- 'Nine global trends in religion' (Sellers, 1998)

- 'Food scarcity: a global wake-up call' (Brown, 1998)

- 'Future View: why we simulate' (Branch, 1998) – the author of this article about one aspect of forecasting is a Distinguished Emeritus Professor of Urban and Regional Planning at the University of Southern California.

- 'The future that has already happened' (Drucker, 1998) – an article by the famed management 'guru', excerpted from Stone (1998).

We are less convinced of the reliability of prognostications by the self-named Dr. Tomorrow (www.drtomorrow.com) . Nor do 'pop trendy' or 'dire warning' books generally convince us that their authors have done their homework, or indeed that they have taken their medication before writing! Even serious attempts, such as *The Great Reckoning* (Davidson and Rees-Mogg, 1991) can be too much in love with their own predictions. Minkin (1995) at least has the sense to simply ask prominent industry players for their opinions. Thus, the 'forecasts' are focused on particular areas of technology or industry. Another credible effort is that of Foot (1998; the comments here relate to the first edition), who applies demographics to forecasting economic and social evolutions. We caution, however, that Foot's book does not give full details of how demographic methods are applied quantitatively rather than qualitatively or semi-quantitatively. Indeed, it is a book for a general readership and the methods are relegated to two appendices:
Appendix I: Product and Activity Forecasting – How to forecast just about anything.

Appendix II: Demographic Forecasting – How demographers think about the future.

Nevertheless, long-term forecasting is important and we will look at some of the methods used for such forecasting. Our own experience is that, in business and administration, this sort of forecasting is rarely formalized. We do not believe that managers should necessarily practice the detailed application of the methods discussed in this chapter as a regular activity. However, we strongly recommend that managers regularly and consciously set aside time to consider the evolution of their business, its technology and culture. Changes of a qualitative nature are often quantum jumps. You do not want to be a manufacturer of buggy whips when everyone is driving an automobile unless you have a very specialized and secure market for your product. In a very general way, of course, the modern obsession with a 'business plan' is closely linked to medium- and long-term forecasting.

Our own categorization of the methods of forecasting, and indeed the way we have chosen to group methods into quantitative, qualitative and semi-quantitative methods, clearly involves personal choice and taste. Other workers, with different backgrounds and experience, have different viewpoints that are no less valid and useful than ours. In the arena of long-term forecasting especially, considering different approaches and viewpoints is important. We recommend, for example, the Principles of Forecasting web site directed by Scott Armstrong at www-marketing.wharton.upenn.edu/forecast/. (*www[hyphen]marketing…*) It contains a decision tree for choosing forecasting methods that is quite different from any we would draw, yet we can also say that there is absolutely nothing 'wrong' with it. Such is the field of forecasting.

Expert opinion

One of the most common approaches to qualitative forecasting is to ask someone who is (supposedly) knowledgeable on the subject of interest. The difficulty, of course, is that nobody knows the future. We depend on the knowledge, experience and, most of all, wisdom of the experts we consult. How can we judge their competence?

- Has the 'expert' a documented record of success in providing forecasts of the type we seek? Are these documents open to public scrutiny and/or peer review?

- Does the 'expert' have knowledge and experience in the particular situation in which we are interested? Industrial patterns, business cultures, laws and regulations differ from jurisdiction to jurisdiction. Agribusiness is quite different from semiconductors.

- Does the 'expert' provide arguments that pass tests of reasonableness? That is, are the scenarios suggested realizable?

- Besides subject knowledge, is the 'expert' familiar with forecasting and its processes?

Our opinion is that in most industries, the 'experts' consulted are almost always those who are currently regarded as successful. They do not necessarily have experience with forecasting, but have in some way (usually money) been in the right place at the right time. Possibly they have managed this by good forecasting, but since huge numbers of people are actively seeking business success, it is very difficult to distinguish good planning from good luck.

We further believe that any credible expert can provide not only a forecast of future situations, but also the pathways by which they are reached, along with the key events that will promote or hinder developments. We return to these ideas under Cross-Impact Analysis below.

As a cautionary note, we do *not* believe that even the best experts can reliably predict when their forecasted events will occur. Timing of events is, as we have indicated before, extremely tricky to predict with any accuracy in most applications. We should still press our experts to provide such estimates, but consider them with great care.

Panels of experts

If one expert is good, two or more must be better! That is the motivation for using not one, but a panel of experts. Frequently, conference organizers will arrange for such a panel to present their views on the evolution of the subject of the conference; for example, 'The evolution of the role of statisticians in government policy making'. There are even whole conferences devoted to examining the future. Agriculture Canada used to run an annual Agricultural Outlook Conference in the late Fall of each year where various commodity experts, individually and in panels, would present their projections for prices and production for the coming year.

The Outlook Conference, however, had a rather interesting motivation. As one colleague said to us once: 'I give my forecast for the coming year and hope that I'll be wrong'. What he meant was that if he anticipated that too few producers were going to plant crops or breed animals, he would project a high price to encourage more activity, which would actually lower prices and make his forecast wrong. If too many producers were in the market, he would predict a very low price, some producers would change to different products or stay out of the market and again he would be wrong.

Panels usually offer more than one opinion, as they should if they reflect a range of expertise and interest. However, they have their own difficulties:

- Personalities may become more important than ideas. Strong-willed panellists, or those with power over others on the panel can lead the discussion away from the central purpose of the panel or focus it on entirely minor issues. A case in point involved a Deputy Minister and a Research Scientist under his authority on the same panel, seen as a concern in the 1998 Canadian Senate enquiry into the safety of rBST, a hormone to promote milk production in cattle. (Ref. www.sierraclub.ca/national/genetic/index.html)

- If there is an attempt to summarize the deliberations, the result may be a set of inconsistent ideas because the panel may simply contribute snippets of scenarios that are pasted together as a collage rather than an orderly and reasoned scenario. This is the problem of committee-based design, where there is insufficient direction and purpose to maintain a unity of vision.

- Panels often have very limited time to present ideas, so the output is necessarily truncated and incomplete.

Nevertheless, panels generally avoid the potential extremity of just one expert view.

Delphi techniques

In an effort to overcome the issue of personality in panels of experts, the Rand Corporation developed the *Delphi Technique*. We still have a panel, but their contributions are made anonymous by a moderator. Each panellist provides an initial forecast. These are 'anonymized' (names and any references that would divulge the author's identity are removed or edited) and distributed to the other panellists who can then comment on them and/or revise their own forecasts. This process can be repeated. The hope is that a consensus can be reached, or at least a reduced set of scenarios, each of which is hopefully self-consistent. That is, we want each scenario to make sense and be potentially realizable.

Computer technology has made Delphi-like panels fairly easy to organize. Some *groupware* allows for online panel discussions where the participants are automatically made anonymous. Of course, they have to be able to use a keyboard, so good typing skills may render some panellists more 'powerful' than others in the discussion. Internet 'chat' technology, while not directly intended for Delphi forecasting activities, does facilitate very similar discussions. Our own preference is that there should be a moderator, but electronic mail does make his/her job much easier. The moderator can and should intervene to focus the discussion and ensure that all relevant issues are addressed. Purely technological methods such as groupware and Internet chat do not (yet) provide such guidance and discipline to the forecasting exercise.

Example: The following is a quotation from the introduction of a report on an example of a Delphi exercise conducted by Japanese and German investigators, found at
http://www.atip.or.jp/public/atip.reports.94/j-g-tech.94.html:

'Growing competition on the world market and increasing technological change are forcing economies and organisations to concentrate their research and development (R&D) activities on selected areas. In order to identify those technologies which will have the greatest impact on economic competitiveness and social welfare, several new studies on critical technologies have been published in the United States, Japan and Europe. All these studies are written with the more or less express objective to sort out those technologies which are considered most important for the respective countries. They differ considerably in terms of size, disaggregation, methodology and relevance.

'Among them is the Japanese Delphi which includes a comprehensive survey over two rounds with more than 1,000 technological topics included. The Delphi is considered to be highly oriented towards conformity though the huge statistical data base created does not automatically yield evaluations and recommendations. Based on the Delphi data pool, holistic assessments seem to be possible and they are provided within this report. The Japanese Delphi survey puts an established and validated methodology into practice and stresses the power of new technologies to remedy important societal and ecological problems'.

Cross-impact analysis

A method that may allow the forecaster to provide some quantitative information to long-term technological development is cross-impact modelling. This can be both a conceptual and a calculation tool. The principal idea is that the development of one technology either enhances or diminishes the likelihood or rate of development of another. For example, the development of electronic cash registers halted development of electromechanical devices.

The main weaknesses of cross-impact analysis are that the user may fail to include a relevant technology or may provide poor estimates of the relevant weightings. Nevertheless, this analysis does improve on 'guessing' about the future, and is the main form of analysis that will be used here. However, we consider the present form of this approach, as used below at least, to be primarily a tool for structuring one's thoughts. In preparing a technological forecast of the impact of microcomputer technology on sample survey methodology (Nash, 1989), the development of a program to extend the ideas discussed was attempted, but we feel there are so many choices and assumptions made in the implementation of such a program that its use is constrained by too many cautions.

We will present cross-impact analysis by using the example in Nash (1989). This was an attempt at a technological forecast for the impact of personal computers and workstations as they may be used to carry out and analyse statistical surveys. The time horizon for the forecasts was the year 2000, that is, 12 years from the time of the forecast.

The cross-impact analysis was not the only approach to this exercise. By examining ongoing developments and flows of technology, it was moderately straightforward to forecast most of the direct impacts of advances in workstations and related technologies on survey methods.

'The steady, possibly inexorable, march of progress in speed and capacity of workstations will naturally make it possible for a very small statistical team to carry out quite large surveys which do not require personal interviews (i.e. those done by mail or electronic communications or which do not otherwise require human labour).' (This and other quotations in this section are from Nash, 1988)

It was the other issues that concerned us:

'The questions which remain would seem to be those of the integration of people, software, hardware and operating procedures into systems which operate smoothly. That is, I perceive the major obstacles to be those of incompatibilities in the requirements and capabilities of surveys and the means to execute them.

Less easy to predict are the more generalized impacts of these developments.'

To organize our thoughts, we listed five classes of 'technologies' we believed were relevant to forecasting the evolution of survey methods and their application. See Nash (1989) for more details and justification.

Table 5.1. Technologies or situations considered

1.	Computer improvement -- More powerful hardware and software with better user interfaces
2.	Dispossession: homelessness, unemployability, illiteracy, de-skilling of jobs, etc
3.	Human rights and privileges -- growth in privacy, personal security, individual freedoms, anti-discrimination and environmental concerns
4.	Number dependency -- legal requirement or other need for the results of statistical surveys
5.	Communications enhancements – easier transfer of information, compatibility of data systems, standards, etc.

We then developed a table of the 'impact' of each of these technologies on the others and on itself. We can formalize such impacts in terms of probabilities. If we divide our time scale into suitable blocks, such as 1, 5, 10 or 100 years depending on the situation, we can, by use of experts, panels or 'judgement', assign a probability to the realization of technology i in the next time block. A probability of 0 means that the technology cannot be realized, while 1 means that it will be realized. Our cross impact table then reflects the likely change in the probability of realization of the technology between time blocks. A ++ score implies a large increase, a -- a large decrease, while 0 means no effect.

Clearly we need to quantify how the cross-impacts will apply. One technique is to take the vector of 'realization probabilities' and make the cross-impact scores into multipliers, for example, ++ multiplies the probability by 1.3, + by 1.1, 0 by 1, - by 0.9 and -- by 0.7. Those familiar with Markov chains will see a resemblance. There are some critical difficulties, however, in particular if the realization probabilities compute to numbers that are too big or too small. And there is no special reason why the impact should be in the form of multiplication by a factor. We are simply trying to provide a crude model.

For 'technologies' that are already part of the scene, such as political or social opinions, we would want to treat the cross-impacts as modifying the proportion of society that holds the opinions or concerns. We again have to decide on the time block length, initial estimates of proportions, and transition mechanisms that govern how we model the changes in the proportions by use of the cross-impacts.

For our example, we did not attempt to model and iterate, but used the cross-impacts as a way to modify our thinking and forecasts obtained by considering technological flows. Our final forecast statement reflected the concerns raised in the cross-impact table (Table 5.2).

'In summary, statisticians should expect to have an easier time planning, analyzing and reporting surveys, and a much tougher job actually getting hold of the right data. When quality data is collected, however, the tools should allow for better information to be extracted from surveys, with increased timeliness of reporting, precision and reliability of estimates, and finer grain of analysis in terms of time periods, geographic areas or other categorizations. Furthermore, the reporting can and will include better graphical aids and should avoid typographical errors by having tables directly typeset into the report from the most current analysis.'

Systems analysis and modelling

Throughout this book we have developed models and extrapolated them to forecast. This process is also applicable to long-term forecasting, but the types of models are generally different. In the next section we will look at simulation and gaming, both of which are also forms of modelling and extrapolation, although the extrapolation phase is not simply advancing the time variable. An engineering viewpoint, particularly that applied in large computer systems, gives rise to models via *systems analysis*. We define this to be a careful study of the objects, processes and interactions in a complex system of interest so that an understanding is achieved of how the system will respond to various inputs. This was a popular subject in management in the 1960s and 1970s but has been less prominent in recent years. For example, this approach to forecasting is mentioned in the book by Makridakis *et al.* (1983) but is dropped in the 1998 edition.

If we have sufficient understanding, we can codify the results of our systems analysis, which gives us a model. Then we can formulate the input conditions that correspond to future time periods and derive forecasts from our model. This presupposes that our analysis has been done correctly and also that the understanding gained has been correctly codified. Given time and resource constraints and the possible evolution of the system itself over time – an evolution that may escape our analysis – the forecasts are unlikely to be accurate. Nevertheless, this approach to forecasting does have some useful applications, particularly in energy and transportation systems, where interaction and substitution between different modes of supply and demand are

Table 5.2. Cross-impact of different 'technologies' on each other on a scale -2 to 2 (--, -, 0, +, ++). The scale of the effect of the technology at the left on the technology at the top of the table is given in the entry in the body of the table

	Computer improvement	Dispossession	Human Rights	Number Dependency	Communi- cations
Computer improvement	++	+	-	++	++
Dispossession	0	+	+	– (Note b)	- (Note b)
Human Rights	-	- (Note a)	+	-	-
Number Dependency	+	0	-	+	+
Communi- cations	+	+	0 (Note c)	+	+

a   this element is negative, since improved concerns about human rights and privileges will REDUCE dispossession.

b   inability to place people in a sampling frame may make it very difficult to include them in a survey. However, a view may be taken that those who cannot afford any of the products of society do not need to be considered. Here, I have taken the view that all citizens are part of the society.

c   communications technology may have profoundly negative impacts on rights and privileges, but at the same time may be used positively. The author has personal knowledge of several instances where serious human rights violations were avoided or minimized by Amnesty International members and supporters using communications technology intelligently.

important. For example, people can commute by several forms of transport, sometimes in combination. If the effective price of one mode rises, commuters will change behaviour, sometimes in ways that lead to very unfortunate congestion. It is important for forecasters to anticipate and hopefully model some of the substitutions and interactions if their forecasts are to be useful.

Simulation and gaming

If we have a very good understanding of a situation, we may be able to *simulate* its behaviour. Historically, simulations were made with physical models. For example, an *orrery* is a model of the solar system that puts the planets in the correct relative positions so that one can work out the past or future alignments of the planets and other objects. Such models were usually masterpieces of the clockmaker's art, providing quite accurate representations of reality. We include simulation and gaming under qualitative forecasting since some quite crude models can be used profitably to make useful predictions. Moreover, the ideas of simulation are linked naturally to scenario writing. However, the ideas of this section can be made very detailed, often in ways that are well beyond the scope of this book.

Today we generally use mathematical models – which is what two-thirds of this book is about! However, up until this point, we have generally been satisfied with quite simple models of time series. Simulation, as currently practised, attempts to model systems in much more detail, with multiple equations or computer routines to extrapolate deterministic processes and a variety of pseudo-random number generators to provide for stochastic variation. By making many 'runs' of the computer program, the pattern of behaviour of the system can be generated. We have used simulation to analyse and forecast a number of situations, usually quite technical in nature. However, none of the examples in our experience relate easily to typical business or administrative situations. Simulations generally require time, effort and skill to build and analyse properly. Small errors in their construction can cause the results to be very wrong. We do not recommend this sort of simulation unless it is really needed.

Since many business and administrative situations involve competition, or even conflict as in military contexts, such simulation has been extended to *gaming*, which allows several participants to interact with a simulation of the system of interest. This may be mediated by an umpire, who decides the merit of the players' actions, or by a computer program. In either case, some mechanism is generally used to provide for the element of chance, or 'luck', and the 'rules' of the game set the conditions of the simulation. We have little experience of this approach to forecasting. For our students and readers, we believe that gaming will require too much effort and too many resources to be helpful in any but the most specialized of forecasting activities.

Scenario writing

An approach that is much less demanding of resources is *scenario writing*. We simply write down, in a few sentences, what *might* happen or be the state of affairs at a given point in time. Then we attempt to list the consequences of this state. That is, we examine the implications of the situation on resources, people, or related events. We will want to describe possible ways to arrive at the future state to see if there are pathways or *trajectories* that allow us to arrive at the future state from the present one. We will also want to provide a critique of the future situation. Is it desirable or undesirable? Are there features we would want to enhance or avoid?

Clearly this is a technique that can lead to much fantasy and little worthwhile forecasting unless we are very disciplined. Nevertheless, the act of writing down ideas and then exploring their consequences, especially with group involvement, may be helpful in formulating goals and directions. It may be more useful in 'forecasting' what we do *not* want to happen than in making reliable predictions. The effort to make a consistent scenario that is *realizable* by evolution from the present state can also be helpful as a first step in building models for better forecasting.

Most scenario writing is informal, including our own experience. As an example, we juxtapose an extract from an opinion article one of us published in the (Toronto) *Globe and Mail* in 1991 with a recent advertisement from Bell Canada.

Page A19, Globe and Mail, Monday 22 July 1991

TELEPHONES / *Why not do away with long distance charges?*

Just dial a number and shrink the country

BY JOHN C. NASH

Ottawa

...

How would a single Canada-wide calling zone be implemented? ...

Perhaps all calls would cost a few cents -- in most countries of the world, local calls are individually charged.

The advertisement (Internet Web page, TV commercial) downloaded from the Internet in November 1998.

Introducing Bell's new Advantage Per Call™ savings plan for small business. Call anytime, anywhere in Canada, and talk for up to 10 minutes for just 35¢ a call. That's right. 35¢ per call. So you can call more, sell more, and save more.

Exercises

E5.1. For classroom discussion: set up and run two or three rounds of a Delphi forecast process for one or more of the following topics, which should be restricted and defined before starting.

The class performance in the current (or other) course;

The outcome of the peace process in Northern Ireland (or similar political activity);

The nature and timing of settlements in copyright issues for Internet file 'sharing', or similar intellectual property issues;

The uses and abuses of genetic engineering.

E5.2. Define the 'technologies' of importance in one or more of the topics of Exercise E5.1 and conduct a cross-impact analysis.

E5.3. Discuss and/or carry out systems analysis and/or scenario writing forecasts for one or more of the topics of Exercise E5.1.

6

Semi-quantitative methods

Market versus situational forecasting

We have entitled this chapter 'semi-quantitative methods' rather than 'market forecasting' or 'qualitative methods' because there are a number of forecasting tasks that do not fit nicely into the two latter categories. The examples and approaches we will consider are not only drawn from the problem of predicting market acceptance of a new product or service, nor are they purely qualitative in nature, although we are often interested in predicting qualitative outcomes such as consumer behaviours.

We will consider *market forecasting* to comprise activities directed toward predicting the success (or lack thereof!) of new ventures that require consumer acceptance. Such forecasting makes up a large part of the commercial forecasting industry, especially if we include the activities of political polling agencies and their associated analysts. On the other hand, we may wish to understand the overall social impact of such products or services on a community. For example, we may want to know how access to the Internet has changed children's study and recreational habits, or how cell-phones are being used, and how these phenomena are evolving over time. We call such activities *situational forecasting*, although we caution readers that this term is our own and not in general use. Clearly, the aims and the types of information sought are different in market and situational forecasting. However, the methods used are often similar.

Technological forecasting

When forecasting for the long term, situational forecasting is usually referred to as *technological forecasting*. Here we try to predict what technologies will exist at some point in the future. Clearly, forecasting just one technology is likely to be unrewarding. If people could, as in the well-known *Star Trek* television and movie series, be transported by radio beam, automobiles would be redundant, as would many of our roads, traffic signals etc. Parking meters and fines would also be historical artefacts.

Technological forecasting methods have already been considered in Chapter 5, but we note that the forecasting time horizon is not rigidly divided, and methods discussed there may also be appropriate, in many cases, to the situations of interest in this chapter. That is, we want our readers to recognize that our division of the material between the chapters, while we believe it to be a logical one, should not be considered as prohibiting application of the methods wherever they work.

Surveys

One of the major tools for market and situational forecasting is the sample survey. This is a statistical method. While the title of this chapter is 'semi-quantitative methods', we must admit that the methodology can be highly mathematical. The 'measurement', however, is generally of human behaviour, so that our forecast is directed not immediately toward 'We expect to sell X Widgets next year' but to '3 out of 10 consumers will prefer the new Widget design, while only 1 out of 10 has a definite preference for the old model'.

The usual steps of a survey are (Tryfos, 1996; Bailar and Lanphier, 1978; Satin and Shastry, 1983):

- Decide the question(s) to be answered by the survey exercise, and the time interval for which they are to be valid. We caution that the references above give limited guidance on this topic, and we have not found a source to recommend.

- Cast these questions into a form that can be posed to survey respondents in a questionnaire (the sample instrument). This is a critical part of getting valid results from a survey, as illustrated graphically in the Canadian Broadcasting Corporation's *Witness* programme 'Ask a silly question'(Kastner, 1998).

- Develop a sampling frame, that is, a list or description of the population of possible respondents.

- Develop a sampling plan that will select appropriate members of this frame. Typically, students have been made aware of Simple Random Sampling, where each member of the population has an equal chance of being asked to respond to the questionnaire. However, in the real world we can save time and money by using other schemes without necessarily compromising the statistical validity of our results. Stratified and cluster sampling schemes are widely used by both official and commercial statistical agencies to good effect. These topics are discussed in a number of good books on survey sampling, such as Barnett (1974).

- Carry out the sampling plan as faithfully as possible. This is *always* difficult. People are unavailable or refuse to answer questions for surveys and censuses, even under threat of legal sanction. Survey staff are often paid per respondent, so there is an almost overwhelming urge to use the path of least resistance, even if this biases the survey results.

- Using the sample data, compute estimates of the quantities of interest for the population. Also compute measures of reliability for these figures, that is, standard errors and possibly measures of potential bias. In serious studies, attempts will be made to understand and even model the non-response, that is, the people who did not or would not answer the questionnaire. An obvious example is that telephone surveys conducted during daytime working hours will find few people who have 'regular jobs', at home. (We make no effort here to define 'regular'.)

- Report on the findings of the survey in a fair and comprehensible manner.

Leading indicators

Leading indicators are variables that our experience and, hopefully, analysis have shown us provide a good prediction of later behaviour of the quantity we are interested in forecasting. Typically, we use quantities that are easily measured and that have a structural linkage to the forecast. An example is deliveries of capital equipment to manufacturers, which is often used as an indication of manufacturing output several quarters later. Similarly, the number of 'help wanted' advertisements for particular types of workers are used as an indicator of subsequent activity in the relevant fields. However, the indicator variables could be unrelated. There is a persistent – and we cheerfully admit no attempt at serious verification of this – statement that the height of Paris fashion hemlines presages the level of the stock market. Measurement issues affect both the indicator and forecast here, as they do even in situations where we may expect a linkage.

Leading indicators are most likely to be used in multiple regression models in order to generate forecasts. If we want long-term forecasts, we are forced to predict the indicators. This is clearly a nuisance, and leading indicators are generally useful only for a fixed time horizon that is imposed by the nature of the linkage.

A variant on this approach is sometimes useful. We may wish to forecast population flows in and out of an area (e.g. our population of Calgary project assignment). Many cities experience an 'average' net flow of people, in addition to births and deaths. This may be shifted upward by 'good' economic times, and downward by 'bad' ones. We may not be able to find a specific variable to use for 'good' and 'bad' economic times. 'Domestic product' or 'expenditure' figures may be unavailable or unreliable for geographic areas of the size of interest, and they may be too specific, since salary and cost levels are likely to be of more interest to people thinking of moving in or out of the city. However, from a mixture of quantitative and qualitative information, including media reports, it is usually possible to assess 'good', 'bad' or 'average' times, or even a five-point scale. We code this 'economic well-being' as a three or five point variable that has zero as its central value e.g. -2, -1, 0, 1, 2 for a five-point scale. This is then used in our modelling, and kept if it actually improves our model, with the usual tests for significance if these can be made. For more on this example, see Chapter 19.

We do, of course, need to predict the 'economic well-being' in order to forecast. However, we may simply skirt the issue by providing a set of forecasts for each future time period conditional on the state of this variable. Some workers prefer to use a variable that is a continuous index, rather than our 3- or 5-category indicator. It is our opinion, however, that we can get knowledgeable observers to provide us with forecasts of economic activity in terms of 'bust', 'slow', 'average', 'good' and 'boom', and that these are often sufficient to provide for improved forecasts. Note that we are using these more or less like seasonal shifts (and here we must refer the reader forward to Chapter 12).

Market penetration models

We have not yet encountered the mathematical models needed for extrapolating market penetration models. These are the S-curves of the Logistic and Gompertz functions, the estimation of which will be discussed in Chapter 20 (Nonlinear regression modelling), although other functional forms can and should be used in special situations. For the moment we will not worry about the technicalities, but look at the forecasting concepts. The essence of our activity will be to take data for market penetration, for example, the percentage of a given market that a product of interest commands at different time points, and fit an S-curve to this data. We then extrapolate the curve by substituting future time values into the function. Another name for this is a *diffusion model*.

Stated this way, the process seems straightforward. Of course, reality is never so kind. The nonlinear estimation, as we will see, is often sensitive to small perturbations of the data. Was that 'blip' in the first quarter simply due to the Great Ice Storm of 1998? Moreover, nonlinear modelling adds another layer of approximation to the estimation of the standard errors of the model parameters and hence to the potential range of values the extrapolated model is likely to cover. Let us consider the logistic growth curve in three parameters (b_1, b_2, b_3) and time t

$$y(t) = b_1 /(1 + b_2 \exp(- b_3 t))$$

over the time points $t=0$ to $t=45$. The graph in Figure 6.1 shows various logistic curves that arise when we perturb the parameters singly and in some combinations of up to 10% of their magnitude from the 'usual' values of $b_1=100$, $b_2=10$ and $b_3=0.1$. This graph is too 'busy' to illustrate more than the general flavour of these perturbations.

If one examines the changes in the logistic in detail, the sensitivity to modest *relative* changes in the parameters is not too severe, but the b_3 parameter needs to be very much smaller than the b_2 parameter in magnitude to get the rise in the function to occur over the full range of the data. Readers may find it helpful to experiment with spreadsheet software that allows the function values to be changed quickly. *A warning:* use absolute column and row addressing to ensure that the formulas get the correct input if they are copied between cells. It is imperative to have correct cell references to the time period (t) and to the three parameters.

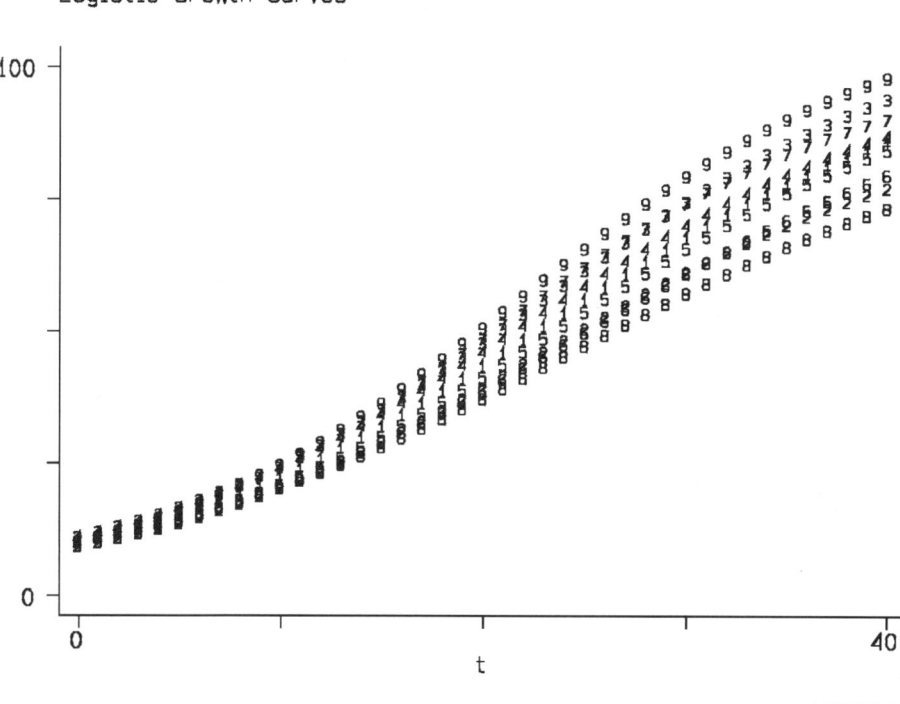

Logistic Growth Curves

Figure 6.1. Sensitivity of sigmoid models to parameter changes. The parameters for the series (labelled by the numbers 1–9 in the graph) are in the table below.

Series	1	2	3	4	5	6	7	8	9
$b1$=Asymptote	100	90	110	100	100	100	100	90	110
$b2$=startpoint	10	10	10	9	11	10	10	9	11
$b3$=inflexion	0.1	0.1	0.1	0.1	0.1	0.09	0.11	0.09	0.11

To quote from a study by Meade (1984):

> There is ample evidence that the use of growth curves for forecasting is widespread. However, there is less evidence that the applications are theoretically sound and statistically valid.

We concur with this opinion, even though we have ourselves used growth curves in forecasting applications (Nash and Teeter, 1975; Nash, 1977b). Several examples of use are described in :

- Blackman (1971), for the rate of innovation in the commercial aircraft jet engine market;

- Mahajan and Peterson (1978) who consider models for adoption of United Nations membership and of washing machines (clearly the methods are flexible!);

- Merino (1990), who looks at tire cord textiles;

- Easingwood *et al.* (1981), who use a modified ('nonsymmetric responding') logistic model to analyse hospital adoption of four 'new' technologies; namely, ultrasound, computed tomography (CT) head scans, CT body scans, and mammography;

- Wang and Kettinger (1995), who look at the adoption of cell phones (see Chapter 20);

- Rai *et al.* (1998), who consider the growth in Internet hosts (also Chapter 20).

Other extrapolation approaches

The ideas of the previous section can be applied more loosely to forecasting diverse situational and technological changes. Here, we are re-treading the path of scenario writing, simulation, gaming and systems analysis of Chapter 5, but taking a more quantitative viewpoint. The difficulty is to prescribe how to do this. That is, there are examples of forecasts prepared by making various extrapolations, some of them well done, some exasperatingly poorly prepared, but it is not easy to codify the process, given that different authors sometimes use different tools or vocabulary. Foot (1998) takes a demographic perspective, while Hughes (1985) and Bloom (1995) look more to trend lines and simple sums and differences of key quantities.

While details are elusive, we can underline a few general points.

- Many of the types of long term semi-quantitative forecasts of interest are for the entire world, so it is very important to check that units and scales are correct. Production and consumption figures are worth reducing to a per capita basis to see if they make sense.

- It is almost always worthwhile to draw a (good!) graph.

- Despite the common practice of not doing so, we feel it is important to try to provide some form of interval for forecasts, especially best/worst case scenarios.

Media analysis – the Bhopal example

Sometimes it is not necessary or possible to carry out the surveys or modelling of the methods above. Indeed, the data we seek may be unavailable to us directly. As an example, consider the problem of getting information on homelessness in a particular city. It is very hard to find people who have no home via a household survey, which is one of the favoured approaches of statistical agencies. In attacking this problem, some of our students have used numbers of persons sleeping in shelters as a proxy for homelessness, although it is reported that many 'street people' refuse to use such shelters.

Another proxy is available in the news media. We can look at the number and type of stories about homelessness and 'street people' over time in media reports. Particularly for newspapers and magazines, there are paper and even machine readable indexes that allow us to get a 'profile' of what is going on fairly quickly. We must, of course, be aware that the media has fashions; when newspaper A is writing a feature on homelessness, magazine B may feel they also should 'do a pièce on it'. Moreover, figures quoted in articles are frequently 'borrowed' widely. We provide an example.

The Methyl Isocyanate (MIC) gas leak at the Union Carbide plant in Bhopal, India, on 3 December 1984 has become an icon of the dangers of technology. The magnitude of the disaster, in terms of death and injury reported, puts the Bhopal incident among the worst peacetime industrial accidents on record. What actually happened, and how many people it would kill or injure, were not immediately obvious. The unfolding of the story coincided with the first course in Managing Technological Risk taught by one of us (JCN). The course participants were motivated to produce the limited-distribution monograph *The Bhopal Incident: Origins, Issues and Impacts* (John C. Nash, Joseph R. Charlton, Stephen J. Delaney, and Alain Lahaie, 1985). This relied on contemporary news reports, including magazines and newspapers from the Indian sub-continent as well as some industrial trade magazines that were available to the authors. Five years later, J-L Hotte, another MBA student, assessed the success of this analysis based on subsequent substantive reports. He wrote 'the purpose of this study was to investigate if it is possible to find a reasonable approximation to the truth concerning a technological disaster story directly from early, publicly available sources. Is it realistic to think that stringent questioning of the reasonableness of each report and diligent cross-checking is sufficient to find the truth?'

Hotte's conclusions subdivide the analysis and assess the results as mostly accurate or at least satisfactory where they are applicable. For example, some court actions were and are still ongoing, so predictions for their outcomes cannot be checked. His overall reaction is 'From these results it appears that it is possible to find a reasonable approximation to the truth concerning a technological disaster story directly from early, publicly available sources.'

To illustrate the process, we consider the problem of estimating the number of deaths due to the MIC leak. This is not a simple matter because many of those in the path of the gas cloud were squatters. One report we found suggested, for example, that it was 'illegal' for people to live within 2 km of the MIC plant. Furthermore, deaths due to MIC can superficially resemble those due to respiratory illness, of which any night in Bhopal would be expected to have a fairly sizable number. Moreover, once financial compensation for injury or death became a possibility, false claims became an attraction. Deaths may or may not be properly recorded – one body could be counted more than once, particularly if it is possible to get the death registered as two or more family members.

To visualize the distribution of the reported number of deaths, we elected to use a variant of the Tukey boxplot. We modified this by plotting a letter for the month in which the number of deaths graphed was published. We also deleted observations that we deemed to be simply repetitions of the same wire-service report.

Although we did not do so here, we could have sorted the symbols to put all the data for similar publication date together. With some notable exceptions, the later data suggests a figure of 2500 deaths. The particular value of 1431 is, we believe, reflecting the number of 'official' or documented deaths. The figures of 2000 and 2500 deaths are clearly the most common, and we would likely accept that a figure in the range 2000–2500 is a reasonable approximation to reality. There are, however, continuing court actions and controversies concerning Bhopal. For example, Leonard (2000) cites a figure of 3800 deaths.

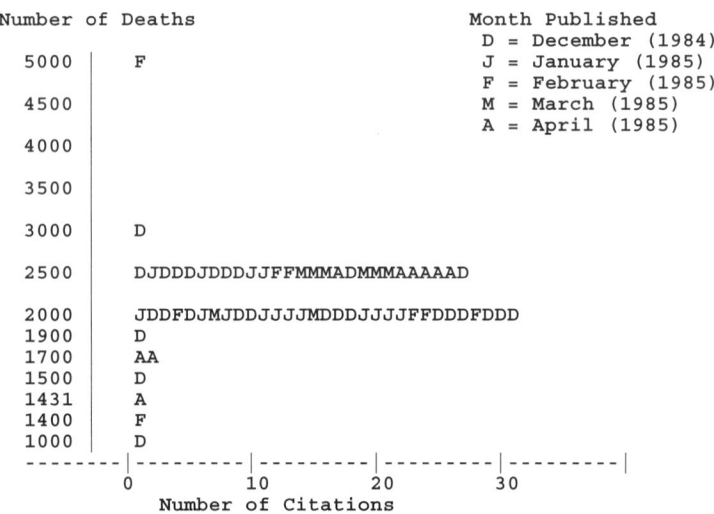

```
Number of Deaths                        Month Published
                                        D = December (1984)
    5000    F                           J = January  (1985)
                                        F = February (1985)
    4500                                M = March    (1985)
                                        A = April    (1985)
    4000

    3500

    3000    D

    2500    DJDDDJDDDJJFFMMMADMMMAAAAAD

    2000    JDDFDJMJDDJJJJMDDDJJJJFFDDDFDDD
    1900    D
    1700    AA
    1500    D
    1431    A
    1400    F
    1000    D
         --------|---------|---------|---------|---------|
            0        10        20        30
               Number of Citations
```

Figure 6.2. Distribution of the number of deaths reported due to the gas leak at Bhopal, 1984–12–3

Exercises

E6.1. Discussion: What types of forecast may be carried out using surveys?

E6.2. Find books or other sources of information on sample surveys. Prepare a short checklist of steps in conducting a survey. Try to estimate costs for each step and the overall exercise. This is a useful group discussion project/topic.

E6.3. Based on E1.2 and E6.1, select some forecasting project topics that might be suitable for resolution by survey. Develop the survey instrument (questionnaire) for each of these.

E6.4. Using a spreadsheet, try several sets of parameters in different growth curve functions and examine the shape of the resulting functions. [Note: It can be surprisingly difficult to choose appropriate values.] Are there any obvious effects of changes in the parameters; in which parameters in which models?

E6.5. Select a recent major event with widespread media reporting. This may be a disaster with deaths and injuries, an election, or a debate over costing of some large public expenditure. Perform a media analysis such as that suggested in this chapter. Discuss the results and the success or failure of the approach.

7

Forecasting, risk, and strategic management

The role of forecasting in strategic management

A few years ago, a colleague had students survey local high-tech firms to ask how they used forecasting. Most responded that they 'do not use forecasting'. If we took this response at face value, we clearly would know which firms we should definitely *NOT* invest in. Any firm that is not forecasting its own activities and those of its main partners and competitors, as well as generally predicting the environment in which it will operate, is not going to flourish.

The most likely explanation of the survey results is that managers generally take forecasting to be a specialized and formal activity. The ongoing short and medium term *planning* that they undertake presupposes some sort of forecast. Many such forecasts are 'bought' or 'found' from general or particular trade publications or communications. A senior executive with a major actuarial firm told us that the life tables used for developing and managing pension funds are developed by the Society of Actuaries, and indeed we have found some of these resources to be available over the Internet. Magazines such as *The Economist* or *Fortune* include various economic forecasts, as do the business pages of major newspapers. A variety of companies, trade organizations and government departments commission studies which they publicize, as and when it suits their objectives.

Internal forecasts of revenues and expenditures are prepared from simple aggregations and extrapolations of accounting information. However, the manager may not regard the activity of preparing these as 'forecasts'. Long term forecasting is, we believe, quite rarely practised. Even when it is, we suspect that the task is undertaken informally by the Chief Executive Officer or equivalent, who then communicates the resulting decisions to the rest of the organization. Because the inputs to such decisions are not available to others, the cross-checking and validation of the ideas is usually omitted.

Our view is that forecasting should be a more organized activity, with opportunity for input from the entire organization. The hints and whispers of change are likely to be heard first by those least likely to communicate them to 'the boss', especially if there is no structure to allow such communication. Moreover, if the message is one of bad news, the messenger may get shot! And if the word is of opportunities, it will be said out loud only if there is appropriate reward.

Incorporating a view of the future is an important aspect of strategic management. For many organizations, the prediction of possible misfortune is an important tool for survival. We can all name companies that have disappeared because they did not see the future quickly enough or did not act upon available forecasts. Equally important is being ready to capitalize on opportunities. This is also true for governments, since lack of available infrastructure in the form of transportation, communication, living accommodation for workers, educated citizens, and so forth will cause business to move elsewhere and result in lowered living standards and lower relative prosperity.

Risk anticipation

The *risk* associated with a particular hazard is technically defined as the product of the *hazard magnitude* and *probability of occurrence*. This is the cost of having the hazard actually hurt us multiplied by the chance that it does. When there are several possible outcomes, we work out the *expected value* of the hazard costs. A simple and artificial example illustrates these ideas.

Example: Damage to automobile

Hazards & attributed hazard cost

Collision	$ 3,000
Fire	$10,000
Theft	$10,000
Ice cream + 2-year-old	$ 30

Probabilities (est. for a 1 year period)

P(Collision)	= 0.05
P(Fire)	= 0.0005
P(Theft)	= 0.001
P(Ice cream + 2-year-old)	= 0.1

Risks

Risk(car damage in year)   = $168
= 0.05*3000 + 0.0005*10000 + 0.001*100000+ 0.1*30
= 150 + 5 + 10 + 3

Note that the individual products of hazard magnitude * probability, that is, the *components* of the risk, are useful in ranking the overall importance of different hazards. This is important as a way to overcome the errors humans make in their *perception of risk*. Indeed, we tend to overestimate the importance of rare but 'nasty' or 'dramatic' hazards, such as air crashes, and ignore commonplace ones, such as driving while intoxicated, or smoking, even though the risks as we have defined them above, are far greater for the latter type of activities.

This is not the place for a treatise on the management of risks; we have done that elsewhere (Nash and Nash, 1995, 1997), but we will provide a brief summary of the major approaches to the estimation and management of risk.

First, we note that taking risks is really about gambling, and that the most formal of gamblers are insurance underwriters. They employ specialists – actuaries – to study the mathematics and statistics of the hazards of interest and to compute the risks. This usually goes by the name *risk theory*. However, there are also people interested in safety or reliability. They are usually engineers, and *risk analysis* is the name generally given to such studies. As is common in human activities, one could be tempted to believe that practitioners of these arts never talk to each other.

To estimate and manage risks, we will be concerned with estimating and hopefully controlling either or both hazard magnitudes and probabilities of occurrence. A hazard that costs us nothing is no longer a risk; one that cannot occur is not at issue either.

Typically, we estimate hazard magnitudes by reference to costs, e.g. to build or acquire structures or equipment or goods. The most difficult magnitudes to estimate are those associated with human suffering or death, particularly if we must combine such estimates with those relating to property. We want to *quantify damage*. This cannot be done as a generality, but only in reference to one or a set of *stakeholders*. There is always plenty of controversy to go around in any such exercise. Nevertheless, we have found that even quite crude estimates of hazard magnitudes help us to sort out which hazards are important for a given stakeholder.

When we want the probability of occurrence – the chance a hazard is realized – we must consider the *sample space of possible events*. Failure to include events invalidates the whole exercise. The *Titanic* was unsinkable! Because human perceptions give rise to very poor judgemental estimates of the probabilities we want, it is important to try to provide at least order of magnitude bounds to them. Engineers in the nuclear and chemical industries frequently make use of *fault trees* and *event trees*. These are tree diagrams that trace possible events and attach conditional probabilities to each node of the tree so that composite probabilities can be built up by multiplication. Fault trees work backward from a realized hazard event, while event trees work forward from some incident, e.g. worker cuts wire accidentally, to possible outcomes. The difficulty is including all the possibilities, and such exercises are rarely suitable for regular use by non-specialists.

Statistical tools offer more help. Simple data analysis involving tables, graphs and summary statistics, is likely to be our best hope. Carried forward on a formal level, this mirrors actuarial and epidemiological methods where tables are built up from experience data, but we need to have the records to make use of these. For the more adventurous, Bayesian statistical methods may be helpful, although we do not feel they are generally suitable for managerial application, and the assumptions embedded in choices of prior distributions can be important.

When we lack data, we are forced into more subjective techniques, as we have seen earlier, in Chapter 5. Scenario writing, simulation, and the use of opinion, expert or otherwise, are tools we can attempt to use here.

Fortunately, when it comes to risk management, that is, the application of our risk 'forecasts' to planning, we may be able to prune our possible set of choices by simply eliminating or controlling some hazards. If we choose not to participate in certain activities, such as Internet

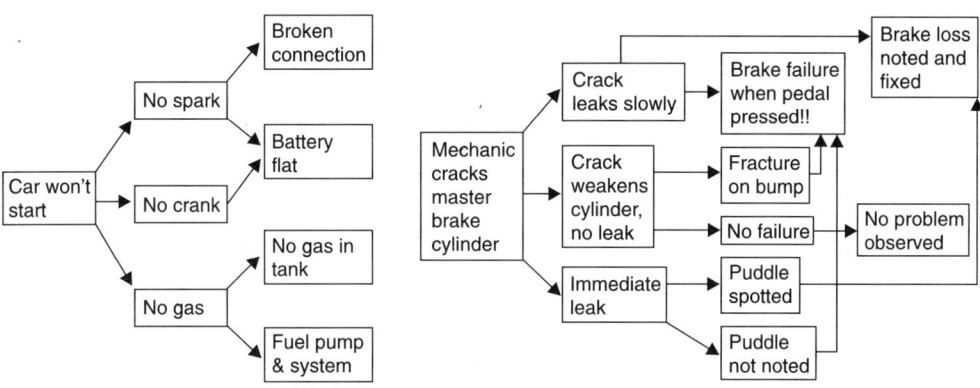

Figure 7.1. Examples of a Fault Tree (left) and an Event Tree (right).

financial transactions, we eliminate the risk of fraud from that particular direction. Insurance limits our costs due to fire, for example, to the annual premium. We can then focus on the risks that are associated with our particular business activities.

One important message we feel deserves stating about risk is that *any* formal attempt to estimate hazard magnitudes and probabilities of occurrence is worthwhile compared to managing by the 'seat of the pants' approach. By simply organizing our thoughts, we are going to clarify our view of the situations we face. Moreover, we eliminate a great deal of the perception bias once we start listing the hazards and dealing with them one at a time.

Predicting opportunities

Opportunity is the converse of risk. We can apply very similar techniques to place a value on each of a set of opportunities and on a portfolio. We can, moreover, mix in the risks. We should, of course, always do this, but the emphasis on 'safety' in our society often pushes this out of view.

People have, in our opinion, a more active imagination in proposing possible hazards than possible opportunities. In forecasting the occurrence of both risks and opportunities, we need to have a sample space of what may happen. If we leave out elements of this sample space, we do not attribute any probability and therefore forecast no occurrence. How can we develop a list of opportunities?

Let us assume that we are working on risks and opportunities for a business organization. Furthermore, assume we have a reasonable knowledge of what it has become fashionable to call the *core competencies* of our business sector. First, then, we should list what we feel are the ongoing or natural outgrowth activities of the business as it is currently organized. This should include activities we think are going to disappear along with why and when.

Second, we should list activities that we feel the business could undertake in the future by using existing resources or by modifying the resources in fairly obvious ways. That is, we list what are, in a reasonable sense, understood as potential avenues for evolution of our organization.

Third, we try to consider the 'off the wall' ideas that provide directions for future activities that are less well-understood in our organization. These are the items that will be most difficult to develop, as they are not part of our regular scene. To develop items to list in this category, we will need to make some special efforts. Suggestions include:

- 'Brainstorming' sessions with staff to elicit potential items to consider and discuss;

- Review of magazines and journals to pick out possible technologies and developments that may affect our business activities to provide potential items for discussion as in the first item;

- Encouraging staff participation in conferences and trade meetings with instructions to gather ideas for future business directions.

The ideas collected in such exercises tend to be an amorphous, disorganized mass of sometimes conflicting concepts. To provide some organization, we suggest a table such as Table 7.1. This includes the regular and easily-predictable ideas, as well as the wilder ones. In this table, we want to include some measure of the benefit or risk posed by a particular activity or idea, along with likelihood of occurrence and time frame. There may be multiple estimates of the magnitude of benefit or risk, of likelihood of occurrence or time frame, and we should provide references or reasoning to support or at least document each. The purpose of the table is to set out possibilities. At this stage, we have not decided *what* we will undertake in the future, but we want to be able to visualize the range of possibilities. Our measures of magnitudes, likelihoods and times do not need to be precise, nor even in any conventional units, but they should be consistent. That is, they should scale to costs, probabilities and periods of time if they are to be comparable and useful to us in aiding decision making.

Once we have the table of possibilities, we should subject it to scrutiny. By examining the 'Reasons and References' column and exposing the estimates in the magnitude, likelihood and time frame columns to review and comment by staff and possibly outside consultants, the table can be revised and edited. Some activities will be discarded, while others will be added.

Deciding a direction for actions

Given a table of possibilities that management accepts as reasonable, the activities can be assigned weights according to their importance to the organization. Clearly, a part of this weight will come from the product of the magnitude of benefit or risk (with appropriate sign + or -, respectively) and the likelihood of occurrence. However, the time frame and the organizational interests will have a strong influence on the choices of which items we select as important.

Table 7.1. A partial table of possible evolution of the organizational activities of universities

Activity	Income (yearly)	Costs (yearly)	Likelihood of change or occurrence	Time frame (years)	Reasoning / references
Undergraduate teaching					
Traditional	$8000	$7500	90%	15	Making money on u-grads Will likely continue
Internet methods	$6000	$5100	25%	5	Pressure to be 'modern' and to have alternative products
Self-learning tools	$0	$500	20%	5	Students register elsewhere, but often use facilities here
Graduate instruction					
Traditional	$12000	$25000	75%	10	Need grants, or else!
Internet methods	$8000	$5000	15%	5	Profitable, but finding suitable projects so research and supervision 'work' may be difficult.
Professional instruction					
Traditional	$500	$400	50%	5	Much commercial and academic competition
Internet methods	$250	$200	20%	5	Even more competition. How can we be 'better'?
Self-learning tools	$0	$25	40%	5	Students use our facilities/ bandwidth to take competing 'courses'
Continuing Education					
Traditional	$100	$80	80%	5	Partly promotional
Internet methods	$50	$45	50%	5	ditto
Self-learning tools	$0	$25	40%	5	Students use our facilities/ bandwidth to take competing 'courses'

Similarly for **Grant Research**, **Contract Research** and **Donation seeking**

Rearranging the items according to the agreed importance ranking gives us a Pareto diagram of activities that the organization takes to be within its interests. What is done about them moves us from forecasting to planning. However, the action of planning will increase or decrease the likelihood of occurrence of some activities. If we plan to do them and we have the knowledge and resources to do so, then they will become reality. The time frame will also need revision.

Feedback from management to forecasting

This last point suggests something we have only vaguely mentioned elsewhere in the book. Our actions and decisions as managers may put in train events that alter the future. This is especially true in the case of large organizations, governments, or where we are at the centre of innovation.

As forecasters, we therefore need information from management concerning their decisions. If we have provided forecasts to aid our management in decision making and planning, we need to know what was decided and planned in order to update our forecasts. Indeed, not knowing will lead to confusion within the organization. In some organizations, this is a normal state of affairs. Large organizations, in particular, are often found to suffer from extreme lack of communication between those who are supposed to provide information and senior management. Moreover, management may 'inform' staff of decisions, but not communicate with them. Scott Adam's *Dilbert* has real-world antecedents, and most of our readers will be able to identify plenty of models for the different characters in the comic strip.

Exercises

E7.1. Develop a fault tree for 'my computer will not boot up'.

E7.2. Develop an analysis like Table 7.1 for electronic books (or a similar topic of your choice).

E7.3. Build a spreadsheet for the risks of flood for a grocery supermarket in your area.

8

Measuring how well forecasting goals are met. Part 2

How good is a model? Part 2

In Chapter 3, we saw how 'fit' is defined and quantified in a several ways. Recall that we want a model that exhausts any time pattern in the data. There should only be 'noise' left in the residuals; that is, the deviations between model and data. Once we have the residuals, we usually want to look at their distribution so that we may attempt to apply statistical methods to our model and forecasts. Remember to be careful in assuming normal (Gaussian) distributional properties, but also that samples from normal populations may still show patterns quite different from the classic 'bell' curve. We now extend the ideas that were introduced in Chapter 3 to some very useful statistical measures of the quality of our models.

R_squared – a unitless comparison of model fit

A useful tool for comparing the fit of different regression models (or other models for that matter) is the coefficient of determination or *R_squared*. This statistic has a notation based on the fact that for simple linear regression *with* a constant term, this statistic is the square of the correlation coefficient, usually denoted by *r*. [We, too, are annoyed that many concepts in statistics use the same letter of the alphabet!]

R_squared uses sums of squared deviations as its main computation. While R_squared is traditionally associated with regression models, to the extent that few textbooks talk about it outside that context, we believe it is useful as a comparison tool for all the forecasting models you wish to examine.

R_squared considers the sum of squared deviations for a particular model and compares them to the special 'simple' model that is a single number, *Y*_bar, that is, the mean of the data. We compute, in words,

R_squared(for present model)
   = 1 - (residual sum of squares) / (sum of squared deviations from *Y*_bar)
   = 1 - (residual sum of squares) / (total sum of squares)

For a *linear* regression model *with* a constant term, the residual sum of squares is *always* less than the total sum of squares, i.e. the sum of squared deviations from the mean Y_bar, or $(n-1) * VAR(Y)$, so that for such models $R\_squared > 0$. Clearly it cannot exceed 1. For such models, which are the vast majority of multiple linear regressions,

$$0 \ <= \ R\_squared \ <= \ 1$$

For nonlinear models or models with *no* constant term, this inequality need not hold, and $R\_squared$ can actually be negative. That means that the mean Y_bar is actually a better model for the data! Clearly, we cannot get $R\_squared$ larger than 1, so it is still a good way to check the fit of a model. Just keep in mind that nonlinear or no-constant models can do worse than the mean. We would not want to use them in such cases for forecasting, but some students waste a lot of time trying to find the 'mistake' when they compute a negative $R\_squared$ for such situations.

$R\_squared$ *must* increase with the number of predicting variables (equal to the number of coefficients) for *linear* multiple regression models with a constant term. To compare between models with different numbers of coefficients (or parameters), we use the Adjusted $R\_squared$, which for a model with k parameters ($(k - 1)$ predicting variables) is

$$R\_squared\_adj \ = \ 1 \ - \ (1 - R\_squared) \ (\ (n - 1) \ / \ (n - k) \ )$$

This is *smaller* the greater the number of parameters. It is primarily intended for comparisons of linear regression models *with* a constant term, but is still a useful guideline for nonlinear or no-constant models.

Generally, we recommend only using $R\_squared$ for deviations computed over an estimation period (test or full), and not those for validation test periods. The reasoning behind this recommendation is that for the model estimation periods we are usually 'fitting' the model, so expect the deviations to be relatively 'small'. In the validation period we are allowing our model to advance in time and comparing to actual data that has been set aside. Moreover, as $R\_squared$ is based on squared deviations, we may miss features of the comparison, such as all deviations being of one sign (a consistently 'high' or 'low' forecast). In fact, we try to avoid single number measures of the quality of validation forecasts in favour of graphs and other ways of describing the behaviour of the forecasting model over the validation period.

Serial correlation

Serial correlation is a linear relationship between data elements a fixed number of periods apart in time. We measure the strength of association by computing the autocorrelation and partial autocorrelation functions. A related measure, useful mainly to detect first-order autocorrelation (relationship between data that are just one period apart), is the Durbin–Watson statistic.

Recall that correlation is a measure of linear association between two variables, say W and Z. We compute the covariance of W and Z as

$$COV(W, Z) = (\sum_{t=1}^{n} (W_t - \overline{W})(Z_t - \overline{Z}))/(n - 1)$$

and the correlation is defined as

$$CORR(W, Z) = COV(W, Z) / (SD(W) * SD(Z))$$

where $SD(Z)$ is the standard deviation of Z, which is the square root of its variance.

The autocorrelation is very similar to the correlation. Let us define the variables W and Z from the time series variable Y. Suppose that the time series Y has values $y(1), y(2), ... y(n)$. The mean of this series is y_bar. Then let W be the variable defined by starting at time point 1, so $w(1) = y(1)$, $w(2) = y(2)$, etc. while Z is the variable defined by starting at time point $k+1$, so $z(1) = y(k+1)$, $z(2) = y(k+2)$, etc.

In other words, W is Z *lagged* k periods. Clearly, with finite data, we are going to have to deal with running out of data at the 'ends' by using the first $(n - k)$ data elements of Y for W and the last $(n - k)$ for Z. Furthermore, we can use all of the Y data to compute the variance, rather than compute standard deviations of the two parts of Y that are in W and Z. So our formula for the kth order autocorrelation coefficient, or autocorrelation of lag k is defined, for this sample time series data, as follows. Note that we use subscripts here rather than the $y(..)$ indexing. Readers should learn how to read either form for the elements of a variable or time series.

$$r(k) = (\sum_{t=1}^{n-k} (y_{n-t+1} - \overline{y})(y_{n-k-t+1} - \overline{y})) / TSS$$

$$\text{where} \quad TSS = \sum_{t=1}^{n-k} (y_{n-t+1} - \overline{y})^2$$

Note that we must be careful how we define our formulas. In a spreadsheet the absence of the observations is obvious, and programs such as Minitab 'know' how to compute the autocorrelation. However, one of the most troublesome feature of all time series methods, and autocorrelations in particular, is the need to include the correct elements in the summations.

The collection of the pairs $(k, r(k))$ is taken to form the autocorrelation function or ACF. These *sample* autocorrelations, called $r(k)$ here, have population equivalents rho$(k) = \rho(k)$. These are the true but unknown correlations that we would like to know and that the $r(k)$ estimate, just as X_bar estimates mu $= \mu$ and s estimates sigma $= \sigma$.

Partial autocorrelations

If we suppose that there is an autocorrelation of lag 1, then

$y(t) = \text{rho}(1) * y(t\text{-}1) + \text{error}(t)$

Hence,

$y(t) = \text{rho}(1) * (\text{rho}(1) * y(t\text{-}2) + \text{error}(t\text{-}1)) + \text{error}(t)$

and there is a relationship between $y(t)$ and $y(t\text{-}2)$ *through* the lower order autocorrelation. However, there may be a relationship between $y(t)$ and $y(t\text{-}2)$ directly. We try to uncover this using the partial autocorrelations, which attempt to discount the contribution of lower order autocorrelations to the relationship between $y(t)$ and $y(t\text{-}k)$. We will not attempt to even give the formulas, as they are not terribly important to us (DeLurgio, 1998, Appendix 7-G gives a fairly algebraic treatment). We assume programs such as Minitab compute the partial autocorrelation correctly. The partial autocorrelations are grouped together in pairs $(k, \text{pacf}(k))$ as the partial autocorrelation function or PACF. In this book we will treat them like the autocorrelations (ACF).

One way to view the partial autocorrelation coefficients is as the coefficients of a regression model (Chapter 12) of $y(t)$ on $y(t\text{-}1)$, $y(t\text{-}2)$,..., $y(t\text{-}k)$. If we label these coefficients as b_0 (for the constant) and b_1, b_2, ..., b_k , then the partial autocorrelation of order j is given by b_j . This viewpoint is suggested by Makridakis *et al.* (1998). While it helps as a way to view the partial autocorrelations, we are not convinced that it is an easy concept to understand. Some authors, e.g. Tryfos (1998), choose not to mention it. However, it does aid in identifying some types of models that we will encounter later in Chapters 15 and 16 (ARIMA models).

Use of autocorrelations

The autocorrelations $r(k)$ graphed against k are a common tool to detect time patterns in data (sometimes called the *correlogram*). Note that correlation – and by extension, autocorrelation – is a property of a *linear* relationship between variables. There may be more complicated relationships the ACF does not reveal.

For a truly random series, samples may *still* have non-zero ACF (or PACF) values for various lags. This is the same as having a difference between the population mean mu and the mean X_bar of a sample of n values from that population. Accessible yet rigorous references to the sampling theory for the PACF have so far eluded us.. However, for the ACF, it is well-established that samples of size n from a 'white noise' set of disturbances (errors) will give rise to ACF values for all lags k that have mean 0 and (approximate) variance $1/n$. (The standard deviation is therefore $1/\text{sqrt}(n)$). Moreover, the distribution of the sample values around the mean is

Gaussian (normal). Clearly we can then compute confidence intervals for the autocorrelations. If z is the standard normal variate for a symmetric (1-alpha)*100 percent confidence interval, then the (1 - alpha) * 100 percent CI for $r(k)$ is (from Bartlett, in Kendall and Stuart, 1968)

   0   +/- z/sqrt(n)

Example: 95% of $r(k)$ values from a time series of length 400 will be smaller in absolute value than $1.96/20 = 0.098$.

Wilson and Keating (1998, p. 73) point out that we should be using a t–test (rather than the Gaussian z) when the sample is 'finite', but often this is overlooked in practice. They suggest using a hypothesis test for H_0: $\rho(k) = 0$ versus H_1: $\rho(k) <> 0$ using the test statistic

   $t_{calc} = (r(k) - 0\ )* \sqrt{(n - k)}$

which we compare to Student's t for $(n-k)$ degrees of freedom for $\alpha/2$ where α is the significance level of our test, that is, α is the probability we wrongly conclude that there is autocorrelation at lag k when in fact there is none. We can analyse partial autocorrelations in the same manner.

Ljung–Box Q statistic and its use

Of course, we want *all* the $r(k)$ to be zero at once, and can develop a test to check them, although we have to decide how many lags is 'all'. Typically, we choose a limit such as 15. Minitab, in its ARIMA estimation output, provides tests using the limits 12, 24, 36 and 48 where there are enough data points that the autocorrelations are available. Tests that simultaneously check several statistics are called *portmanteau* tests (Makridakis *et al.*, 1998, p. 318).

The Box–Pierce Q statistic, Q_{BP}, is the simplest of these test statistics, and is computed

$$Q_{BP} = (n - d) \sum_{k=1}^{m} r(k)^2$$

that is, the sum of squares of the first m autocorrelations, where

   n = number of elements in the original time series

   d = degree of differencing to obtain a stationary time series. This relates to the particular use of the Box–Pierce statistic in the Box–Jenkins methodology for estimating ARIMA models. If there are both seasonal and non-seasonal differences used, then the total level of differencing should be inserted for d. These details are not clear in the texts, but clearly will make only very small changes in the value of Q_{BP} for most data.

   k = lag of autocorrelation
   $r(k)$ = sample autocorrelation for k lags based on residuals

m = maximum number of lags to be included in the test Typically, Minitab uses m = 12, 24, 36, 48. We like to check the highest m first, then use smaller values if it seems that the model only fails to satisfy the test for high lag values.

Although not in the above formula, we let b = the number of parameters estimated in fitting a model from which the data (generally residuals) have been drawn to compute the autocorrelations.

The Q_{BP} has a sample distribution that is approximately chi-square with $(m - b)$ degrees of freedom. This allows us to carry out a *hypothesis test* for the presence of non-zero autocorrelation for any lag up to order m, where m is the specified number of lags in the computed Q_{BP} statistic. Let us give the formal structure of this test:

H_0: All autocorrelations $\rho(k)$ are zero up to order m, or $\rho(k) = 0$, $k=1, 2, ..., m$
H_1: At least one of the $\rho(k)$ is *not* zero for $k = 1, 2, ..., m$

Assumptions: we need to assume that the series for which we are computing the autocorrelations, is distributed according to a Gaussian distribution with mean zero and variance σ.

The significance level of the test is α (Greek alpha), which is the probability we reject H_0 when it is *true*. You should set this in advance of carrying out a test. However, we must admit that most practitioners compute the *prob-value* or *p-value*, which is the probability we would reject H_0 if it is true.

The test statistic is Q_{BP}, defined above. We compare its computed value to a Chi-squared statistic with $(m - b)$ degrees of freedom, where b = (number of parameters in model). Thus, if we have three variables plus a constant in a multiple regression model, $b = 4$, and for $m = 24$ lags, we would use 20 degrees of freedom. There are tables of Chi-squared statistics, but we find that we rarely use them, since packages such as Minitab now offer ways to compute the numbers we need. We illustrate via an example below.

Note that we now have a test for the 'randomness' over time of our residuals from a model by using either ACF or PACF. This is a desirable property, although it depends on another desirable property of the residuals, namely their approximation to a Gaussian distribution. While it is nice if our forecasting models have these properties, we first and foremost want them to have small residuals over the estimation period and the validation period. As mentioned in Chapter 3, students will sometimes comment at length on the 'bad' distributional or serial correlation properties of very tiny residuals.

Before we move to illustrating the Box–Pierce statistic, we note that while Box and Pierce (1970) stated that Q_{BP} could be compared to a Chi-squared statistic, one of the authors returned to the subject later, claiming in Ljung and Box (1978) that the alternative *Modified Box–Pierce*, or *Ljung–Box* statistic was better. This is,

$$Q_{LB} = n\,(n+2) \sum_{k=1}^{m} r(k)^2 / (n-k)$$

We feel that, for practical forecasting, the difference between Q_{BP} and Q_{LB} should not be critical to decision making, and while we document which statistic is used, we use them more or less interchangeably. Minitab, in the ARIMA functions, uses Q_{LB}.

Example of use of the Box–Pierce statistic

To illustrate the use of the Box–Pierce statistic, we will actually start with numbers that should be Gaussian. These are generated pseudo-random numbers which are supposed to come from a source that has mean zero and standard deviation of 1.

```
MTB > note Example showing Box-Pierce Q statistic
MTB > note we generate some "random" data (Gaussian noise)
MTB > random 120 c1;
SUBC> normal 0 1.
MTB > desc c1
               N       MEAN     MEDIAN     TRMEAN      STDEV     SEMEAN
C1           120     0.0385     0.0377     0.0462     0.9437     0.0861

             MIN        MAX         Q1         Q3
C1       -2.2667     1.9763    -0.6527     0.7830
MTB > name c1 'gnoise'
MTB > note now put 48 acf coeffs into c11
MTB > acf 48 c1 c11
```

The ACF values are drawn by Minitab in Figure 8.1. They are, in fact, small, as we would hope. However, they are *not* zero, because we are *sampling* the values, that is, taking a finite number of them at random from a population.

Recall that the standard error of the ACF is $1/\mathrm{sqrt}(n)$, in this case 0.0913. An approximate 95% confidence interval is twice this (1.96 times it, to be correct) or approximately 0.18. In Figure 8.1, Minitab has drawn lines at roughly +/- 0.18 on the ACF graph, and most of the ACF values are within the lines. These lines are not perfectly straight; we believe this is due to the use to $1/\mathrm{sqrt}(n-k)$ for the standard error of $r(k)$ as in Wilson and Keating (1998, p. 73).We could just look at the ACF numbers, although this is, in our view, more tedious. Generally we use the character graph of the ACF since this is much easier to include in log files and to transfer to reports. (Figure 8.1 cost us several hours of work.)

To perform the hypothesis test for the Box–Pierce statistic, let us choose to use $m = 48$ lags, which is a quite severe test. The hypotheses are

H_0: data has true population autocorrelation coefficients all zero up to lag 48
H_1: at least one of the underlying autocorrelations up to lag 48 is non-zero

We set $\alpha = 0.05$, that is, we are willing to reject the null hypothesis, *even if it is true*, if there is less than a 5% chance of so doing.

The Q_{BP} can be computed from the saved autocorrelations in column C11. We find $Q_{BP} = 39.7394$. This value is to be compared to a Chi-squared statistic (χ^2) for 48 degrees of

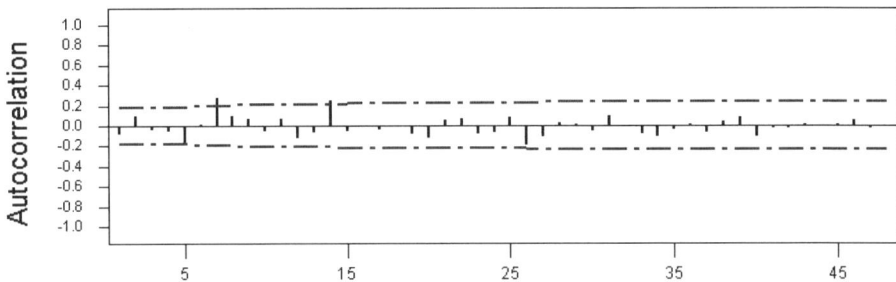

Figure 8.1. ACF of a pseudo-random Gaussian sequence using Minitab 'graphic' ACF.

freedom, since we have no parameters here. (This will be unusual in our later work.)

One way to do this is to look for the value of the Chi-squared statistic for 48 degrees of freedom that will be exceeded 5% of the time. In tables – for example, those in the back of many texts – we usually do not find 48 degrees of freedom listed. Aczel (1996, p. 828) has Chi-squared$(50, 0.05) = 67.5048$. This means a Chi-squared(50) statistic will exceed 67.5048 one time in 20. Most people, ourselves included, have to work at understanding the tables.

We find an easier approach is to ask software such as Minitab to compute the probability our Q_{BP} will or will not be exceeded. This is accomplished with the commands (there are also menu choices; note that the semicolon and period are important.)

```
CDF 39.7394;
chisquare 48.
```

This gives the result

```
39.7394      0.2039
```

meaning that the probability a Chi-squared(48) is less than or equal to the computed value 39.7394 is only 20.39%. There is an almost 80% chance of getting a bigger value.

Using any of these approaches, we conclude that the Box–Pierce Q_{BP} is not 'big' and is consistent with our null hypothesis. The autocorrelations up to lag 48 can be taken to be all zero.

If you prefer to find the 'table' value of Chi-squared(48) that will be exceeded 5% of the time, we need the INVCDF function of Minitab or equivalent software. Here is what the output looks like.

```
MTB > invcdf .95;
SUBC> chisquare 48.
Inverse Cumulative Distribution Function
Chi-Square with 48 DF
  P( X <= x)          x
    0.9500        65.1708
```

Since 39.7394 is less than the output value, we do *not* reject the null hypothesis. Note that the cumulative and inverse cumulative distribution functions work with the values of statistics that are *not* exceeded.

You will rarely see situations where the autocorrelations are so 'nice'. Therefore, we will give an example where there *are* non-zero autocorrelations. We create a sine wave pattern, then add the noise series above. We then compute the Box–Pierce Q_{BP} and test it. The $Q_{BP} = 273.7107$, so we can use the results above to conclude that there really is at least one non-zero autocorrelation here. Of course, we could see this clearly from the ACF plot below, even if we do not bother to draw in the confidence interval lines. To underline this result, we compute

```
MTB > cdf 273.7107;
SUBC> chis 48.
   273.7107    1.0000
```

This implies 100% of Chi-squared(48) statistics are less than the computed Q_{BP}. There is no chance we would observe this value if the null hypothesis is true and the assumption of normal disturbances holds.

```
MTB > note now create a series WITH autocorrelation
MTB > let k50=4*atan(1)
note this gives pi = 3.14159
MTB > set c2;
DATA> 1:120;
DATA> end;
MTB > name c2 'count'
MTB > let c3=2*sin(c2*k50/24)
MTB > name c3 'sinwave'
MTB > let c4=c3+2*c1
MTB > name c4 'sindata'
MTB > note c4 has sine wave plus noise
MTB > acf 48 c4 c21
```

It is straightforward to use Minitab and similar packages to experiment in this way with the Box–Pierce statistic and related tests. If you need to use these ideas in your forecasting, we recommend such experiments as an aid to understanding.

```
ACF of sindata
            -1.0 -0.8 -0.6 -0.4 -0.2  0.0  0.2  0.4  0.6  0.8  1.0
            +----+----+----+----+----+----+----+----+----+----+
  1    0.303                              XXXXXXXXX
  2    0.376                              XXXXXXXXXX
  3    0.201                              XXXXXX
  4    0.359                              XXXXXXXXXX
  5    0.201                              XXXXXX
  6    0.230                              XXXXXXX
  7    0.110                              XXXX
  8    0.224                              XXXXXXX
  9    0.192                              XXXXXX
 10    0.153                              XXXXX
 11    0.027                              XX
 12   -0.017                              X
 ... etc.
```

The Durbin–Watson statistic

The Durbin–Watson (DW) statistic for a time series Y is given by

$$DW = \frac{\sum_{t=1}^{n-1} (e(t+1)-e(t))^2}{\sum_{t=1}^{n} e(t)^2}$$

where $e(t)$ is generally the residual at time t. The DW statistic is used mostly in conjunction with multiple regression models. The Durbin–Watson test for serial correlation is based on the statistic given above. This tests *only* first-order autocorrelation. The model for the *theoretical* errors eps(t) upon which the test is founded is

eps(t) = rho * eps(t-1) + $v(t)$

where $v(t)$ is some uncorrelated error. We want to test whether rho can be taken as zero versus the possibility it is not zero.

The DW statistic takes values between 0 and 4. When the deviations $e(t)$ vary slowly, then from the formula above we can see that DW will be small. But when the $e(t)$ vary slowly, we have positive first-order autocorrelation. Conversely, if the $e(t)$ vary very quickly, indeed more quickly than expected, we have negative first order autocorrelation. If the $e(t)$ have a saw-tooth pattern, then the difference between adjacent deviations will be twice their magnitude (i.e. from +1 to -1). Squaring gives a value of 4 as the most extreme DW statistic.

The steps in the test are as follows.

1. Compute the multiple regression model in p-1 independent variables plus a constant (that is, the dimension of predictor matrix is n by p), and compute the residuals.

2. Calculate DW, if this has not been computed within some other procedure. Note that we could compute DW for any set of residuals, but will rarely do this, since we might as well use the ACF and Box–Pierce or Ljung–Box Q.

3. If DW > 2, then use DW' = 4 - DW for the rest of the test. This implies that we suspect negative serial correlation. Otherwise, we are testing for positive serial correlation.

4. Compare DW or DW' with tables of critical values for the Durbin–Watson statistic for the appropriate n and p-1. There are two values L and U for each set of significance level, n, and p. We should find $0 < L < U < 2$.

    a) If DW > U, there is no first-order serial correlation;

    b) if L <= DW <= U, the test is inconclusive;

c) if DW < L there is first-order serial correlation. This is positive serial correlation if we have used DW, negative serial correlation if DW' was used.

Note that the significance of the test is decided by the tables. We have noted some possible inconsistencies between the published tables. We believe that readers and students should know about the DW test and more or less how it is performed, but we generally prefer the (Modified) Box–Pierce Q statistic or other portmanteau test, since these require no tables if we are using Minitab or a similar package and deal with several orders of autocorrelation. We will reserve our example of the use of the DW statistic until Chapter 12. For more discussion of the DW statistic, see Draper and Smith (1981, section 3.11).

Desiderata and parsimony. Part 2

Our list of desiderata for forecasting models is getting longer, and is evolving with our managerial goals:

1. Small residuals over the estimation period, whether for the trial or main model

2. Zero mean residuals over the estimation period

3. If the residuals are not very small, then they are distributed in a well-behaved way that allows us to perform statistical tests, as in items (4) and (5) below

4. Small autocorrelations or partial autocorrelations over the estimation period

5. Model parameters that are significantly different from zero

6. Small errors for validation forecasts

7. Zero, or at least small, mean for validation forecast errors

And again, on top of all these, we would like

8. A parsimonious model, or one that can be easily explained

All of the above are coloured by the amount and quality of the data available to us for forecasting, by our forecasting needs in terms of the time-frame, nature and precision of forecasts required, and the time and resources that we have at our disposal.

Exercises

E8.1. There are many possible measures for the 'fit' relative to how many parameters we use. Investigate the meaning and motivations of the following measures:

• Theil's U statistic

• Aikake Information Criterion

• Bayesian Information Criterion

E8.2. Try to re-create the example of the use of the Box–Pierce statistic. Note that because the numbers used are *generated*, your results may be different from those we obtained. Modify the calculation to use the Ljung–Box statistic rather than the Box–Pierce Statistic.

E8.3. Re-examine your answers to Exercises E3.1 and E3.2. What changes, if any, would you make in your answers?

9

Preliminary data analysis
for forecasting

The purpose of this chapter is to show how to carry out a preliminary analysis of a real data set. Because there is such a variety of different possible situations, we must necessarily present more tools than the reader is likely to apply in any single problem. The purpose of the preliminary analysis is to understand as much as possible of what has been, and is, happening so that we can proceed to choose an appropriate method for the forecasts we desire. The preliminary analysis will almost always lead to some preliminary forecasts. In fact, the generation of such forecasts is often a very natural extension of the graphs and summary statistics we will prepare. In this book, however, we will postpone the preliminary forecasts to Chapter 10.

We assume in our presentation that we have already sought and gathered at least some data, and that we understand the nature and time horizon for the forecasts that we wish to prepare (Chapter 2). Now we dive into the detail. Students have specifically mentioned the usefulness of this detail in working through their own problems. However, we have tried to put the tedious nuts and bolts in the Appendices.

Graph the data!

If we can persuade you of just one rule in forecasting, it is

GRAPH THE DATA!

Sometimes a simple graph is sufficient to tell us all we need to know for forecasting, even if we decide to pursue more sophisticated approaches to confirm what the graphs reveal by simple observation. By 'graph the data', we do not mean a single display, but a set of suitable graphs that help us to understand what may underlie the situation at hand. We believe that pictures, providing they are faithful to the situation under study, carry more weight than words with the audience for our forecasts. Bad graphs are a waste of time and effort. To learn about graphs we recommend the books by Cleveland (1985) and by Tufte (1983). While different in style, both have much to offer anyone who wishes to prepare or study graphs.

What should we choose to graph in order to help us forecast? First, we need to decide on the particular variables we will graph. If we have data, over time, of the variable(s) we want to forecast, then these are obvious choices. From our understanding of the situation under examination, we may be able to suggest other variables that are related to the one we wish to forecast. We will also want to look at various *derived variables*, namely variables that arise from transforming, adjusting, aggregating, disaggregating, or otherwise combining or manipulating our data. We discuss such derived variables below.

Next, we need to know what graphs can tell us. The two main functions are:

- They show the general movement of data over time, or the general 'shape' of data in the form of a distribution, or the overall 'size' of errors or deviations. In this role, they present what is usual.

- They help us spot 'outliers' or obvious deviations from usual behaviour.

That is, graphs help in establishing what is normal and in highlighting what is different from the norm. In helping us perceive the abnormal however, there is little we can do unless we have good documentation of the background to our situation. What does a day of poor sales mean unless we know, for example, that there was a blizzard so our store was closed. This documentation, as we have already mentioned, is often called *metadata*.

It is not our intent here to provide a treatise on how to graph data, but it will help to review a few key principles.

1. Graphs should be scaled so that they reveal clearly the variability of data.

2. Despite the previous point, the starting level and scale of data should be clear.

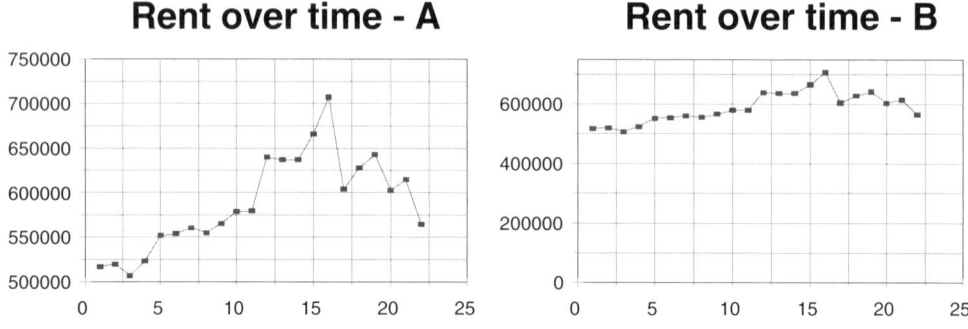

Figure 9.1. Illustration of axis scaling to show variability. Note that the graph on the left does NOT have the y-axis starting at 0, so the variation is more pronounced. In relative terms, the variation is not as great as we may be led to perceive. Note that the *x*-axis labelling style depends on the choice of the 'type' of chart, in this case a 'Specialty' chart. Such details may seem unimportant, but can waste a lot of time when we seek consistency of appearance.

Figure 9.1 presents some data for rent paid over time in two different scalings to illustrate our concerns here. Some practitioners draw graphs such as those in Figure 9.1 so that the y axis has a clear interruption, such as a zig-zag, to indicate that the bottom of the graph is *not* zero. We know, however, of no commonly available software for such graphs. The main issue is clear documentation, so that the reader may recognize that the y axis does not start at zero.

Continuing our suggestions regarding graphics:

3. Graphs should be 'information rich'. That is, they should convey a lot of information, although not overwhelm the reader. Pie charts, in particular, convey very little information but occupy a lot of display space. By contrast, it is often straightforward to draw multiple graphs on the same set of axes, for example, the separate quarterly data for the Quarterly Traffic Fatalities in Canada, as in Figure 9.2. This is based on data subsets, discussed later.

4. Where possible, we should annotate the graph to point out key information. For example, the introduction of new rules or regulations at some point in time may be expected to alter outcomes, so the time of introduction should be noted on a time graph. Figure 9.2 shows such an annotation.

Fatalities by quarter over time

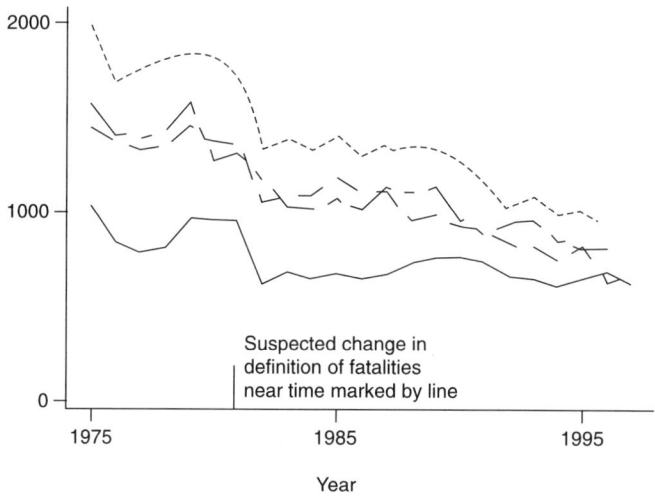

q1 solid, q2 dash, q3 dot, q4 dot-dash

Figure 9.2. Multiple plot: individual quarters vs. year for Quarterly Traffic Fatalities in Canada.

When drawing graphs, computers can be a great help. They can also be a huge hindrance to our analysis. In particular, computers are often programmed to scale the axes automatically so the data fills the space nicely. This can be disastrous if we are trying to compare data. For example, we frequently want to look at residuals, that is, the deviations of data from models. When we prepare a time plot of a two sets of residuals, we must be careful to specify the scale for the residuals, or we will not see which of the sets of residuals is larger. We urge you to keep in mind the importance of the choice of scale of axes, and especially consistency in this choice.

We now describe some types of graphs we will use for our analysis and forecasting.

Time plot

The *time plot*, also called a time series plot, is simply a graph of our data against time. It is one of the most common graphs for displaying social, economic and political data. The graphs used in describing the Pegels' classification are time plots, and we clearly will use a time plot to help us decide if we can classify our data following such a classification. Sometimes the points are joined, while other choices are to use symbols, particularly if more than one series is drawn on the same set of axes. The choices should, hopefully, be aesthetically pleasing and be readable without a magnifying glass. That is, the scale should be suitable for human viewing. Modern laser and ink-jet printers have presented forecasters with the difficulty that time plots for long series must be 'stitched' together. See the Appendix for more details.

Figure 9.3 lets us see what our Quarterly Traffic Fatalities in Canada data look like. Evidently, we have a quite strong seasonality with a modest downward trend, although the sudden drop at index point 28–29 suggests that something may need our attention. We suspect, based on conversations we have not been able to support with documentation, that a change in the definition of a traffic fatality was introduced around 1981.

The time plot is also the basis for what we call the 'ruler forecast' where we place a ruler on the time plot and extrapolate to get a very crude forecast as a guide to the scale of numbers we may expect. Two refinements to the simple 'general trend' are worth mentioning:

- The ruler can also be used to draw 'envelope' lines by setting the ruler to positions that have all points above or below it. (In the case where there are suspected 'outliers', one may relax the principle.)

- By use of an acetate sheet or similar tool for capturing pattern, one may forecast seasonality.

More details concerning the drawing of the ruler forecast appear in Chapter 10 and the Appendix.

We caution that it is important to get the time points correct in time plots. Data for a particular time period may not be reported until much later. We want to graph it at the time for which the data applies in almost all cases, not the time of publication. An unfortunate aspect of the time plot as presented in Figure 9.3 – and we would point out that this figure is typical – is that most humans do not easily relate the index of the time period (in this case the quarter number) to actual time periods. The quarters in the Quarterly Traffic Fatalities in Canada data run from 1975-1 to 1997-2, that is, 22.5 years or 90 quarters. If we continue the data to period 93, we get quarter 1998-1. This will have particular reality for people living in eastern Canada and parts of New England as the time of the 'big ice storm' and its aftermath. 'Period 93' just does not convey this reality. Should we decide to drop data prior to 1981 as 'too old' to be relevant, the period numbering will change. Thus, we would like to have software that would provide a better time axis. This is not a trivial task, so the alternative is to include information in the caption of the time plot so that readers have some information with which to position their view.

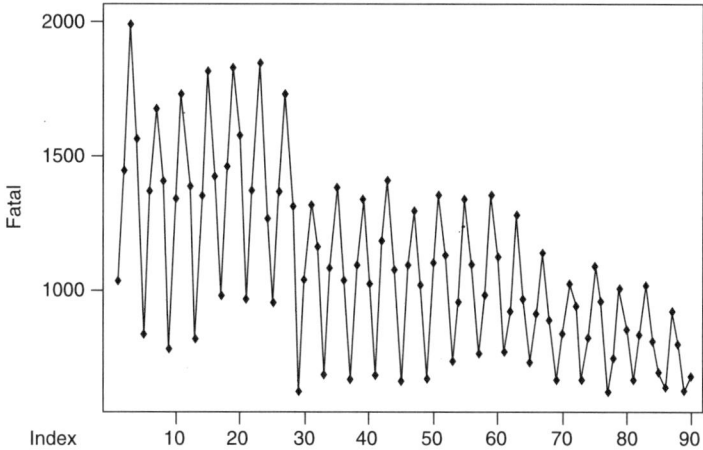

Figure 9.3. Time plot of Quarterly Traffic Fatalities in Canada. The time index is the number of the quarter (three-monthly calendar period) with the index=1 for quarter 1975-1 (Jan – Mar. 1975).

Although we were concerned above that the scale of the time plot may make it difficult to distinguish the individual time points, sometimes the time scale may be too expanded to see long-term trends in the data. It is often useful in our preliminary analysis to use tools that prepare scatterplots rather than time plots, since we can then control the axis scaling to provide the compression or extension desired. With some software, the aspect ratio of the graph may be controllable by 'dragging' the corner or other control points with the mouse or other pointer. We strongly recommend viewing time series data at several scales while exploring it.

Histogram or other distributional plot

The distribution of data is important in showing us the relative likelihood that different values of a variable are actually observed. This can be important for forecasting the likelihood of observations in the future if the distribution is expected to persist. For example, for a stationary series (no trend), and especially one without seasonality or cycle, we can use a histogram to predict the probability that observations greater than some stated value will be observed.

```
Stem-and-leaf of oct        N  = 106
Leaf Unit = 0.10            N* =   1

   16      0 0000111223334444
   45      0 5555555555666666666777789999999
  (27)     1 000000000001111112223333444444
   34      1 555677889
   25      2 112234444
   16      2 556777
   10      3 14
    8      3 67
    6      4 2
    5      4 67
    3      5 34
    1      5
    1      6 0
```

Figure 9.4. Stem and Leaf diagram of Boulder precipitation data.

For example, we have some data on precipitation (in inches) observed in Boulder, Colorado, for the month of October from 1894 to 1999 (some observations for 2000 are in the file we used, but October is missing). There are a total of 105 usable observations. See the PFM Web site for details. A time plot of this data shows some ups and downs which we will ignore for the moment, but we assume there is not any major trend; that is, that the data is stationary. We draw the Stem and Leaf diagram of this data as Figure 9.4.

Figure 9.4 makes it quite easy to predict that there is a rather small probability that there will be 6 or more inches of precipitation in a month. In fact, based on past experience from the 1894–1999 period, we can say this has occurred in only 1 out of 105 months, or slightly less than 1% of months. Such information is important for planning storm sewer capacities, for example.

Another use of distributional graphs is to show us that data (or residuals or deviations of data from some model) have a particular distributional shape. It is an important assumption of many statistical tests that data or deviations be Gaussian (normally) distributed, yet in practice it is very common that this is not the case. Quantile plots, mentioned below, allow departures from a particular distributional shape to be visualized and interpreted more precisely.

In Figure 9.4, we have used a Stem and Leaf diagram. Histograms are generally a more common and traditional form of distributional plot in the business world, and are also better suited to large data sets (including the present precipitation example). For modest data collections, the Stem and Leaf diagram has the advantage of avoiding some of the information loss when we *bin* the data to draw the histogram. *Dotplots*, a character-based graphic (Wilkinson, 1999), or similar constructs also serve to display distributional information. (Minitab can stack dotplots for comparing distributions; we can also stack boxplots, as discussed below.) For the heavy-duty user, one could also consider kernel density plots. Some distributional information is also revealed by *boxplots*. Most modern textbooks in applied statistics, such as Aczel (1996), present reasonable descriptions of histograms, Stem and Leaf diagrams and boxplots. See the PFM Web site for more examples. Minitab offers character-form histograms as well as a bar-chart form. The character-form histogram has the count of observations in the bin (or bar), which can be used in the same manner as the Stem and Leaf diagram to compute probabilities that observations will be realized.

Boxplots

The boxplot, or box-and-whisker plot, was introduced by John Tukey (1977) to display summary information about one or more variables. *Multiple* or *stacked* boxplots have become a common tool in modern data analysis to allow a quick comparison of a group of variables, possibly subsets of the data selected on the basis of different characteristics or time periods. An example is Figure 9.12.

There are various extensions and variations to boxplots, e.g. Rousseeuw et al. (1999). Figures 9.5 and 9.6 show two forms of the boxplot. The brackets '(' and ')' are *notches* that provide an approximate 95% confidence interval for the median of the data. In Minitab, notched boxplots are available only in the character-form graph.

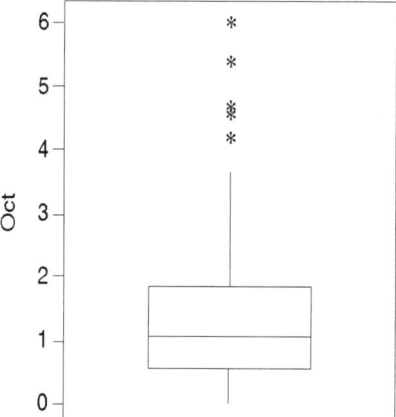

Figure 9.5. Hi-res form of the individual boxplot for the precipitation data for Boulder Colorado, month of October.

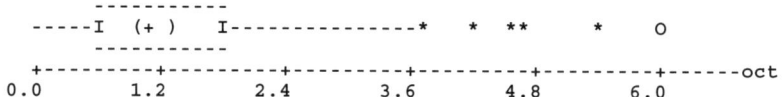

Figure 9.6. Character form of the individual boxplot for the precipitation data for Boulder Colorado, month of October.

Scatterplots for possible relationships

The graph of one or more variables against another is called a scatterplot. These simple graphs are among the most widely employed in data analysis. The time plot is clearly a special case where the horizontal axis is the time index of the data. Scatterplots help us to discern possible relationships between variables, but with equal importance they point out potentially significant departures from 'usual' behaviour.

As an example, we graph the maximum monthly temperature at the Bundaberg Post Office in Australia against the minimum monthly temperature. Generally, this follows a linear relationship that we could use to provide a fairly good forecast of maximum temperature from the minimum. However, there are a few outliers that it would be useful to identify. *Point identification* is not available in Minitab, but Stata, DataDesk and other software do permit the outliers to be identified. We have used Stata to draw Figure 9.7, using the row index of each point as its plotting symbol, which is easy to do with the Stata `graph` command. However, while it is not particularly difficult to do so, obtaining 'nice' axis divisions and labelling only the obvious outliers in this graph is fairly tedious. For example, point 104 is August 1915, while point 379 is July 1938, which would be a better labelling except for the length. Spreadsheet software is not very helpful in preparing scatterplots useful for data analysis and forecasting.

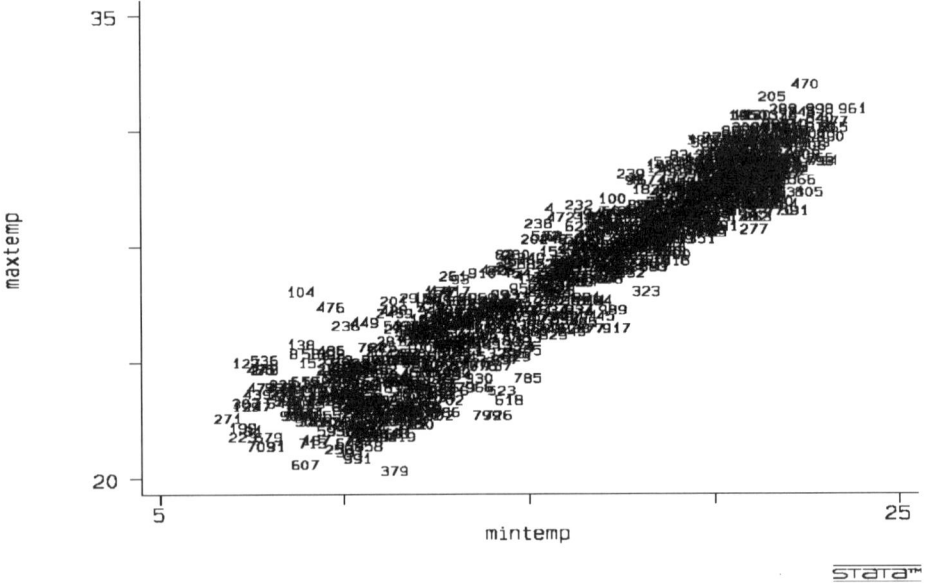

Figure 9.7. Scatterplot of maximum versus minimum monthly temperature in degrees Celsius at Bundaberg Post Office (Australia) 1894–1990, with points identified by their observation number.

Quantile plots

The pth *quantile* of a set of data is the value below which fraction p of the data lies. This corresponds to the $100*p^{th}$ *percentile*. Quantiles and percentiles are based on ranks, so data that follow a known distribution can be characterized by its quantiles. The standard Normal (or Gaussian) distribution having mean 0 and standard deviation 1 has quantile $Q(0.5)$ at 0 (the median as well as the mean). The Empirical Rule uses the approximate quantiles from Table 9.1.

Table 9.1. Standard normal (Gaussian) quantiles

Cumulative probatility	0	0.0228	0.1587	0.5	0.8413	0.9773	1
Quantile		~= -3	~= -2	~= -1	0	~= +1	~= +2	~= +3

In fact, we can draw the graph of the standard Normal as $Q(z)$ versus z from the usual tables given in statistics texts. It looks like a stretched out 'S'.

Given an observed column of data, we could sort it, so $X(i)$ is the ith observation. Following Chambers *et al.* (1983, p. 12), we then could draw the quantile plot by graphing the quantile fractions $q(i) = (i - 0.5)/n$ versus $X(i)$. This adjustment by 0.5 is intended to correct for the fact that the observed data are sampled, supposedly from an infinite set of possible values that follow the governing probability distribution. Our quantile plot can then be compared with the 'theoretical' graph we draw from the tables of the distribution, such as the standard Normal.

Figure 9.8 presents such a quantile plot for the Boulder precipitation data for October. This is the graph of $q(i) = (i - 0.5)/n$ versus the sorted data $Xs(i)$, where $n = 374$ is the number of observations. For comparison purposes we have also plotted $q(i)$ versus the *normal quantiles*, $Xn(i)$, where we compute the $Xn(i)$ as follows:

Let μ = mean of precipitation (we actually use the mean of the data) and σ = standard deviation (again from the data), and $F(z)$ is the cumulative distribution function of the standard Normal (Gaussian) distribution. Thus,

$F(z) = P(Z <= z)$ where $Z \sim N(0, 1)$

i.e. Z is distributed as a Gaussian with mean 0 and variance 1,

Then $Xn(i) = \sigma F^{-1}(q(i)) + \mu$

That is, we compute the *inverse cumulative standard Normal function* and adjust to the mean and standard deviation of the data.

Judging how closely a curved line approximates its theoretical ideal is not a talent many of us possess. Therefore, we make the task simpler by transforming the data or else the plotting scale so the 'ideal' or theoretical graph should be a straight line, since most of us are quite good at detecting deviations from straight lines. This skill can be improved – and we strongly recommend this approach – by the use of a small ruler. Our preference is the 6–inch clear plastic variety that can be easily carried and which allows graph features that are under the ruler to be seen.

Minitab and other software include tools for drawing quantile plots, in particular the Normal Probability Plot. There are several styles for drawing such graphs. Novice data analysts may find these differences unsettling. Look for the common idea, that is, a graph that draws rank versus value and adjusts the scale so that the ideal pattern is a straight line.

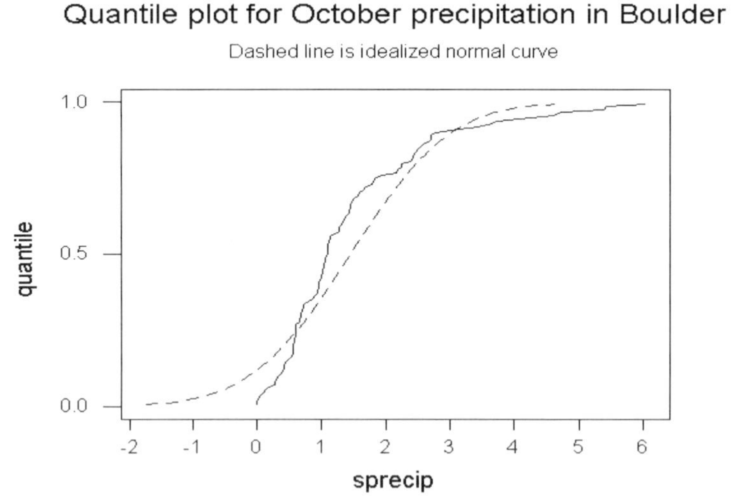

Quantile plot for October precipitation in Boulder
Dashed line is idealized normal curve

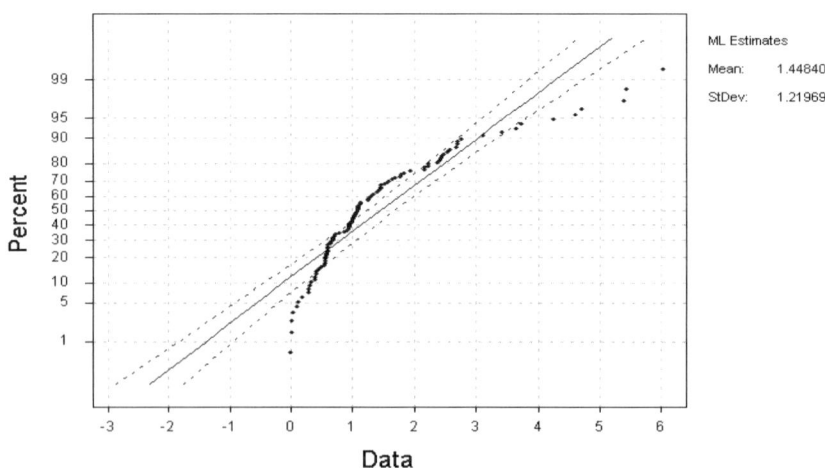

Figure 9.8. Quantile plot in linear scale with superposed idealized curve. We need to transform the axis scaling to make the deviations from normality easier to visualize.

Figure 9.9 shows a Normal Probability Plot for the Boulder precipitation data. We already suspect that this is *not* Gaussian from earlier distribution plots, but this tool makes it very clear.

Normal Probability Plot for precip

Figure 9.9. Normal probability plot of the Boulder October precipitation data. The title is a result of using 'precip' as the variable name and the Minitab automatic title generation.

Colour, patterns, symbols and shading

We can use colour, special patterns of lines or special symbols, or shading of different parts of a graph in order to include more variables on the same plot. In this book, and generally, we avoid colour, even though it can be useful for display and presentation purposes, because it is still relatively awkward and/or expensive to print. This is a choice in approach to forecasting (Chapter 11). Even where colour printing is easily available, the colours seen on a screen may appear different on paper or overhead slides or a projection screen. If our clients need to print or photocopy our work, colour may add nuisance and expense. In such cases, using different line types, shading, or hatching patterns may allow our work to be presented clearly.

When our work is to be presented via a colour screen, e.g. a Web site, then colour is a definite help and should be used. We advise against using a dark background and light coloured type, however. Sometimes printers ignore the background and do a wonderful job of 'printing' white on white! We also do not want to annoy people by forcing them to use up lots of ink or toner for the dark background. Note that colours display somewhat differently on different systems; see www.efuse.com/Design/web_color_basics.html.

For internal analysis reports and student assignments, a simple and inexpensive method for using colour is to highlight manually with a coloured marking pen the lines on a graph, using a small ruler for neatness. This is surprisingly effective; it also saves hours of 'fiddling' with special options in software packages that are supposed to make coloured graphics easy. For machine readable material, it may be possible to add coloured lines or highlighting with a drawing or painting program, although this is seldom as easy as we believe it should be.

Adjustment of the data

The data for forecasting is almost never immediately suitable for our methods. Or perhaps we could better say that our methods need adapting to the situations for which they are used. Such 'adjustments' may be due to special events that do not correspond to the assumptions we wish to make about the 'usual' course of events. For example, severe weather may cause loss of business or unusually high accident rates. Other adjustments are used to render data consistent between periods. For example, a city may annex a neighbouring suburb, so water consumption and similar figures will rise accordingly. We will likely want to use consumption per household or per capita to render data consistent over time. Finally, we may need to impute data that is missing for some reason. Let us consider these different types of adjustments.

Special adjustments for unusual events

The need for such adjustments may be the easiest to justify, provided that we have the ancillary information about the special conditions or events that have affected our data. Making the adjustment is more difficult, and falls into the same sand-trap as missing data. For example, failure to incorporate the fact of the great ice-storm of January 1998 into many types of data for Eastern Canada will certainly distort forecasts.

Imputation for missing or known incorrect data

Imputation is generally difficult and should be left to specialists in statistical agencies, for example, see Sande (1982). In principle, we wish to adjust to conform to our expectations. For example, we may expect that if sales one month are 25 units, and two months later they are 35 units, that it may be reasonable to suggest that for the month in between they might be 30 units. Of course, if the 'months' are really 'days' and the middle day is a Sunday, then we could be totally misled by this simple interpolation. Nevertheless, interpolation – use of the average of the adjacent values, or a smoothing of nearby values – is a common method for imputing missing data or for substituting for values that are known to be inappropriate because of special events.

Scaling of data

We often scale data when forecasting to avoid masking behavioural changes in general growth of the population of interest. For example, when an automobile maker ships a new model, there are initially very few cars actually available to break down. If the rate of breakdown is proportional to the number of cars delivered to customers, then we can forecast breakdowns by forecasting deliveries. Of course, manufacturers would really like to know if they are actually managing to reduce the proportion of breakdowns. Therefore, it may be better to work with 'breakdowns per 1000 cars delivered'. If the data can be linked to the date of manufacture, then it is even more useful because we know which cars are giving trouble.

Trading day adjustments

A very common scaling of this type is a *trading-day adjustment*. This adjustment divides monthly sales or similar data by the number of days that business was conducted. That is, we work with, and forecast, not our total sales etc., but the rate of sales per day or other unit of time. For modern retail businesses, we may need to refine such adjustments to count hours or possibly even a weighted average of hours, since there are 'prime' and 'non-prime' times for shopping. Another possibility that arises quite often is the number of weekends in a month, clearly an important factor in tourism and recreation industries. This is an area of forecasting where an intimate knowledge of the background and situation are extremely important, and a strong reason why forecasting should not be delegated to those whose only function is forecasting. More details are given in the Appendix. First, we may want to examine how the data are collected and the meaning of the numbers. If we are working with sales data, different periods of the year may imply different conditions under which data are collected or applied. For example, many retail stores open extended hours for different seasons of the year: summer vs. winter, Christmas or holiday hours, school term vs. break. For insight, we may examine sales per hour or sales per $100 of salary paid.

Typically, we make such adjustments to render the basis of measurement the same for different time periods. A more difficult adjustment of this type occurs when the time scale is 'wrong' for the forecasting task. The classic example here is a data series of airline passenger arrivals in Israel. These are given by month on the usual Gregorian calendar. Pope Gregory XIII commissioned the work on the calendar scheme bearing his name, and, in 1582, ten days were 'dropped' to realign the calendar with the astronomical seasons, and leap year rules were revised. Other jurisdictions adopted this idea much later: in 1752 in England and not until the 20th century in regions under the Eastern Church. Other calendars still have uses. The Jewish calendar is approximately two millennia older than the Gregorian form. Its use is crucial to the understanding of airline passenger arrivals data for Israel, which is presented as an exercise for this chapter.

First differences

First differences are rates rather than totals. They aid insight by highlighting changes or differences in data. As an example, consider the changes in the closing price of a stock or commodity from one trading day to the next. (Note that we now explicitly look at trading days. Would trading days preceding or following holidays or weekends be important?) A time plot of the *first difference*, that is, the data at a given time point minus the data at the preceding time point, shows changes or patterns in such day to day changes. However, in the case of our Quarterly Traffic Fatalities in Canada data, a time plot of first differences will be confusing because we have obvious cyclic patterns in the numbers of fatalities.

Error checks

Sometimes scaling is a helpful tool to discover reporting errors. Some years ago, a newspaper reported the dire news that North Americans threw away 200 million *tons* of aluminum pop cans every year. A simple consideration of the numbers shows the report must be in error. Pop cans are quite light; let us assume there are about 10 cans per pound of aluminum (actually an overestimate). This suggests that there are $200 * 10^6$ tons $* 2000$ lbs / ton $* 10$ cans / lb = $4 * 10^{12}$ cans thrown away each year. If we generously include Mexico, the population of North

America is approximately $4 * 10^8$ implying a consumption of 10,000 cans of pop per year per person! One suspects that the unit should have been pounds or kilograms rather than tons in the report.

Transforming data to help us understand it

Apart from such adjustments, we may look at derived variables to help understand the variable(s) that we wish to forecast. We have already used a transformation in drawing quantile plots. Statisticians often lump scaling and transformation together, but most managers view them (we believe correctly) as different operations. Scaling can be viewed as changing the numbers on the graph axes – the picture does not fundamentally alter. A transformation actually changes the shape of the graph. Transformations may allow methods to be used outside their usual zone of applicability. For example, a method that extrapolates a linear model – a straight line – will not forecast the growth of the amount of compounded principal of an investment. Recall that the amount is

$$\text{Amount} = \text{principal} * (1 + \text{rate}/100)^{\text{periods}}$$

which is an exponential growth curve. However, if we take logarithms (base 10 or base e) we have

$$\text{Log(Amount)} = \text{log(principal)} + \text{periods} * \text{log}(1 + \text{rate}/100)$$

which does follow a linear form (Figure 9.10). Note that we must back-transform when we want the forecasts in a form we can use. That is, we cannot spend log($), only $. Students often know how (and even why) to transform data. Back-transformations seem to give them trouble.

To spot changes in behaviour, we can use first differences (see above) of data series $x(t)$,

$$d1(t) = x(t) - x(t\text{-}1)$$

Series $d1$ has one less element than the series x. The distribution of the first differences, e.g. a stem-and-leaf plot, shows the historical frequency of large changes in a single period. If we are told that a particular stock is going to increase $5 tomorrow, but it has only ever achieved such a change once in the last ten years, the tip has little historical support. However, if something that underlies the stock price has changed drastically, we cannot trust in a principle of continuity, and this approach will not help us.

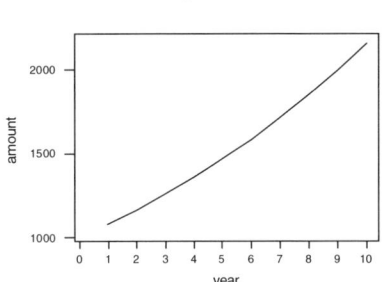

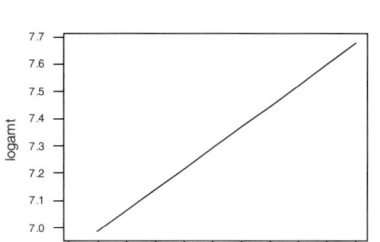

Figure 9.10. Illustration of logarithmic transformation. The data are the amount of compound interest on $1000 at 8% per annum. Natural logs (base e) are used in this transformation in the right-hand graph.

Level and variability

The Pegels' classification is a form of description of our data in terms of level and variability. That is, the *level* of our data gives us an idea 'where' the numbers are found in a general way. This is like saying that an average family in some moderately developed nation has 2.45 children. No family has this number of children. We are likely to want to know the variability. For example, are there a great many childless families and a few with more than a dozen, or is the pattern mostly 2-children families, with some single and 3-children ones? Such *variability* (or spread or dispersion) will be of interest. The boxplot is a good way to see both level and spread. Histograms or stem-and-leaf diagrams can also be used, but only if we are careful to set the scales to be the same. Software generally does not make this very easy, while multiple (or stacked) boxplots are commonly available. In the video series 'Against All Odds', a nice example of a 'back to back' stem and leaf diagram is presented to compare the rates at which hysterectomies are performed on patients by male and female gynecologists. Has anyone seen software for such an effective graph?

Often the variability in our data will depend on the level or on the time index. This can be seen fairly quickly from the ruler forecasting lines used to form the upper and lower envelope to our data (see Chapter 10). If these envelope lines are parallel, then the seasonal variation is *additive*. If we have a seasonal variation that increases (or decreases) with the level, we have a *multiplicative* seasonality. The Quarterly Traffic Fatalities in Canada data have a spread decreasing with time. Since the level of fatalities is also decreasing with time, the spread is still roughly proportional to the level. Spread-level (or spread-location) graphs help us see such relationships (Cleveland, 1993). To prepare such graphs we need to subset the data according to some criteria, e.g. years, and examine the yearly ranges versus their means – hence spread versus level. To provide a visual reference, we usually draw a regression line for the spread fitted to the level on the same frame. Figure 9.11 shows two spread-level graphs of the trial data for the

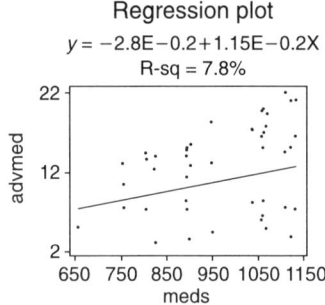
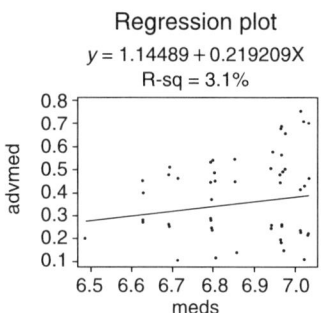

Figure 9.11. Spread-level graphs (raw and log data) for the Quarterly Traffic Fatalities in Canada.

fatalities problem. The first of these uses raw data, the second the logarithms of spreads and levels. Here we have applied our new 'tool' – the spread-level graph – with the ideas of transforming data to help us to understand it and possibly to detect additive or multiplicative seasonality. (If the logarithmic graph gives a better fit, we lean toward a multiplicative model for our seasonality.)

In this case, there is little to choose between the two graphs in that the deviations from the fitted lines are very similar. This implies that we will have difficulty distinguishing additive or multiplicative seasonal models, which turns out to be the case. If the underlying situation were truly multiplicative – with spread increasing in proportion to the level – we would expect the logarithmic spreads and levels to be closer to the reference line. We prepared these graphs quite easily in Minitab, although there is no built-in procedure for drawing them. See the Appendix.

The reader may have noted that we did not yet mention the case where the spread declines as the level increases. Such cases are not common in our experience. They would require specialized models to describe the relationship, if any, of the spread to the level or the time index.

Managing the graphs we draw

Readers will notice that a number of the graphs in this book use 'old-fashioned' character graphs, also known as printer-plots or typewriter graphs. We have software and hardware to prepare high-resolution graphs and have avoided colour only because it is expensive to print. There are several reasons we prefer to use character graphs, at least in our preliminary analysis. We do not prefer character graphs to more precisely drawn ones, but they simplify the overall forecasting task .

First, character graphs are plain text files. They can be sent as part of the message in e-mail. Their computer files are compact. There are multitudes of excellent text editors for all computing platforms so that annotations can be added (cautiously so as not to disturb the graphic information!), and they can be printed on all the printers in existence, including Gutenberg's original – and we have even observed that famous machine in action! We can import them easily into reports without fear of losing the image. This is not the case for many high-resolution graphics. We have on one occasion spent more than three days' worth of effort playing with high-resolution graphics for the Pegels' classification so that WordPerfect (version 5.1) could print it. Several times, the printer preview screen showed delightful images that were rendered totally blank by the printer driver. Forecasters usually want the graph of the forecast before the actual event arrives! Where character graphs are adequate for our needs in forecasting, we advise their use to save a lot of time, effort and frustration.

The only caveat is that to display character graphs correctly, a *monospaced* or typewriter-like font is needed. Students have, in our experience, a very poor understanding of how typography works. They can 'see' the difference between Times New Roman and Courier, but do not understand that one is proportionally spaced and the other is not. Make sure you know how to choose a monospaced font if you intend to use any plain text files that require spacing for purposes such as a graph or table. A related issue is the presence of tab characters in the plain text file, as these may be interpreted in different ways by different editors, word processors and printers or printer drivers.

Second, in Minitab and some other packages, the character graphs can be dumped into the log file (initiated with the OUTFILE command). As we usually create a number of graphs in any forecasting project, keeping them in sequence in the log file is one way of maintaining control of them. With many software packages, we must be careful to make notes of the filenames of each of the graphs we draw in high-resolution form. Few of us – and we include ourselves – do a good job of this. *DataDesk* is one package that does allow a non-text audit trail. See the comments by Paul Velleman in Goldstein (1993). Indeed, it is a surprise to us that more packages have not followed the lead of *DataDesk*. Furthermore, at the time of writing (December 2000), there are many examples of HTML, XML or similar forms of output from software (but as yet not much statistical software) that suggest how a sensible log file with embedded graphics and other information could be prepared.

For presentations, we recognize the need for high-resolution graphs, although again we note the difficulty of extending them over multiple pages to get long time scales (see the Appendix). Our experience is that it takes considerable time and effort to adjust scales, symbols, line types and other features of graphs so that they accomplish our purposes. An additional caution regarding high-resolution graphics concerns the use of a number of them in a single report. On one occasion, a student had many graphs in his report. He decided to bring them all into his document rather than use a manual 'cut, paste and photocopy' approach. Unfortunately the file he created was bigger than his entire disk allocation! When he tried to save it, it caused a 'disk full' failure. There was, however, worse to come, since the fullness of the disk precluded updating his account information, so he was unable to log into the system to do any work at all, and this was three days before the end of term! The 'lock' was so bad that it took the Faculty's most senior systems manager more than three hours to resolve. Such issues are gradually disappearing, but problems always arise that are 'larger' than some of our resources.

Numerical descriptive statistics

Besides graphs, we also use numerical descriptions of our data. The mean and median of a data series, for example, give us immediate measures of the general level or 'size' of the data – its *location*. The standard deviation, the inter-quartile range, or the range (the largest number minus the smallest number) provide us with a scale to the variability of the data – its *dispersion*. This is another way to consider level and variability. We can obtain an understanding of the structure and evolution of the process under study by applying these concepts of location and dispersion to our data and to subsets of this data. This understanding can be made visual if we also use graphical displays designed to elucidate similar facets of our data.

We caution again that forecasting exercises often report the 'mean square error', a concept closely allied to that of the variance of a set of numbers. Unfortunately, such measures – and we cannot avoid using them ourselves because they are so ingrained in the subject – have units in the square of the natural units of our problem.

Using data subsets

The amount of data available to forecasters varies greatly with the problem at hand. In many instances, there is so little data that dividing the main data set into trial and validation subsets is just not feasible. In other cases, we may have plenty of data to work with and can exploit this luxury to experiment with different forecasting methods and to test their capability by setting aside portions of the data for validation.

The amount of data to set aside is always an issue. As we have indicated in Chapter 2, the sub-sample set aside for validation should correspond to our desired forecast horizon. There, we presumed that the rest of the data would be used to estimate the trial model. This is not, however, always appropriate. Indeed, for Quarterly Traffic Fatalities in Canada, we discard the first seven years of data because this period seems to be shifted significantly upward from the rest of the data, possibly as a result of regulatory or data collection changes.

With sufficient data, the idea of testing subsets of the data can be expanded. We can check if a given model for the phenomenon under study is stable over the separate subsets. We are looking for consistency, or at least for predictable changes in the model. The justification for subset selection in the trial/validation approach is that this mirrors the actual forecasting process. This is a pragmatic and empirical viewpoint of forecasting. If it works, use it.

The essential element of the multiple time graph or stacked boxplot is the comparison of subsets of data. We will often want to draw these and other graphs of subsets of our data. For example, we may want to compare June sales figures with December ones. Obviously, we do not expect to sell many Christmas trees in June, at least not at the retail level. On the other hand, our students are constantly surprised that July and August are the months when Canadians most frequently manage to kill themselves on the roads. The lowest months are those when ice and snow make driving 'treacherous'. Such artefacts in the data may be useful for forecasting, planning, or development of regulatory policy. This has already been made clear by both the multiple season time plot for the Quarterly Traffic Fatalities in Canada data as well as from an analysis of where the traditional time-plot (Figure 9.3) has its peaks.

The essential work in drawing Figure 9.2 (multiple season time plot) is separating the data into several series, each relating to one identifiable phase of a repeating cycle. In our Quarterly Traffic Fatalities in Canada case, we broke the data into four series, one for each quarter. We advise against drawing *separate* time plots of each of these. Minitab, along with most graphing software such as spreadsheets, generally will scale graphs automatically. This is bad news for us! We want to compare the different quarters or seasons across time, so need to graph them on the *same* scale. Figure 9.2 does this, representing different seasons (quarters) by different line types.

Note that we now have clear evidence that the 'summer' season, July through September, has almost twice as many fatalities as the 'winter' of January through March, with the other two quarters in between. We have chosen to force the lower limit of the y-axis to be zero so we can see the true scale of the number of fatalities. (This was one of the points made in our discussion of graphing data.) We have also been able to get the time axis to be the calendar year. However, the version of Minitab we used gave some trouble in adding the legend to identify the quarters.

We now have good evidence that our fatalities are declining over time and that the spread of the seasonal variation is also declining. This could also be seen from looking at boxplots (or other distributional plots) over both quarters and years. Figure 9.12 shows the boxplots of Quarterly Traffic Fatalities in Canada by quarter number. We could also prepare a set of stacked boxplots by year to compare the level and variability of the data over each year. For monthly data this is recommended, but for quarterly data we must recognize that each 'box' is drawn with only 4 numbers, so is not a good application of the boxplot, which is often introduced as a way to visualize a 5-number data summary (min, quartile-1, median, quartile-3, max).

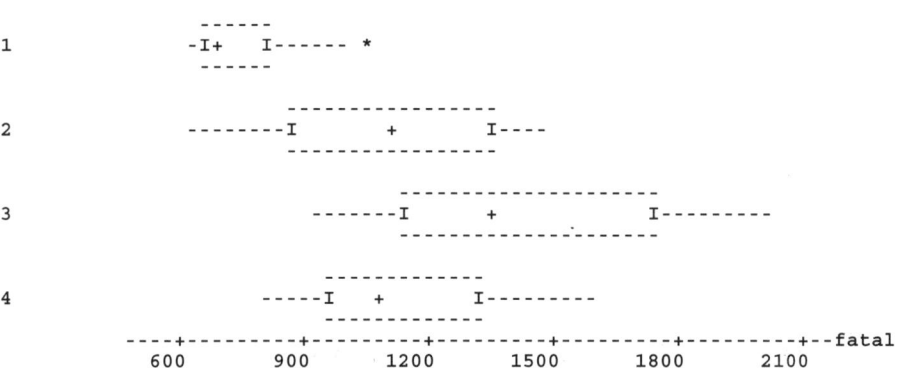

Figure 9.12. Boxplots of Quarterly Traffic Fatalities in Canada by quarter num , 1975–1997.

Subsetting the data can be a lot of work, and we recommend, where possible, automating the task by means of program scripts. Truthfully, we do not maintain fully automatic scripts to carry out these manipulations in Minitab or other packages. Instead, we try to keep well-documented examples of the operations as scripts related to specific forecasting problems. We then edit these as needed for a new problem.   The reason for *not* fully automating is that data often have odd starting or finishing points that are a nuisance to deal with automatically in every case. Also, each problem is different. We may be happy to use January as the start of many annual series. However, because its main academic period runs from September to May, the University of Ottawa, among others, uses a working year that is July 1 to June 30. For such applications, we would be better to align our monthly data series to start in July. Another reason for avoiding automation is that we want to name our data series sensibly. Computers can do a good job of many things, but they do not (yet!) automatically document our data.

We can also use our manipulated data to compute summary statistics of subset data. Clearly this should mirror the graphical analysis of our data. Once again, we note that subsetting quarterly data into years will compute statistics from only four observations per year. The quartiles and standard deviation should not be taken seriously. Tabulated statistics are rather dull reading. Moreover, formatting the tables to fit available page space and pagination can be more work than actually computing the statistics. We recommend that the statistics be computed and reviewed during a preliminary analysis, but generally not included in reports. As an example, Table 9.2 gives summary statistics by quarter for the Fatalities case. These were computed by the Minitab DESCRIBE command, with the output dumped into a log file and then (manually) edited to give just one line of figures per quarter. (Those with a good knowledge of PERL programming could likely automate this.) We chose to name the quarters q1 (Jan – Mar), q2, q3 and q4, while Minitab uses Q1 and Q3 for quartiles 1 and 3 of the data, an unfortunate collision of naming conventions.

In reviewing such statistical data, it is helpful to convert the numbers to measures that humans can understand. For example, we note that even at the minimum of 618 fatalities per quarter, there are approximately 7 Canadians killed each day on the roads, while the maximum of almost 2000 in a single quarter implies that approximately 22 died per day on the roads. Later, we may want to compare the numbers of fatalities with the population of Canada, or with 'amount of driving' measures such as sales of gasoline, number of licensed vehicles or drivers, or other data we may be able to find that may enhance our ability to forecast the number of fatalities.

Table 9.2. Summary statistics for Quarterly Traffic Fatalities in Canada by quarter

```
Descriptive Statistics

Variable   N   Mean    Median   TrMean   StDev   SE Mean   Min     Max       Q1       Q3
q1        23   754.9   697.0    747.7    125.9    26.2     618.0  1043.0    666.0    821.0
q2        23  1075.7  1095.0   1078.0    253.2    52.8     643.0  1460.0    841.0   1353.0
q3        22  1408.3  1350.0   1402.8    314.8    67.1     926.0  1999.0   1133.5   1732.7
q4        22  1133.7  1092.0   1127.3    233.4    49.8     813.0  1582.0    960.2   1338.7
```

Summary and application to a real data set

This chapter has considered the preliminary analysis of data from which we intend to prepare a forecast. The three main themes were:

- Typical patterns expected in forecasting data, as illustrated by the Pegels' classification

- Use of graphs and summary statistics to find patterns and exceptions from them in our data, with particular focus on the general level and variability in the data

- Facets of rendering and managing our data to make our task easier

Use of these tools allows us to characterize our Quarterly Traffic Fatalities in Canada data. Clearly we have a *declining trend* with *quarterly seasonality* that has declining magnitude. This is suggestive of *multiplicative seasonality*. We may wish to question whether the trend is linear or some form of exponential decay. This is *not* obvious from the graphs. It can (and has) been argued that the data show two or even three 'plateaus'. In Figure 9.3, we see an apparent period where the data are more or less 'flat' from quarter 1 (1975-1) to quarter 28 (1981-4). Then the data apparently stays relatively flat or slowly declining until quarter 68 (1991-4), when some viewers claim they see another drop. The separated quarterly plots in Figure 9.2 show the 1981/82 drop, but the 1991 transition is far from obvious, at least in two of the quarterly series. We will choose to look at the time periods 1975–1981 and 1982 to the present as distinct. We have been told, but never provided with documentation, that at some time early in the 1980s the definition of a 'traffic fatality' was changed for purposes of official statistics. Apparently, until the change, anyone who died within six months of a road accident was classed as a traffic fatality, while afterwards, a period of two weeks was used. This would certainly account for some decline in the figures. As, however, we have not been able to document this, it remains an open question. The issue is muddied by the occurrence of a number of changes in legislation and improvements in safety design of vehicles. Various provinces in Canada passed seat-belt laws around the start of the 1980s, and enforcement and promotion efforts were starting. The authors would be very glad to receive documentation relating to this issue. (We thank Stacey Muise for giving us a copy of the statute for the Province of Ontario, which received Royal Assent 7 July 1982.)

Given the characterization above, we propose two possible adjustments of the Quarterly Traffic Fatalities in Canada data.

- For much of our work, the early 28 quarters may suggest a level of fatalities that is too high. Therefore, we simply remove these from our working data. This is easily done in Minitab using the DELETE command. Clearly, it is generally easy in almost all data analysis software. We caution that it is easy to make small but dangerous mistakes. Document your work! The deletion of the first 7 years of data is the approach used throughout the rest of the book for the Quarterly Traffic Fatalities in Canada data.

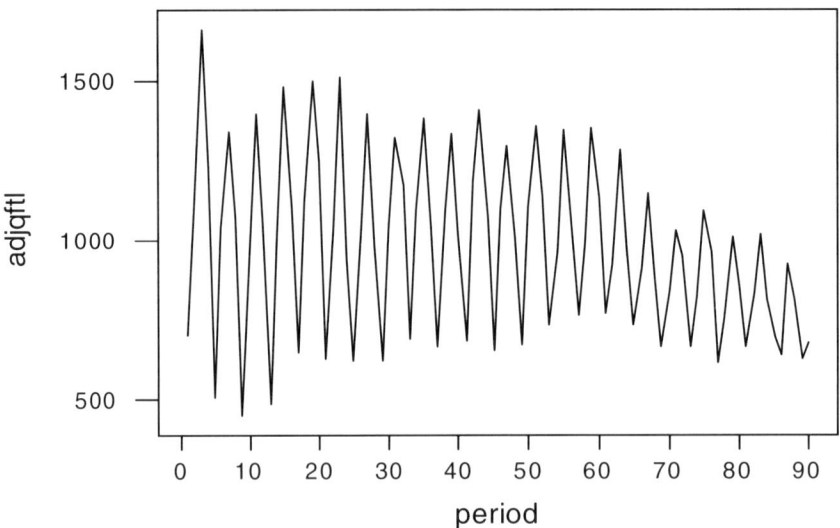

Figure 9.13. Quarterly Traffic Fatalities in Canada adjusted during the first seven years using script qftladj.mtb.

- When we are interested in the seasonal pattern of the fatalities, it may be useful to make an adjustment for the change in level at the start of 1982. One possibility – and we stress that it is just one of several choices we could make – is to average observations 1 to 28 and average observations 29 to 56. We label these avg(1:28) and avg(29:56). Then subtract avg(1:28) – avg(29:56) from the first 28 observations. We show how to do this in script qftladj.mtb presented in the Appendix. The result of this adjustment is shown in Figure 9.13. We will only wish to use the adjusted data if we feel that it is important to model the seasonal behaviour. That is, we are concerned that the seasonal pattern may be changing and wish to observe this without the change in level (or trend) interfering. Generally, we discourage such adjustments to the data, which are made in a very *ad hoc* fashion, as they may hide real features in the problem at hand.

Exercises

E9.1. Consider the following 'trading day' adjustments and provide a one-paragraph critique of the good and bad aspects of each.

a) For a business that is primarily a *weekend* activity, but for which data are monthly, we work out the number of weekends in a month and get the sales/weekend and forecast these. We then transform back to sales for a month using the number of weekends in the month, which may be fractional. How should we deal with 'long weekends'?

b) For businesses that can be considered roughly scalable by the number of hours of operation, such as a physician's or dentist's office, we could work with the volume of work/hour. Again we rescale back to get the total 'work' or whatever we are trying to forecast.

E9.2. The following situations suggest the difficulty, both technical and political, of imputation. Suggest whether you feel we should impute data or otherwise make adjustments and provide a one-paragraph critique of the good and bad aspects of such adjustments.
a) The number of black males of military service age counted in the 1940 US Census turned out to be fewer than the number from that category who actually showed up to join the military after Pearl Harbour. More recently, this issue, called the 'undercount', has been a political controversy because funds transferred to large urban centres are based on population, and these are the very places where undercounting has been shown by ancillary statistical work to be the most troublesome. Note that, in Canada, it is estimated that males aged 20–29 are undercounted by 10% in areas such as Vancouver.
b) Again with the Census, what should one do with figures indicating a small but definite number of reported cases of women who are reportedly less than 14 years old yet have recorded more than 3 children? This looks like a recording error, but the Census legislation requires that the recorded data be taken as fact!
c) In Canada, the legislation setting up the Egg Marketing agencies (there is a Canadian Egg Marketing Agency but there are also parallel provincial regulators) 'grandfathered' those producers already in the system at their production average for a short period preceding the inauguration of quotas. To establish quota, producers over-produced, but reported fewer hens than they really had. As a colleague working in the area once remarked: 'Some producers have hens that have, on paper, exceeded the biological capability to lay.' (This limit is about 1 egg every 2 days.) Similar comments have been raised about the reported population of Nigeria, where a perception that the census was to be used for taxation led to severe undercounts on one occasion, while belief that higher populations meant more representatives and grants led to serious double counting in another.

E9.3. Documentation of *how* data is recorded is important. Consider that technology for measurement changes. Also units of measurement change and 'old' measurements then need to be transformed. There are many cases where conversion factors have been misapplied. Properly documented data at least allow us to try to figure out what has happened. Find an example of measurement change and discuss its nature and consequences.

E9.4. Run an analysis of the Dry Cleaning data, which is provided below and on the PFM Web site (Minitab script). Use a two year forecasting and validation period.

```
note data in $1000s for volume of drycleaning in a province
note by month, Jan 1970 onward to Dec 1981
note
set c1
 2905 3677 5756 5041 4423 4885 5345 6172 6320 5468 4027 3744 3340 4001
 6036 5845 5041 5622 5831 6878 7079 6165 4434 4632 3607 4679 6739 6505
 5818 6093 5897 7138 7502 6511 4837 4558 3958 4874 7127 6908 6213 6600
 6512 8127 8016 7020 4780 5270 4444 5398 7739 7213 6694 6706 7190 8760
 8834 8288 6129 5523 4838 5480 8497 7858 6942 6863 7550 8921 9055 8739
 6320 5815 4806 5597 8328 8548 7614 7832 8225 9625 10018 9403 6686 6116
 5088 5840 8864 8709 7762 8078 8374 10266 10494 9257 6596 6646 5385 6190
 9534 9069 8139 8735 8868 10946 10976 9724 7236 6790 5646 6494 9982 8946
 8891 9088 9281 11236 11261 10371 7555 6920 5855 6830 9811 9761 9075 9695
 9964 11573 11761 10670 8207 7565 6279 7229 10247 9876 9713 10186 10009
 11399 12044 10544 8648 7744
end
name c1 'drycln'
```

E9.5. The following is the famous Tourist arrivals in Israel data set. Try a preliminary analysis on this data to see how subtle the calendar issue can be. (See the PFM Web site.)

```
Note      TOURIST ARRIVALS BY AIR FOR ISRAEL, BY MONTHS, 1956-1976.
Note      TAKEN FROM RAPHAEL RAYMOND V. BARON, "THE ANALYSIS OF SINGLE
Note      AND RELATED TIME SERIES INTO COMPONENTS: PROPOSALS FOR
Note      IMPROVING X-11", IN SEASONAL ANALYSIS OF ECONOMIC TIME SERIES
Note      (PROCEEDINGS OF THE CONFERENCE ON THE SEASONAL ANALYSIS OF
Note      ECONOMIC TIME SERIES, WASHINGTON, D.C. SEPT. 9-10, 1976.)
Note      Data entered by Carl Yue, Faculty of Administration student.
name c1 'ARRIVE'
set c1
   924  1384  2547  2646  2448  1822  2514  2019  1876  1962   864  1063
  1151  1293  1767  2911  2660  2103  3177  2416  3317  2389  1349  1491
  1466  2014  4524  8003  4096  3772  4819  3288  2599  2610  1929  2074
  1526  2218  4922  6832  6632  4101  6495  3903  4496  3935  2842  3012
  2143  3744  6709  9896  6971  5370  9480  6320  5022  6420  4309  5487
  3655  6327  9288  9834  8084  7369 14310  9701  7041  7145  4788  5015
  3972  5987  9167 13001 10731  8405 14944  8565  7584  7226  5167  6818
  4268  6790 10646 16803 11010  9732 18454 11690  9090 10919  6178  9169
  5670  7989 16205 14599 12926 11759 19460 13235 10331 10727  7729 10265
  5720  8142 12089 21055 17780 13880 22209 17290 12741 12336  7053 12695
  7159 11689 17984 22199 16417 16729 28072 17443 13692 15919  7319 11027
  7295 10311 17752 19817 15222  9868 30427 22790 18147 22896 12841 24220
 12641 18402 27585 47020 27869 30269 57170 39132 28527 31584 19125 27200
 14284 16538 34474 32496 25236 31978 60411 40099 24170 26821 17412 29200
 17682 22987 35850 32338 29115 34456 64519 46843 25907 21765 17023 33370
 20308 26414 44344 57840 39998 51294 87832 64013 43094 53703 31277 46208
 29482 39182 72632 59304 58557 54678 83326 55635 47129 58466 30179 38543
 27648 35019 48774 72864 54622 53095 84695 58486 51665 16991 20599 37058
 24304 34303 52551 59562 41172 41899 67668 52789 37486 47071 29685 37633
 21844 27238 49300 39941 36983 40420 65322 51421 41860 50924 31195 51948
 28079 37503 62344 78792 55087 51165 80145 64743 50699 61261 44111 57668
end of data
note      M E A N 22994.5754
note      S . DEV 20599.4874
```

10

The preliminary forecast: concepts and examples

Ruler forecasts

Time plots let us prepare *ruler forecasts*. These will serve as a safety check on the other methods we will use. We create these forecasts by simply drawing lines on time plots that – in our judgement and eyesight – represent the trend and envelope of the data pattern. Forecasts are obtained by extending the lines to future time points in a simple extrapolation. We will have lines for the 'centre' of the data, which is our estimate of the average level of the quantity we wish to forecast, as well as for the upper and lower limits of the envelope. The envelope is not widely used by other practitioners, but we feel that it provides a useful guideline to the limits of future behaviour of a quantity if we assume that current trends and patterns continue.

Figure 10.1 shows the application of these ideas to the Quarterly Traffic Fatalities in Canada data. It also shows how we can forecast seasonal patterns by tracing the ups and downs of our series onto an acetate sheet (such as those used to present information on an overhead projector) or else tracing paper. We can then slide the sheet along the trend line, using registration marks such as the line shown in the figure, and 'read off' forecasts from the graph axes.

Using paper and acetate, this is remarkably easy. Preparing Figure 10.1 was almost a day's work! We took our Minitab output and imported it into Corel Draw, Version 6. Adding the lines and patterns was easy. The difficult part was transferring our results – which printed beautifully – to the WordPerfect file for this chapter. We tried importing the Corel Draw file (.cdr), Adobe Illustrator (.ai), and several versions of Computer Graphics Metafiles. Sometimes the Minitab graph was rescaled, sometimes the lines would not appear, other times we were 'given' huge top and bottom margins. After some playing with the pen nib-size parameters, an HPGL (.plt) file was used, but the results were less than fully satisfying. More recent updates to this diagram have used a Windows MetaFile (WMF) format, which appears to work as well as the HPGL form and requires less 'fiddling' with parameters. However, we made some patches to the WordPerfect software in the meantime, first using a free 'replacement' CD and later installing an upgrade. We believe problems of this sort will continue to occupy forecasters' time and effort in matters that are not central to forecasting, but important to its presentation.

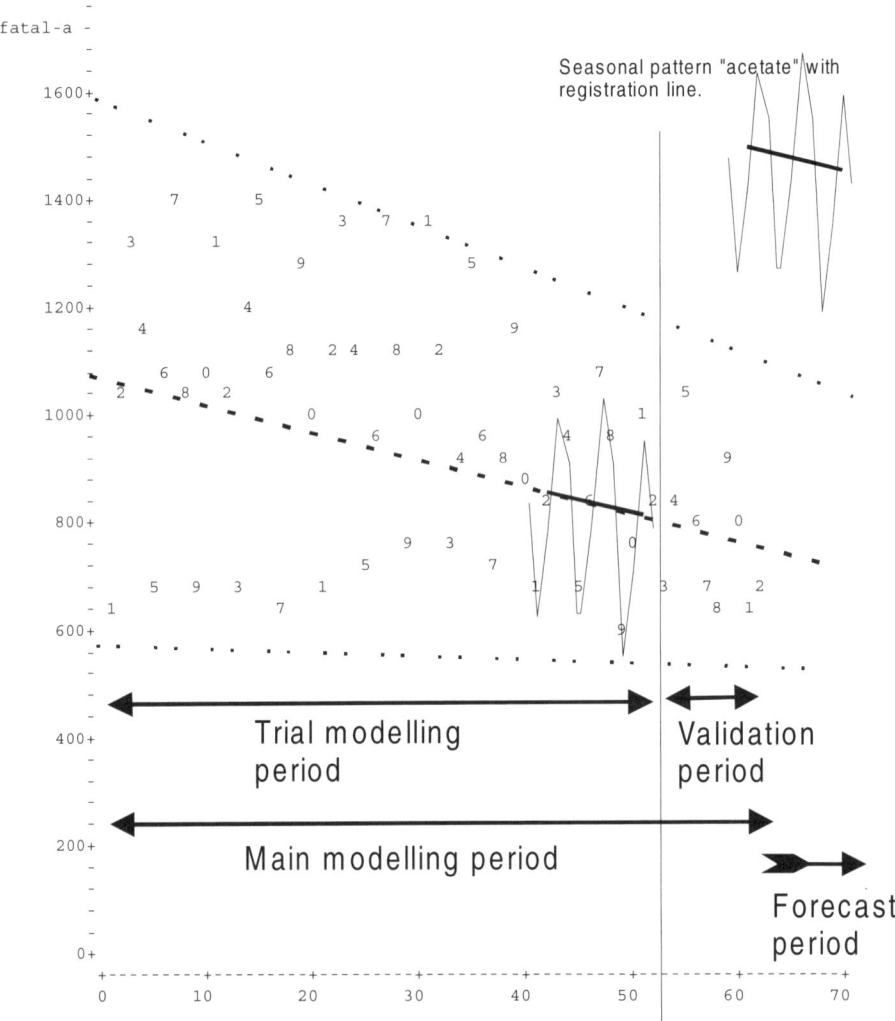

Figure 10.1. Illustration of the creation of ruler forecasts.

Besides acetate or tracing paper, we have also used a lead ruler, a bar of lead encased in rubber or plastic. It can be bent to the shape of the pattern in the data and then moved forward in the time domain and re-drawn. Truthfully, while this is fine for smooth, very simple patterns, it does not work well for any realistic data we have encountered. Indeed, the lead ruler may be more useful for drawing shaped curves such as the *sigmoid* or S-curves that are often applied to market penetration or technological change, as we shall consider in Chapter 20. Other tools for such smooth curves exist: the French curve and the spline. These have largely fallen out of use as mathematical versions (some of them called splines also, as well as Bezier curves) embedded in drawing software have taken their place.

In summary, we recommend the use of ruler forecasts, with traced pattern extensions, for:

* Obtaining rough forecasts of the central trend of the data;

* Estimating upper and lower bounds for such forecasts via the envelope lines;

* Projecting existing patterns forward in time, if these patterns are stable.

The last proviso is important. Humans often try to see patterns where none exist.

When preparing ruler forecasts, we recommend using graphs large enough to permit easy physical measurements so we can compute approximate slopes and intercepts of the trend and envelope lines and also read off the forecasts. If you use character graphs, be careful not to move the points around when preparing the graphs for printing. We often use landscape mode printing (11 inches wide by 8.5 inches high in North America, 27.94 cm by 21.59 cm), and we set the margins to 0.5 inches (1.27 cm) or less all around. Note that similar ideas apply to ISO paper sizes such as A4. We generally choose a Courier font and where possible vary the point size. This last point can be tricky – you need to have a scalable font *but* it must be a typewriter-like or *monospaced* font. That is, each character uses the same width, otherwise we ruin the effectiveness of the character plot. We could, of course, use other ways to graph the time plot, and in so doing would avoid the issue that the character plots have a built-in 'roughness' when points are forced into the coarse grid imposed by the character sizes. Nevertheless, we and our students have found the character graphs very quick and easy to use, with the added advantage that they can be dropped into an electronic message 'as is' as plain text. Unfortunately, printers may lack the Courier or other monospaced fonts in the sizes we would like. For example, our (rather old) Hewlett-Packard LaserJet III has only 10 and 12 point sizes. We can get around this problem by 'printing' to a PostScript (TM) file and then processing the PostScript to the printer via Adobe Acrobat or the public-licence GhostScript software.

Trend equations

We drew the ruler forecast trend lines 'by eye', positioning them through the middle of the data, across the top of the points, and along the bottom of them. You may wish to 'fudge' the position for the upper and lower envelope lines if you suspect a point or two are outliers or otherwise unrepresentative of the behaviour of the series. You may also wish to question our positioning of the upper line in particular, since it is teetering rather dangerously on a single point, marked with a '1' that tells us it is at time period 31. Once we have our lines, we draw a new y-axis. Note that the character plots and many other graphs, not just in Minitab, have the *y*-axis line offset from the time=0 or *x*=0 position. We often also draw one or more vertical lines at convenient time positions along the time or *x* line to aid in reading off *y*-values.

In Figure 10.1, we have used data from script FQ98TSV2.MTB that drops the first 7 years' data. However, since we have used a character-form TSPLOT (time plot) of the data, the time index runs from 1 to 62. These represent 1982, quarter 1 to 1997 quarter 2. We use the first 13 years, or 52 periods, for our trial modelling.

We now read off the intercepts of each of the lines. This takes a little experience in estimating the position of the intercept between axis ticks, but given the coarseness of the character plots and the intended usage of the ruler forecasts as a safety check, we should not be too worried about perfection here. The important issue is to get the scaling and time positioning right. We suggest preparing ruler forecasts for this and other data sets as an exercise.

Sometimes the graph does not start at time = 0. To allow for this, we will develop the equation for a line – be it the central trend or the upper or lower envelope line – using the intersection point with the left axis at ($t1$, $y1$) and the intersection with a vertical line at ($t2$, $y2$). The two-point form of the equation of the line is then

$$y = y1 + \text{slope} * (t - t1) \qquad \text{where} \quad \text{slope} = (y2-y1)/(t2-t1)$$

Thus the intercept at t=0 is at y=($y1$-slope*$t1$)

In the case that $t1$=0, we get the intercept = $y1$, as we should. Note that we want the times $t1$ and $t2$ to be well-separated to allow us to minimize the relative error in our measurements.

Example

We apply these ideas to the trial estimation period of Quarterly Traffic Fatalities in Canada data. For our central model, the intercept is 1120 fatalities. The intersection with the line at t=52 (the end of 1994) is 860 fatalities. Thus, the model trend line is

$$y(t) = 1120 + [(860 - 1120)/(52 - 0)] * (t - 0) = 1120 - 5\ t$$

(We did *not* fudge the data to get such a nice result!) We are able to use this model, substituting the appropriate value for the time period t. For example, the model predicts

$$y(62) = 1120 - 5 * 62 = 810 \text{ fatalities for 1997, quarter 2.}$$

The models for the upper and lower envelope lines, as drawn, are

$$y\_upper(t) = 1640 - 7.31\ t \qquad\qquad y\_lower(t) = 620 - 0.38\ t$$

We clearly have reservations about the reliability of these models.

To get reasonable forecasts, however, we need to apply a seasonal shift. The graph shows how to do this with a pattern traced on acetate or tracing paper (see Chapter 5). We could measure the distance above or below the line in the seasonal pattern over some 'reasonable' time span as in the diagram. This is eminently feasible, but a little messy, and we will not do it here. Note that there will be some uncertainty in our forecast because the distances of corresponding seasons from the trend line is not constant. We could use the mean of several values, or some approximation thereto, or possibly use the middle of three values (the median in this case). Note that there are many choices. We would be uncomfortable if the different choices gave wildly different predictions. We will develop computational methods that mimic the ruler forecast, at least for the central trend. This does not mean that we discard the ruler forecast. Instead we use it mainly as a safety or 'reality' check on our other results.

Data subsetting

Subsetting the data allows us to look at shorter periods and compare them, or to choose all seasons of one type and group them together to see if there are ways to characterize the parts rather than the whole of the data. If the parts can be modelled successfully, then we can make several partial forecasts that can then be combined into a prediction of the full series.

We have already seen, in Figures 9.6 and 9.7, the process of segregating the data by season or year. We can prepare ruler forecasts for each of the seasonal series. Now, however, we should no longer have seasonal variation, although there may be cyclic variation in our quarterly series that can still be 'explained' by advancing a traced pattern. Comparing ruler forecasts made this way with those from the original time plot will indicate the magnitude of errors in our forecasts.

Example

Figure 9.7 is too compact for easy measurements from ruler forecasts (trial period) for the separated seasons. Draw larger graphs for these ruler forecasts. For Table 10.1, we used expanded Minitab character graphs (via `height = 55`) and forced axis scaling. Minitab displays the commands if we use menu selections. We can use the commands in scripts for similar problems.

'Simple' seasonal models

So far, we have used the time translation of traced patterns to forecast seasonal values of traffic fatalities, but note that such 'models' are not easy to convert to a computational or mathematical form, which we now wish to do. The general process is fairly obvious: we decompose the data into parts that we can extrapolate or forecast easily, then put the forecast components together again to make our forecasts of the quantity of interest at appropriate time points.

Table 10.1. Ruler forecasts for individual quarters of the Quarterly Traffic Fatalities in Canada data based on the time period 1982-1 to 1994-4 (`Qruler1. Prepared 1998-11-3`)

	q1			q2		
	lower	centre	upper	lower	centre	upper
int(0)	610	675	770	1120	1240	1350
int(13)	610	675	770	710	780	830
slope	0.00	0.00	0.00	-31.54	-35.38	-40.00
y_hat(14)	610	675	770	1562	1735	1910
y_hat(15)	610	675	770	1593	1771	1950
y_hat(16)	610	675	770	1625	1806	1990

	q3			q4		
	lower	centre	upper	lower	centre	upper
int(0)	1340	1475	1650	1070	1155	1300
int(13)	990	1060	1210	870	900	1005
slope	-26.92	-31.92	-33.85	-15.38	-19.62	-22.69
y_hat(14)	1717	1922	2124	1285	1430	1618
y_hat(15)	1744	1954	2158	1301	1449	1640
y_hat(16)	1771	1986	2192	1316	1469	1663

In this section, we consider conceptually simple methods to provide computational models of the seasonality in data series. Our goal is to describe how a particular season (in our case, quarter) differs from the general trend. There are several ways to do this. While all are simple in their basic idea, they may be tedious in their implementation. This tedium makes such approaches very error-prone. In general, they are *not* recommended unless they can be automated with carefully prepared programs or scripts that allow for careful checking. Nevertheless, we believe that it is useful for those learning forecasting to try one or more such methods as exercises. Methods such as these are often referred to as *ad hoc* methods.

We will divide our methods into two classes: those that *multiply* an 'average' value for the quantity of interest at a given time by a *seasonal factor*, and those that *add* a *seasonal shift* to this 'average', which will itself be some sort of model value of the trend. First let us consider the multiplicative form and compute seasonal factors. Because we will use similar ideas for both factors and shifts, we label our factor approaches as A, C and E, with the corresponding additive shift techniques labelled B, D and F. To provide some notation, in the seasonal factor approaches we want to find a set of L numbers $I(k)$, $k=1,2,3,...,L$, where L is the number of 'seasons' in our cycle (e.g. $L=4$ for quarterly data in a a year) so that the data $y(t)$ can be approximated as the product of a trend and the seasonal factor appropriate to a time period of interest. We write

$$y(t) \sim= trend(t) * I(t \bmod L + 1)$$

but it is common to write simply $I(t)$ instead of $I(t \bmod L + 1)$. (For those unfamiliar with the modulus or mod function, it is simply the remainder, e.g. 4 mod 3 = 1, and (3 mod 3 + 1) = 1.)

Seasonal shifts are handled similarly, but with addition rather than multiplication. If shift(t) represents the shift appropriate for time period t we have (dropping the '$t \bmod L +1$')

$$y(t) \sim= \text{trend}(t) + \text{shift}(t)$$

The trend function need not be the same for the multiplicative and additive forms.

A. *Seasonal factors* can be computed by dividing the average of all observations into the average for a particular season. This is crude, and may not reflect changing patterns over time, but often 'works'. We get the seasonal factors as *ratios of averages*. Once we have the seasonal factors, we can deseasonalize our data and compute a model for the trend component of the data. Let us call this method (A). Other approaches:

B. Similar to method (A), we could compute *seasonal shifts* as differences in the averages. That is, we subtract the overall mean from the individual season means.

C. We compute ratios to annual means. Thus we get a time series for each seasonal factor and can view their time evolution. We get a single factor for a given season by an *average of the ratios* for that season. We deseasonalize and compute the trend component as in (A).

D. Seasonal shifts could be calculated as *deviations from annual means*. Here we are using an additive rather than multiplicative model. Once we compute the seasonal shifts, we subtract these from the original data to deseasonalize it, then compute a model for the trend component. Unlike (B), we now have a time series for each of the seasonal shifts, so can visualize changing seasonal patterns.

E. Rather than use the annual mean, we could use a trend line and get seasonal factors by *ratios to trend line* values at the appropriate time points. Here we are computing the trend component first, then deriving the seasonal factors.

F. Like method (E), seasonal shifts could be computed as *deviations from a trend line*. As in (E), the trend model is found first, then the seasonal shift.

Which of these methods (or, indeed, others that one can devise) should we use? Our view is that we should use the simplest method that suffices.

Table 10.2. Summary of some methods for computing seasonal factors and shifts

Method	Description of computation. '$t == k$' means time period t is in season k
A	$I(k)$ = (average $y(t)$ for $t == k$) / (average all $y(t)$)
B	shift(k) = (average $y(t)$ for $t == k$) - (average all $y(t)$)
C	$I(k)$ = average over all years of ($y(t == k)$ / (average for year including t))
D	shift(k) = average over all years of ($y(t == k)$ - (average for year including t))
E	$I(k)$ = average over all years of ($y(t == k)$ / trend(t))
F	shift(k) = average over all years of ($y(t == k)$ - trend(t))

Trend line calculations for seasonal data

It is temptingly easy to compute trend lines in statistical packages such as Minitab. We simply run a *regression* of the data against time. The graph in Figure 10.2 shows why this may be dangerous with seasonal data that will be used to generate seasonal factors.

In this figure, the particular rising *seasonal* pattern has tilted the *annual trend* upward. We really want to get the trend line of the yearly average data against time, but if we are averaging the data, we should also 'average the time'. That is, the average of a full year's data corresponds to a time point at the middle of the data time points.

Our example, which happens to be quarterly, will have the yearly average positioned in time between quarter 2 and quarter 3, that is, at the average of 1, 2, 3 and 4 = 10/4 = 2.5. In the Appendix we provide more details of such a calculation in the form of a Minitab script.

All the calculations in this chapter relate to a *linear* trend, that is, a straight line. Clearly we sometimes observe or expect our changes to be nonlinear. We present ways to model nonlinear trends in Chapter 20. Unfortunately, many packages, including Minitab, do not make them easy to incorporate, though there are some 'tricks' to approximate the desired calculations that can be carried out fairly easily. We also show that a spreadsheet may sometimes be used to estimate a nonlinear trend in Chapter 20.

Results of 'simple' seasonal techniques

Table 10.3a presents a summary of some of the main quantitative outcomes of the 'simple' methods A, C and E for seasonal factors we have proposed. Table 10.3b gives the equivalent information for the methods B, D and F that use shifts, i.e. an additive model of seasonality. Clearly, the output and its analysis are not simple at all. Note that trial models B and D appear to give the same output which, on checking, appears to be coincidence.

Fit

The Tables 10.3 present several aspects of our forecasting models. From the point of view of the *fit* of the models to the data being used to estimate the models, we see that the R_squared statistics are surprisingly similar for all methods, which 'explain' between 86 and 89% of the variation about the mean of the data for both trial and main data sets. The simplest seasonal factor method A gives the best results for both data sets. Moreover, Method A, for R_squared, is better than the best seasonal shift approach for the trial estimation (Method F) and main data set (Method D). Except as noted, we will use results from Method A as our 'simple' method.

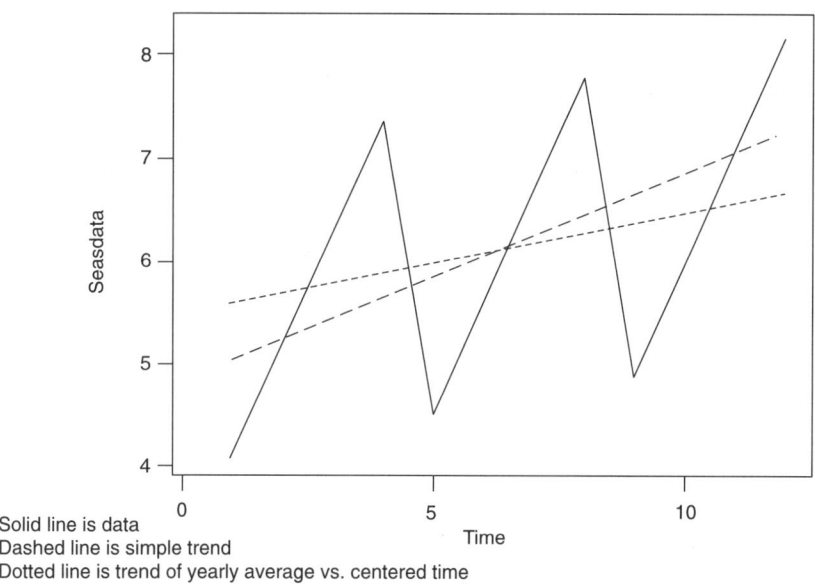

Solid line is data
Dashed line is simple trend
Dotted line is trend of yearly average vs. centered time

Figure 10.2. Potential trend bias difficulties when computing trend lines of seasonal data.

Stability of models

The model parameters for all methods also show similarities, even allowing for the differences in multiplicative (factor) and additive (shift) models. Note the similarity of parameters when we compare a given model as estimated with trial and main data. This supports the necessary presumption of continuity of the underlying structure of the phenomenon we are investigating. We do *not* want drastic changes in our models as we augment our data with recent observations.

The judgment that the model 'is more or less the same' requires a sense of perspective that only comes with experience. Novices may easily fall into one of the two major traps:

•   Disregarding important differences, because admitting they exist requires us to devise better, usually more sophisticated, models that often require more, and costly, data gathering;

•   Exaggerating differences that are really only fluctuations due to natural variation.

Our recommendation to readers is to take note of differences and to try to understand them. If they do appear to be 'real', we may nonetheless decide to maintain the fiction that the models are sufficiently similar in order to calculate and examine the forecasts produced. In other words, we take a 'what if' viewpoint. After all, at this stage the computations for the forecasts are almost free. They can be used for comparisons with other, hopefully better, approaches.

Table 10.3a. Results of *ad hoc* multiplicative methods for seasonal models of the Quarterly Traffic Fatalities in Canada data.

| | Factors | | | | | |
| | Method A | | Method C | | Method E | |
Estimation	Trial	Main	Trial	Main	Trial	Main
R_squared	**0.8985**	**0.8888**	0.8823	0.8588	0.8979	0.8841
Intercept	1220.7	1233.8	1220.1	1243.9	1248	1313.6
Slope	-4.21	-4.83	-4.21	-4.88	-4.71	-6.07
Quarter 1	0.6956	0.7235	0.6997	0.7351	0.6945	0.7263
Quarter 2	0.9976	0.9913	0.9948	0.988	0.9921	0.9835
Quarter 3	1.2715	1.291	1.2694	1.2619	1.2728	1.2778
Quarter 4	1.0353	1.0541	1.0361	1.0335	1.0431	1.0529
Max Res	148.7	167.9	152.2	163.7	148.3	148.9
Min Res	-141.9	-168.6	-144.3	-186.3	-147.7	-202.1
Stdev Res	74.4	77.94	74.6	79.5	73.4	77.35
Mean Res	0.7	0.03	1.2	2.6	0.2	-1.17
Validation						
Max err	97		93.8		108.5	
Min err	-213.4		-210.4		-193	
Stdev err	101.7		99.3		98.4	
Mean err	-61		-60.6		-45.8	
Max % err	13.91		13.45		15.57	
Min % err	-33.19		-32.73		-30.01	

Deviations and errors

The maximum, minimum, standard deviation and mean of the residuals *over the estimation periods* provide us with a scale for the deviations between the models and the data. They tell us how 'big' to expect forecast and validation errors to be, since fitted forecasting models are not perfect. Percentage deviations could be used if we wished. Instead, note that the 1982–1997 data has a mean of approximately 960 fatalities per quarter. That is, the DESCribe command gives

Variable	N	Mean	Median	TrMean	StDev	SE Mean
fqmdata	62	955.9	965.5	950.7	231.8	29.4

Variable	Minimum	Maximum	Q1	Q3
fqmdata	618.0	1411.0	734.0	1106.3

The maximum residuals are thus approximately 15 to 20% of the size of the average quarterly fatalities. If we have roughly equal absolute errors, the percentage errors will be larger in periods when the number of fatalities is few.

Table 10.3b. Results of *ad hoc* additive methods for seasonal models of the Quarterly Traffic Fatalities in Canada data

	Shifts					
	Method B		Method D		Method F	
	Trial	Main	Trial	Main	Trial	Main
R_squared	0.8823	0.8588	0.8823	**0.8609**	**0.8828**	0.859
Intercept	1246.5	1275.3	1246.5	1269.5	1248	1313.6
Slope	-4.69	-5.29	-4.69	-5.27	-4.71	-6.07
Quarter 1	-301.7	-261.8	-301.7	-261.8	-308.8	-270.5
Quarter 2	-2.346	-8.203	-2.346	-8.203	-4.703	-10.86
Quarter 3	269.12	275.42	269.12	256.08	271.47	266.69
Quarter 4	34.962	51.222	34.962	31.883	42.031	48.565
Max Res	142.6	144.9	142.6	150	144.7	170.1
Min Res	-184.9	-236.3	-184.9	-230.9	-178.5	-243
Stdev Res	80.1	86.6	80.1	87.2	79.9	87.8
Mean Res	0	-14.3	0	0	0	0
Validation						
Max err	150.6		150.6		158.4	
Min err	-198.1		-198.1		-195	
Stdev err	123.7		123.7		127.2	
Mean err	-42.4		-42.4		-40.7	
Max % err	21.6		21.6		22.73	
Min % err	-30.81		-30.81		-30.32	

We also note that the minimum residual has a larger magnitude (is further from zero in a negative direction) than the maximum residual. Since the residual is 'data minus model', this implies that our models are overshooting. They are 'high' of the target. Fatalities are actually decreasing faster than our models predict. Note the extent of this in the main rather than trial data sets, indicating the kind of change that suggests we should use a different model.

The mean residual in all cases is small. Note that it is not zero. Students who have been exposed to linear regression calculations sometimes expect that the mean error will be zero. This is, in fact, a necessary outcome of linear regression, but should be regarded as a happy accident in other modelling situations. We certainly want the mean error to be small, or we could apply an overall shift to our model to improve its fit. Zero, however, is not expected.

Validation

The validation results are, of course, critical to acceptance of a forecasting model. Some methods give very good approximations to the data used for estimation of the model, but the resulting model then diverges rapidly from the path of the data outside of the sample range. For all the methods introduced above, we see that the *errors* – and these are truly errors because we reserved 10 observations from our actual data to see how well the forecasting methods worked – are roughly the same size as the residuals obtained in the estimation phase. This is a good sign. We expect a certain amount of 'drift' away from our model, but the results here are 'not too bad'.

We note, of course, that the mean error is negative; our forecasts are generally 'high'. This is underlined by the asymmetry of the maximum and minimum errors around zero. The percentage errors highlight this, and, worse, show that errors are quite large in quarters with few fatalities. We have errors that are of the order of 30% of the number of fatalities in some periods.

Taking the figures for the validation errors together, there is no clear 'winner' among the methods. When the largest error or % error (maximum or minimum) suggests we should avoid a method, its error standard deviation is smaller. Note that if the mean error is large, we should not use the standard deviation alone. Indeed, the root mean square error is the square root of the sum of squared errors, which is

$$\text{sum of squared errors} = (\text{mean error})^2 + (\text{standard deviation of errors})^2$$

Pattern and distribution of residuals

So far, we have only been looking at the quantitative summary of the residuals and errors. We should also pay some attention to their patterns. We use only method A for brevity and look at the residuals from the trial estimation period. Those from the main period are quite similar

```
Stem-and-leaf of trialres
N  = 52      Leaf Unit = 10

     2      -1 44
     3      -1 3
     4      -1 1
     6      -0 88
    13      -0 7766666
    18      -0 55444
    23      -0 33222
   (5)      -0 10000
    24       0 011
    21       0 22333
    16       0 44
    14       0 7777
    10       0 88889
     5       1 001
     2       1 3
     1       1 4
```

Figure 10.3. Stem and leaf diagram of the residuals from the trial estimation period (Method A).

The stem and leaf diagram (Figure 10.3) suggests a moderately 'mound shaped' pattern that may be Gaussian in nature. We could investigate this conjecture further with the Normal Probability plot and related tests, but note that the time plot of the residuals (Figure 10.4) shows a definite pattern, rather like an inverted 'U'. The residuals have a rough

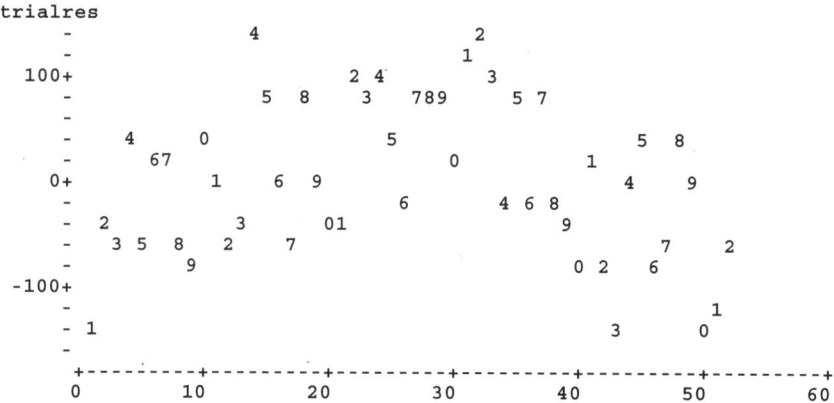

Figure 10.4. Time pattern of trial period residuals from Model A.

negative/positive/negative shape. That is, the model is high/low/high. We may be better with some form of nonlinear trend model. In this case, we would need one that is bowed like an arch, even as it trends downward. Given that the implementation of such ideas may be tricky, we shall not pursue it here. The autocorrelation function also signals time patterns in our data; it does not die out rapidly.

The residuals from the main estimation period are so similar in pattern that we can simply note this fact here and move on to the validation period errors. Since we only have 10 of these, their pattern in a stem and leaf diagram is difficult to discern. The time plot is more informative. To make clear the 'zero' line, we have added a row of equal signs, which is very easy to do with such character plots. Minitab also provides ways to add lines to high resolution graphs.

```
ACF of trialres
              -1.0 -0.8 -0.6 -0.4 -0.2  0.0  0.2  0.4  0.6  0.8  1.0
              +----+----+----+----+----+----+----+----+----+----+
  1   0.346                           XXXXXXXXXX
  2   0.098                           XXX
  3   0.172                           XXXXX
  4   0.495                           XXXXXXXXXXXXXX
  5   0.092                           XXX
  6   0.019                           X
  7   0.131                           XXXX
  8   0.306                           XXXXXXXXX
  9   0.141                           XXXXX
 10   0.055                           XX
 11  -0.081                        XXX
 12  -0.064                        XXX
 13  -0.067                        XXX
```

Figure 10.5. ACF of trial period residuals from Method A.

The Stem and Leaf diagram shows the *distribution* of validation errors. Next, we look at the *time plot* of the validation errors. The added row of '=' signs shows the position of zero on the y-axis, so we can see that more of the errors are negative than positive, implying a 'high' forecast. This is also clear from the Stem and Leaf diagram.

```
Stem-and-leaf of verror
N   = 10    Leaf Unit = 10

   1      -2 1
   3      -1 65
   3      -1
  (3)     -0 886
   4      -0 3
   3       0 4
   2       0 59
```

Figure 10.6. Validation errors from Method A.

The time plot of the validation errors confirms our numerical results in Tables 10.3 and the Minitab DESCribe output. The negative errors (high forecasts) are larger than the positive ones. Moreover, it is clear that only three out of ten errors are positive. The time plot *may*, and we stress that this is only a possibility, indicate that the magnitude of errors is increasing as we get further away from the last available estimating observation. This is clearly not unexpected.

The further we extrapolate from 'known' conditions, the less likely we are to predict well. The early validation errors are clearly somewhat smaller than the later ones, but since the largest error is in the sixth position, the pattern is not one of a rigid increase in size of error.

Evolution of seasonal factors

Method A uses simple ratios of averages for its seasonal factors. Methods C and E allow us to generate 'local' seasonal factors, which are then averaged for our final seasonal factors. These 'local' values may be useful in indicating the changing nature of seasonal variation. For example, method C gives values for the ratios of the quarterly fatalities to the annual mean over the main estimating period, which we graph in Figure 10.8. We can now see quite clearly that the seasonal pattern is *not* stable. We could try to extrapolate the pattern approximately, anticipating the Winters' method of Chapter 13 or the component by season plot of Figure 18.1. The order in which the legend in Figure 10.8 is displayed points out a nuisance feature of the Minitab software, which illustrates many of the minor obstacles to efficient forecasting.

It is also useful to consider the data, the model and the forecasts together. Such a graph is presented in Figure 10.9. This makes it quite clear that we have done reasonably well in following and predicting the fatalities data. All six of the methods presented show a similar form for this graph. Indeed, one of the most annoying facets of forecasting is that many methods work 'well'. The question is whether they work 'well enough', which is a management issue. We can only attempt to provide sufficient understanding of both the capabilities and limitations of our forecasting methods and analytic tools so our readers and students can make rational decisions.

```
                  -                        5
verror            -
                  -      1                                    9
                  -
                0+==================================================
                  -
                  -            2
                  -               3      4                8
                  -
          -125+
                  -                                   7          0
                  -
                  -
                  -                            6
          -250+
                  -
```

Figure 10.7. Time plot of validation errors from Method A.

Main Seasonal Factors vs time: Method C

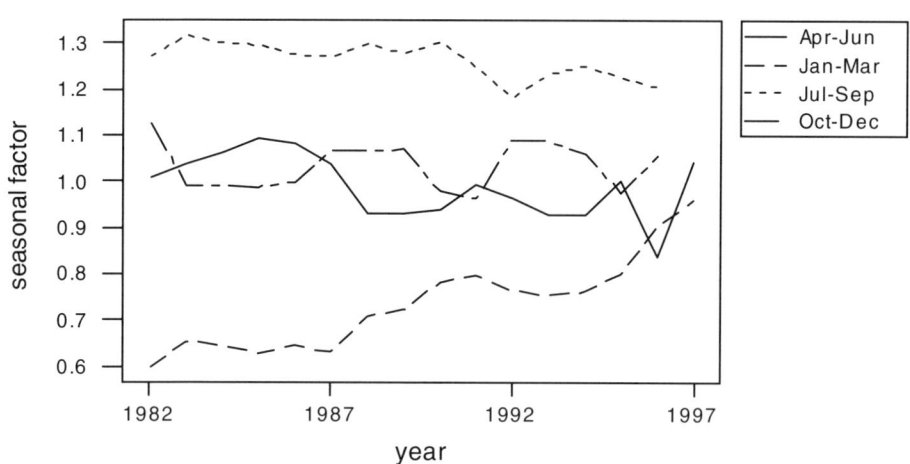

Figure 10.8. Evolution of seasonal factors for Method C over the main estimation period. The ordering of the legend (upper right) is decided by Minitab despite the ordering of the data. Quarter 1 is Jan–Mar. The legend is also confusing in that the last line refers to the dash-dot pattern, but the 'dot' is missing.

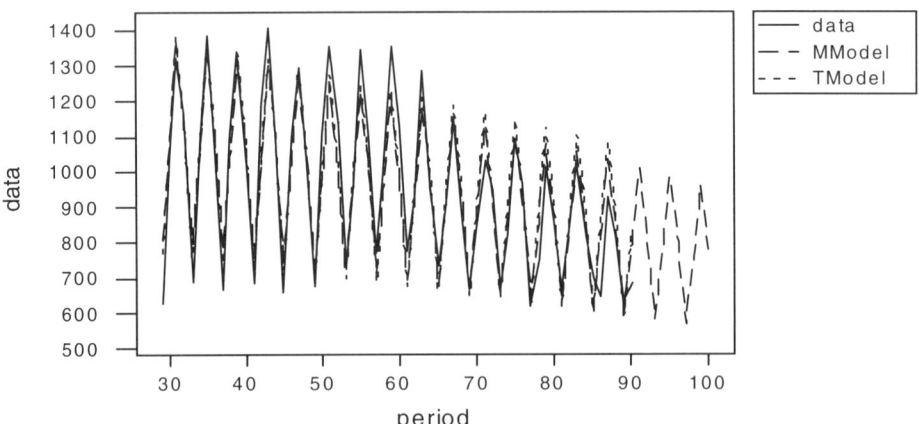

Figure 10.9. Data and both Trial (TModel) and Main (Mmodel) Models for Method C. Forecasts (Main Model) are also drawn.

Housekeeping details

While we started with just a single column of numbers, we now have many columns of data to keep track of. Some are clearly temporary, others we will want again later. Tidiness saves time and effort in the long run. Student editions of software (e.g. Minitab) often limit data storage. Therefore, take the trouble to erase unwanted data. This frees space needed by our statistical package for intermediate calculations and helps us avoid confusion and errors.

Table 10.4 gives a list of columns and constants used in the calculations of this chapter. It is quite long but typical of the amount of data we must deal with in a forecasting project. We performed the computations for the 'simple' seasonal methods A–F via Minitab scripts so we could re-run them if we discovered inconsistencies, errors, or features we felt should be further investigated. To simplify our work, we tried to use certain columns of our Minitab worksheet for specific purposes. A list such as Table 10.4 aids such work, which is still tedious.

As an aside, we note that it becomes quite awkward in Minitab and many similar packages to deal with series that do not start in the first season of a cycle, i.e. do not start in quarter 1 of a year. We therefore did not make our scripts fully general for the six methods presented.   *Reminder warning*: It is, in general, important to be careful if our graphs have been automatically scaled, since the differences in scale may lead us to perceive variability where there really is none. We can usually force the scaling, but often only after an annoying effort. Alternatively, we can use numeric summary statistics to look at the variability of the ratios, or consider boxplots for the same purpose. In a similar fashion, we looked at percentage errors for the validation period, since this is essentially scale-free. The standard deviation (STDEV) of errors and residuals can really only be appreciated in the context of some measure of size of the variable, and relative measures are therefore helpful. For quantities that do *not* have both positive and negative values, we may want to use the Coefficient of Variation (CV) which is the standard deviation divided by the mean (sometimes expressed as a percentage). We could, for example, use the CV to give a percentage variability of the raw seasonal factors about their mean. Seasonal factors are always positive, and 1 is their 'average' value, so we can use the CV, but we should not apply it to seasonal shifts, with positive and negative values.

Table 10.4. Minitab data storage for seasonal factor and shift model computations

Columns / variables		Columns / variables	
C1	our data (drop 1975-1981, 28 periods)	C16	the trial model seasonal factors or shifts (over the whole data period)
C2	subset of data used for trial models		
C4	the time periods of the data in C1	C17	the main trend model predictions (whole data period plus forecast period)
C5	the time periods of the data in column 2		
C7	an index of the quarters for the data	C18	the main seasonal factors or shifts over the same period as C17
C8	the coefficients from the main trend model		
C9	the 4 quarterly seasonal factors or shifts (either trial or main)	**Constants**	
		K3	total sum of squares for the trial model
C10	coefficients from the main trend model	K4	residual sum of squares for the trial model
C12	trial residuals (but over the whole data period, including validation)	K5	R_squared for the trial model
C13	the main model (over the data period extended by the forecast period)	K6	total sum of squares for the main model
C14	the residuals from the main model (over the whole data period)	K7	residual sum of squares for the main model
		K8	R_squared for the main model
C15	the trial trend model predictions	K49	the periodicity (4)

Assessment and validation of forecasting models

Students often worry that they do not know how much analysis of results is 'enough'. This section will summarize our view. We divide our treatment, along with the time axis, into four parts: trial and main modelling periods, the validation period and the forecasts. Overall, we *assess* our forecasts, which are models of the behaviour of our data, by considering how well they perform in 'predicting' the data that has been used to create them. A model must fit the data used to estimate it. The common approach is to 'calculate' residuals, either computationally or, for ruler forecasts, from the visual difference between the forecast lines we draw and the data points on the graph. With equations for trend lines, we could actually calculate the residuals, but this involves a lot of tedious work. Be 'creatively lazy', that is, find the easiest way to do a good job.

Using our residuals, we developed different measures of their 'size', considering both average values, such as the standard deviation, and the largest values (maximum and minimum). Beyond size, we want to look at patterns, both in time (the time plot) and distribution (stem and leaf or histogram). We look for patterns that tell us if we could predict the data better with a different model. We want to find that the time plot of the residuals is simply 'noise'. If we feel the residuals have no pattern over time, then our model is doing a good job of capturing the underlying movement of the data over time.

We can also try to see if the residuals support a hypothesis that the errors around the model have a known distribution, usually Gaussian. If the variation of the residuals around zero is roughly constant (technically no *heteroskedasticity*), there is no time pattern in the residuals such as a sine curve, and the histogram is roughly mound shaped, we could use the Normal Probability plot and associated statistical tests to support a claim that the errors are Gaussian. In any case, we would like our residuals to be more or less symmetrically distributed around zero.

If the residuals are distributed with a shape that is consistent with a Gaussian or t distribution, we can make probability statements about our models and their parameters, but this is not critical here. If our residuals do not seem to indicate that errors are Gaussian, we can still use the model for forecasting. We will just have to work harder to provide confidence bands or other statistical results. We prefer a Stem and Leaf diagram to numerical statistics to get an idea of the spread of our 'errors'. We can see whether the deviations between data and model are mostly small, but with a few large 'outliers', or if the errors are spread out fairly evenly in size. We should get suspicious if the errors are all clustered near just a few positions.

Another view of patterns in our residuals (or other data) is provided by the autocorrelation function. If our errors can be taken to be Gaussian, hypothesis tests on the autocorrelations are possible, e.g. the Box–Pierce or Ljung–Box tests.

Turning to the Validation period, we again use size measures of the errors, and look for patterns in the time plot. As in our example, we are especially interested in whether the errors are symmetric about zero, and if and how fast they grow as we move further from the last data point used in estimating the forecasting model.

Table 10.5 summarizes things to consider in preparing forecasts from quantitative data.

Table 10.5. Considerations in analysis of forecast models

Trial and main modelling periods
Residuals
Measures of size
Mean (should be very near zero)
Standard deviation
Min and Max
R_squared
(Possibly percentage errors)
Pattern
Stem and Leaf or histogram
Time plot
ACF (most useful if errors appear Gaussian)
Models
Stability of trend and season between Trial and Main models
(Possibly use time plots of seasonal factors or shifts, or CV of seasonal factors)
(Possibly consider analysis of deseasonalized series for pattern)
Validation period
Errors
Time plot
Stem and Leaf or histogram
Measures of size as above
Forecast period
(Don't forget to provide the forecasts)

Annual or other long-pattern series

Our example data are quarterly. However, they originated as monthly data, with an annual seasonal pattern. Many real-world series have monthly or even weekly data. Subsetting such data results in many separate series of data. For example, while a quarterly series breaks down into four series, a monthly one gives 12. We get 12 sets of descriptive statistics and graphs. Everything is, in detail, similar, but the housekeeping requires more discipline and data display and formatting in reports is more tedious. The Dry Cleaning data of Excercise E9.4 is annual.

Things get even worse if our data have multiple seasonal patterns. Consider, for example, telephone switch traffic data or patient admissions to a hospital. Here we will have weekly cycles imposed on the annual one. There will be effects due to public holidays as well. In this book we make no attempt to present such cases, since they have a lot of details that relate to the individual situation. Nevertheless, models involving seasonal factors or shifts could be appropriate, although we want to avoid as much of the manual data manipulation as possible.

Forecasts

We have already reported the forecasts for our 'ruler' methods in Table 10.1. The reader may compare these with the results of our six *ad hoc* methods given below in Table 10.6. In this table, the first row of forecasts is for period 92, which is 1997 Jul–Sep or 1997-3.

From the graph of our forecasts given as Figure 10.10, we see that all our methods give quite similar predictions. Method B is generally near the 'top' of the picture, while F is near the bottom. The coding Minitab has assigned is not easy to sort out, and we recommend having a printout of the forecasts to allow for checking. A useful 'trick' is to hand-colour the lines with a coloured pencil or marking pen, since colour printing, while quite common, often requires fiddling with settings and options to get the desired appearance. If you need colour output, take the time and trouble to get it right. For most forecasting needs, we can take short-cuts.

Which of the forecasts should we use? Given the 'fit' of the models, method A is at least a reasonable choice. Moreover, from the graph above, it more or less follows a middle path, so that averaging or otherwise combining the forecasts (Chapter 17) would give results close to these forecasts. Whether this really is a good choice will depend on future events.

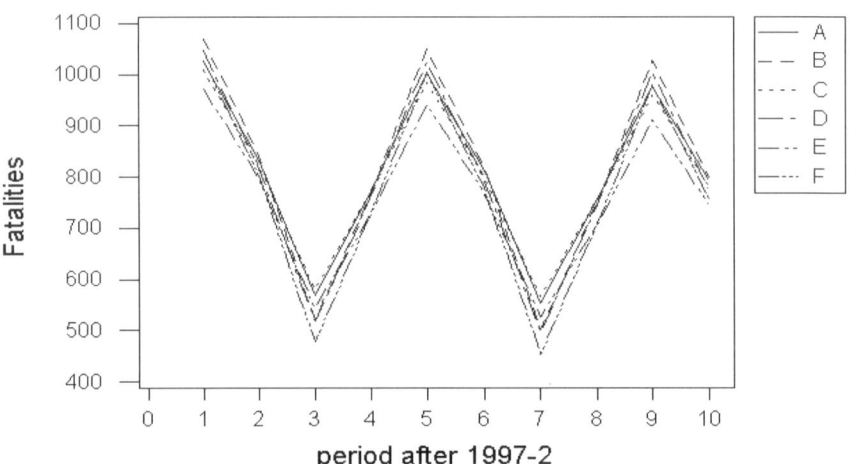

Figure 10.10. The forecasts generated by the six 'simple' seasonal methods.

Table 10.6. Forecasts from the six *ad hoc* methods presented

Row	period	Ammodel	Bmmodel	Cmmodel	Dmmodel	Emmodel	Fmmodel
1	91	1025.73	1069.78	1008.98	1046.00	972.39	1027.69
2	92	832.46	840.29	821.30	816.52	794.88	803.49
3	93	567.84	522.02	580.62	517.61	543.87	478.36
4	94	773.31	770.29	775.50	765.90	730.55	731.92
5	95	1000.81	1048.63	984.34	1024.91	941.36	1003.41
6	96	812.11	819.15	801.11	795.44	769.30	779.21
7	97	553.87	500.88	566.26	496.52	526.23	454.07
8	98	754.17	749.15	756.20	744.82	706.66	707.64
9	99	975.89	1027.49	959.69	1003.83	910.32	979.12
10	100	791.76	798.01	780.93	774.36	743.73	754.92

Exercises

E10.1. Try ruler forecasts on the Dry Cleaning data (Exercise E9.4).

E10.2. Try one or more of the *ad hoc* methods of this chapter on the Dry Cleaning data.

E10.3. *Briefly* try one or more *ad hoc* or ruler methods on the Air Passengers to Israel data (Exercise E9.5). Discussion: What is going wrong here?

11

A strategy for performing forecasting data analysis

Motivations

This chapter concerns the awkward and often 'dirty' matter of actually doing the calculations and making the graphical displays needed for preparing and analysing forecasts and the data that they are based upon. Forecasting often presents us with a lot of data and even more 'output'. Forecasting was exceedingly onerous and tedious before computers became widely available. Moreover, the graphical capabilities of modern personal computers have lifted a huge burden from the shoulders of forecasters and, at the same time, reduced the burden of time, effort and money needed for the preparation of presentation-quality graphics.

Despite the power and availability of personal computers, forecasting still requires software and the user knowledge to employ it effectively. While many software vendors would have us believe that 'you only need software X', where X is their product, we have never found a single product adequate to any broadly based task such as forecasting. It is, in any case, worthwhile to have choices and to be able to make comparisons. This is especially important when we wish to use our results as the basis for business or managerial decisions. We have already made some more general references to the assignment of resources for forecasting in Chapter 2.

Below, we examine some strategies that we believe are viable options for forecasting by managers. These are less recommendations than they are scenarios to allow our readers to formulate their own strategies. These strategies will be dependent on a number of factors:

- The type of situation that is the subject of forecasting;

- The amount and quality of data;

- The amount and quality of meta-data (descriptive information);

- The support that is available to the forecaster in terms of staff, time and budget;

- Our knowledge of the forecasting methods;

- Our facility with the software tools that are at our disposal.

The statistical package approach

In this book, we mainly work with the Minitab statistical software package. We were led to this choice because:

- Minitab is a widely available statistical package with a proven track record;

- It is relatively easy to learn, and is the most commonly used package in teaching statistics, so many students already have some familiarity with it;

- Its student edition includes ARIMA modelling, which is usually available only in 'professional' versions of statistical software, along with a range of other forecasting techniques;

- It provides for a log file, which allows teachers to back-track student work to discover errors.

Many other packages have similar features in their regular (but usually not their educational) versions. We could have worked with Stata, SPSS, SAS, Splus, R, Systat, and a host of other products. In fact, we believe that this book could be used by workers with such packages without difficulty. The output would be similar, although the command syntax is different. Unfortunately, the little details that let a user get the 'right' output are also the ones that cost many hours of frustration at the keyboard. We have tried to include them for Minitab. With the help of readers, we will be happy to prepare a list of parallel 'tricks' for other packages, and invite suggestions.

The principal advantages of using a statistical package are:

- Statistical operations can generally be expected to be properly handled;

- Most packages handle missing values sensibly;

- Most packages offer the option of recording a log file;

- We can generally prepare scripts to perform complex calculations. This is important when we want to change or extend data, or rerun an analysis to replicate our work later.

On the negative side:

- Statistical packages are (generally) expensive to buy and update, although a number of excellent software collections exist that are available free of charge;

- Statistical packages usually take time and effort to learn to use efficiently, and the number of people we can ask for help may be limited;

- Forecasting, as opposed to time series analysis, may be treated as a sideline to 'serious' statistics.

The spreadsheet approach

Most personal computer users have used a spreadsheet package. Since the introduction of Visicalc in 1978, this class of software has become almost as prevalent as word processing packages. Its particular advantages for the forecaster are:

*   Widespread availability, as well as many people who may possibly be able to help and advise on their use;

*   Relatively inexpensive (Star Office is freeware) or available without charge to most users;

*   Reasonably good computational and graphical capabilities;

*   Modern spreadsheet software often incorporates optimization tools (but not Star Office).

The downside of current common spreadsheets (Corel Quattro, Lotus 1-2-3, Microsoft Excel) for forecasting is often not immediately apparent, but a number of workers have expressed dissatisfaction with various aspects of statistical features in spreadsheet programs, ourselves included (Nash and Quon, 1996; McCullough and Wilson, 1999). Some issues are:

*   Missing values are often handled in unpredictable ways. Our experience is that missing values are a very common source of errors and rework, in part because a blank text string and a true 'missing value' appear the same. See Figure 11.1.

*   The effort to prepare scripts to carry out complex calculations is generally beyond the skills or inclinations of most users. On the other hand, it is relatively easy to take an example spreadsheet (forecasting) calculation that has been saved and copy new data into it, then make any needed adjustments for the length or other features of the data series.

*   Spreadsheets do not offer a log file. This makes it very difficult to track down errors.

*   Statistical functions have been a particular concern (Nash and Quon, 1996). Also, ARIMA modelling is not, to our knowledge, available as part of any standard spreadsheet package.

*   Some forecasting techniques may be available in some spreadsheet packages, but otherwise we must seek 'add-ins' such as ForecastX (www.forecastx.com).

*   Forecasting often requires us to graph a lot of data on a single plot. The defaults of spreadsheet graphics, in our experience, lean towards label fonts and display symbols that rapidly clutter the output. Worse, control of these details can be awkward. We must admit that this criticism can also be made of other packages, but a frequently claimed advantage for spreadsheets is the availability of easy-to-use graphics. We have also found it awkward to prepare distributional plots such as histograms or stem-and-leaf diagrams with a spreadsheet, although some capabilities of this sort are becoming available.

- Different spreadsheet programs have different features and mechanisms for their use. For example, Quattro Pro 7 or 8 allow us to export charts to a file directly, while Microsoft Excel 97 does not – at least we could not find such a feature.

An overriding issue for forecasters may be that their data are provided in a particular spreadsheet format, and users of the forecasts wish to receive results in that form. In such a case, and where advanced methods are not needed, a spreadsheet approach is clearly indicated.

Corel Quattro Pro 7/8		<-- empty cell b1; missing	MS Excel 97 / 2000		<-- empty cell b1; missing
@isna(b1)=	0	false=0	isna(b1)=	FALSE	
@isblank(b1)=	1	true=1	isblank(b1)=	TRUE	
isnonblank(b1)=	not available		isnonblank(b1) =	#NAME ?	
@isnumber(b1)=	1		isnumber(b1)=	FALSE	
@iserr(b1)=	0		iserr(b1)=	FALSE	
iserror(b1)=	not available		iserror(b1)=	FALSE	
isref(b1)=	not available		isref(b1)=	TRUE	
@isstring(b1)=	0		istext(b1)=	FALSE	
@isnontext(b1)=	1	<--not in XL			

Figure 11.1. Results of special functions that allow spreadsheet users to determine the properties of cell contents. Note that a 'missing' value or empty cell gives ambiguous results. StarOffice 5.1 gave results equivalent to Microsoft Excel 97.

Special-purpose forecasting software

There are a number of software products that are specifically aimed at forecasting. We have even reviewed some of them (Nash and Walker-Smith, 1989b). As the offerings change frequently, we will limit our discussion to issues that may help or hinder the forecaster. Moreover, the market for forecasting software is both small and fragmented. For example, vendors may focus on a particular application segment, such as investments, cash flow, or sales of industrial goods. Others may offer packages that build on a single class of forecasting methods (e.g. AUTOBOX, which favours the Box–Jenkins methodology). In Chapter 18, we mention some packages that are directed to seasonal adjustment of government or other time series.

The main advantage of the special purpose forecasting packages is (presumably!) that the forecasting methods are set up so that they are easy to use. Output, including graphs, should be more or less automatically produced. This is, in general, our experience, even with packages prepared over a decade ago, where some ingenious program designs were used to overcome the limitations of the computing hardware of the time.

On the negative side, we have almost always found that the issue of preparing data and getting it into the packages is tedious and error-prone. 'Little things mean a lot', especially if they are special characters such as spaces, null characters, tabs, carriage returns and line feeds, or even accented characters in labels or documentation. We also urge our readers to be cautious about vendor claims. Some advertisements for forecasting software are difficult to distinguish from those for psychic telephone hotlines. There is a great deal of very good software, but finding it can be tricky. Some of this software can also be very expensive.

Special purpose forecasting software represents a niche market, and evolves with perceived or actual sales opportunities. We will, as appropriate, put material about such products and their uses on the PFM Web site.

Auxiliary tools

The issue of cleaning and adjusting data to get it into software for forecasting has been mentioned several times. This requires us to have some tools to help us do the job.

Most important among such tools is a *text editor*. We find our students often confuse a text editor and a word processor. The word processor is intended not only to edit text, but also to format it. Formatting output is a very secondary function of statistical software, forecasting software or spreadsheets. Plain text is usually required or preferred for data, metadata, and scripts, such as those used by Minitab and similar statistical packages. Most packages also output files as plain text, including Minitab's log files.

The particular text editor you choose is largely a matter of personal taste, but we can suggest some features that we like:

* Simplicity: We find that we want to be able to use the editor quickly and easily. We will sacrifice fancy functions so that we do not have to remember complicated lists of commands, despite the fact that we came to computers via command line editors that look distinctly medieval compared to a modern text editor with mouse support.

* Cross-program copy-and-paste: Windowed operating systems let us have several programs active. This is very useful when we want to get a few numbers from an electronic mail message and put them into a spreadsheet or statistical program. In such cases, it is useful if the commands for Cut, Copy and Paste are the same in the different packages. We are used to Microsoft Windows where the conventions are Ctrl-X, Ctrl-C and Ctrl-V, respectively. (This means that the Control key is depressed and held while the letter key is pressed. There are similar conventions on other platforms, such as the Apple Macintosh and Linux.) However, several packages use different conventions to cause trouble. Novice users tend to use the menu choice to perform these operations, but this can be slow. Shortcut key combinations can make this sort of work much more efficient, but only if there are common

conventions across programs. Note that spreadsheet programs may give trouble if we Copy cells that are formulas and fail to Paste them as *values*, since we then may get the result of computations with the wrong data.

- The ability to handle large files. The 'default' editors for both Microsoft Windows (NOTEPAD.EXE) and Apple MAC-OS (TeachText or SimpleText) can handle only small files. There are many freeware, shareware and commercial replacements. On Apple Macintosh, we have used BBEdit Lite. On Windows we use SuperPAD, PFE and DeDit.

- A find-and-replace feature: This is now present in most editors.

  Some other features to consider if you want more 'power':

- A few editors allow you to select a set of text columns from a file. This is very handy if you have a text file of output with a table of numbers, but only want the third column. We have used MultiEdit Lite (MEL) in an MS-DOS environment, as well as VIM for DOS. This allows us to extract the numbers, but we lose the cross-program cut and paste. UltraEdit, which is shareware, works in Windows. A note of caution: if you want to do column operations, make sure your editor displays the text in a monospaced font such as Courier. Otherwise, you will find such editing very confusing, indeed practically impossible.

- A number of text editors offer a *macro* facility, as do many word processors. We have used the SEE editor in MS-DOS for many years. It allows us to perform sequences of operations with a single keystroke. This is useful in making adjustments to text, but we do not recommend it to novices, particularly as the most widely available macro editors are based on EMACS, a very powerful editor, but one that requires the user to remember many complicated command sequences.

- Some editors allow the editing of binary files. Thanks to a friend and colleague, Woody Suwalski, we have for many years been able to use WED, which edits large files, including binaries, but is itself very small. In forecasting applications it can be useful to look at files that are giving trouble to see if there are any special characters present and remove them. We have also used it to check for unconventional line or file endings, or for identifiers that may tell us the program that created the file. (WED is a DOS program, but there are some freeware offerings for other platforms on sites such as the Simtel collection.)

Clearly we often use the text editor as a tool for reformatting files, but there are programs that we may call (data) *reformatters*. These take data files from the format intended for one package to that designed for another. For statistical program data, we have used both DBMS Copy and StatTransfer, although not in conjunction with forecasting. Below we suggest an alternative to such programs.

One type of transformation we find that we need to make quite often is between different graphical files. Despite the growing ease with which material can be copi and pasted between applications, there are occasions where this results in graphs or images that are unsatisfactory. We then need to transform the original so that the word processing or publishing package receives an appropriate form to give the results we wish. We would like to say that there are 'rules' we can follow to ensure success. Generally we find we must spend more time and effort than we can afford on such tasks. Some difficulties are due to the nature of the graphic files. We have noted that our students often do not appreciate these differences, so we will briefly explain.

- Most computer images are *bitmaps* (or *raster graphics*). This means that the picture is made up of many *pixels* or dots, each with a colour and intensity. Such image files commonly have filename extensions BMP, TIF, GIF or JPG, among others. While the BMP files are stored as raw pixels, most types are compressed, sometimes via a *lossy compression* that throws away some of the image that we hope is not important. Bitmaps are a nuisance for many statistical graphs because the resolution (dots per inch or dots per centimetre) of one output device may not match that of another. Then we must adjust to the new resolution, often getting 'jaggy' lines or poorly formed symbols. This is especially true if we move or resize the images and store them at each stage.

- Drawing software stores the operations that create an image. We start a line at one point $(x1, y1)$ on the screen and move a 'pen' to $(x2, y2)$. This is a vector, and the resulting *vector graphics* preserve the image across different display devices as long as we keep the original 'instructions'. Unfortunately, word processing and graphics software often does not understand the 'instructions' in our files. For example, Minitab MGF files are clearly vector graphic (they are stored as plain text), but we have had to load them in Minitab, then copy and paste bitmaps for this book. PostScript files are plain text that can have vector graphic instructions as well as bitmaps. PIC, PLT and HPGL files are vector graphic.

Most of the software for reformatting and editing graphics is commercial, although some is distributed as shareware. We rather like the IrfanView freeware package for bitmap graphics.

Besides the transformation and reformatting of graphics files, we may also wish to add annotation, lines, or other features to graphs. Another tool we will want to have, therefore, is one to enable this. We tend to prefer a drawing program such as Corel Draw, but we note that there are some simpler tools built into word processing software such as WordPerfect and Microsoft Word that may suffice. While drawing programs deal with vector objects, other programs, which we will refer to as *paint* programs, use bitmaps. There are more choices here: some examples are Corel Photopaint, Microsoft Paint, Adobe Photoshop, and Jasc PaintShop. We find that it is less easy to edit graphics that are stored as bitmaps. For example, once text has been added, it is difficult to change its position or orientation with a paint-style program.

Finally in our toolkit, we need to prepare reports. Most of our readers will be familiar with word processing software such as Corel WordPerfect (used to prepare this material) or Microsoft Word. These have many features – in fact too many features. We especially dislike the fact that word processors will reformat documents to suit the 'default printer'. This frequently upsets the pagination and location of graphs etc. (an especially annoying matter in the preparation of this book). We are very careful not to save the file in the 'new' format because it cannot generally be changed back successfully. We suggest a way to deal with this difficulty below.

Other strategic and tactical issues

To simplify our (forecasting) life, we use *plain text* files as much as possible. Most statistical and spreadsheet software can deal with plain text data and will output data (not graphics) in this form.

We try to save graphics in a *vector graphic* form where appropriate. Thus, we save the MGF files from Minitab and then reload and transform the resulting bitmap as needed for reports.

File naming can be a nuisance. Long names are helpful descriptively, but not available to users of DOS/Windows 3.1. Moreover, the ISO 9660 standard for CD ROM disks requires 'short' filenames, in fact with tighter restrictions than MS DOS. While such systems are disappearing, we must be aware of them, especially if we aim to share files with others over networks. We do precisely this in our teaching and consulting. Long names also take longer to type. We have so far *chosen* to use 'short' filenames with eight characters in the name and three in the extension. However, we will not be unhappy to be allowed to use a few more characters.

The short filenames are also an advantage in the backup of our files. We have already pointed out that it is important to secure your files by backing them up regularly. We prefer to do this by segregating each project into a separate directory. We have the luxury of high-capacity removable cartridges and writeable and re-writeable CDs, and recommend such devices for anyone who must carry out comprehensive forecasting projects.

Making the choice

We urge you to make a conscious choice of your forecasting toolkit, that is, your approach, software, and practices, because this will enhance your efficiency in doing forecasting.

We find that drawing graphs just as we want them is never easy, no matter what package we use. That is, getting a graph of our data may be easy, but getting one that has the right scale, features and annotation is difficult. Moreover, adding notes to a graph within a software package is never trivial. For graphs we are going to publish, it is likely worthwhile to learn how to do this. For example, Minitab's graphics facilities allow us to add text or lines anywhere on the graph, but there is a lot of busy work entailed in this. When we want to analyse data for understanding, rather than preparing a publication, it is generally much easier to print some fairly large scale graphs and annotate them by hand, in the word processor, or with a drawing program.

Since we may want to prepare multiple graphs such as Figure 9.2 on other occasions and with other data, it is useful to save the command sequence that does the job rather than spend a lot of time (and several tries at least) getting the mouse-click settings. We discuss command scripts for multiple graphs in Minitab in the Appendix. Clearly, we need to learn how to carry out these sorts of manipulations with whatever software tools we choose or are required to use. We believe that facility with such manipulations is important. Because many of the forecasting problems encountered in business and administration are tied to the calendar, it is useful to have a 'toolkit' of scripts or procedures to make common transformations, for example, full series to seasonal series, full series to yearly series, build full series from yearly series, build full series from seasonal series, select trial and validation data from full series.

In one research group to which we once belonged, the instructions for such common tasks were collected in a binder with a large label 'PANIC BOOK'.

Table 11.1. Checklist of strategic and tactical choices for forecasting tasks and tools.

The Task	What forecast do we seek?
	How much data (number of elements, variables, observations, text documentation) do we need to handle?
	What resources are at our disposal?
	How knowledgeable are we (the forecasting team) about forecasting methods?
	Do we know how to use the software available to us effectively?
The Strategy	What software packages are we most adept at using?
	Do we need industrial-strength statistics, that is, are we willing to build arguments based on properly formulated hypothesis tests and checking of assumptions?
	Do we want to 'automate' our process so we can make forecasts on a regular basis, or are we content with a single forecasting exercise?
Tactics	Are we content with 'manual' approaches to documenting our work and work flow? Or do we need a mechanism to keep an audit trail of what we do?
	What do we propose to do about missing values?
	Do we have a naming convention for files we create and use?
	Do we have conventions or standards for file formats we will use for data, text, reports, and graphs?

Exercises

E11.1. Write down a forecasting strategy, including tools and software, that you feel suits your needs. If possible, compare and discuss your choices with those of others.

12

Forecasting trend and season I: multiple regression

Regression – its purposes

Regression, as a statistical tool, is a family of methods for estimating and using models that describe or explain one variable in terms of one or more others. The most common usage of multiple linear regression methodology is to build a (linear) model of some phenomenon of interest. We estimate the parameters of such models by applying the least squares criterion to the residuals associated with the model proposed. That is, we adjust the parameter values to minimize the value of the sum of squared residuals. The parameter values at the minimum are taken as their estimates.

The curvature of the sum of squares surface at the minimum, with a little mathematics, allows us to develop confidence intervals for the parameters. Such measures of precision (or variability) of the parameters help us to understand the phenomenon.

A second usage of regression is simply to'fit' an equation to data. The curve or surface that we have adjusted to the data, regardless of the true underlying structure of the phenomenon, may be extrapolated to provide forecasts for the quantity of interest. This is modelling the pattern, while the viewpoint of the previous paragraphs is to provide a causal or structural model.

When we are trying to develop a causal model, we will want to know that our model is appropriate and that the parameter values are in some way 'correct'. Taking the 'causal model' approach to regression implies that we should attempt to verify assumptions about the model, if possible. For example, if Gaussian errors are assumed, this assumption should be checked with graphs or tests. The absence of serial correlation should be verified, as should any other assumptions of concern to us. The assumptions that regression usually implies in a statistical context can rarely be accepted in full, though we may be able to 'live with' the violations.

Pattern fitting models need not satisfy such criteria. Mainly, we want a very good fit so that we have a clear idea of the pattern to be extrapolated. This is very different from traditional statistical thinking. Thus, we find that regression for pattern fitting is rarely given much space in books on regression modelling. For forecasting, we think it is an important use of regression.The approach is justified or rejected by how well it works, or does not work, in providing suitable forecasts.

Regression jargon

In this section, we will provide a notation for multiple linear regression modelling. We want to provide a model for the data $y(t)$, $t = 1, 2, ..., m$, usually referred to as the dependent or predicted variable, or regressand. The model involves a linear combination of n variables $X(t,j)$, $j=1, 2, ..., n$ that are generally referred to as the independent or predictor variables, or regressors. We prefer to avoid the terms 'dependent' and 'independent', since we may develop multiple equation models where a variable appears on the left of the equality as an 'independent' or predicted variable in one equation and on the right-hand side as a regressor in another. Once we have our data, we want to approximate the y's by the X's, that is, (in two different but equivalent forms)

$$y(t) = \sum_{j=1}^{n} X_{tj} \; b_j = \sum_{j=1}^{n} X(t, j) \, b(j)$$

As an example, we may want to model the monthly consumption of heating oil (variable *oilcons*) of a house in terms of the square footage of the house (variable *area*), and the degree days, that is, the sum of the number of degrees below some target such as 10 degrees Celsius over the days of the month (variable *degdays*). For month t, we then get

oilcons(t) ~ $b(1) + b(2) * $ area(t) $+ b(3) * $ degdays(t)

Of course, few houses actually change area over time, but we could use the same model for many different houses, so the model for month t for house j is:

oilcons(t,j) ~ $b(1) + b(2) * $ area(j) $+ b(3) * $ degdays(t)

which could be a useful forecasting tool for an oil delivery company. Other variables might be added. For example, as children and pets open and close doors, we may include a variable that takes some account of their presence.

To remove the awkwardness of the approximation sign '~', we introduce the *residual $r(t)$* for period t, so that

$$y(t) = \sum_{j=1}^{n} X_{tj} \; b_j + \text{residual}_j$$

This is the actual modelling equation, which is an approximation to the real, but unknown, form

$$y(t) = \sum_{j=1}^{n} X_{tj} \; \beta_j + \text{error}_j$$

Usually one of the X 'variables' is actually the *constant*, that is, a column of ones. We have made it variable number 1 above, and did not even write it in explicitly. Of course, we do not usually call this a 'variable', and just count the number of true variables. Readers should be careful in reading any material on regression to get the correct 'size' of the problem. That is, while we say that a problem has m observations or cases in n parameters, others may use m cases by $(n-1)$ variables, since we have one parameter that multiplies the constant. This parameter is often labelled $b(0)$ and called the intercept. Minitab takes this latter view in the REGRESS command, but if you happen to use Minitab's MATRIX commands, you must remember to put in the constant explicitly.

As indicated, the X variables may be called regressors, independent variables, predictor variables etc. In an econometric model, which may have a number of equations, any variable *not* determined inside the model is called an *exogenous* variable. Those determined from within the model are called *endogenous* variables.

Multiple linear regression generally assumes that the variables, that is, the $X(.,j)$s are linearly independent. This means we cannot find a linear combination of the $X(.,j)$s such that the sum is zero for all index values t. The fact that we want statistical independence is another reason to avoid the jargon 'independent variable'. In exact arithmetic, we almost always find the variables are linearly independent. However, most real data sets have 'near' linear dependencies. This creates the problem of *multicollinearity* (or more simply just *collinearity*), and we say that some of the data are *collinear*.

In causal modelling, we generally assume that the true errors are drawn from a Gaussian (normal) distribution with mean zero and variance σ^2. The sigma is unknown, but by this assumption it is the same for all t. If it varies with t, or with one of the X variables, the data exhibits *heteroskedasticity*. The assumption of homoskedasticity is as heroic as that of independence, but we can check residual plots. Weighted least squares is sometimes used in an effort to overcome this problem (see below).

We also assume that the errors that occur in different time periods are unrelated. That is, there is no *serial correlation*. We have already dealt with how to test for serial correlation via the autocorrelation and partial autocorrelation in Chapter 3. The presence of first-order autocorrelation can be checked with the Durbin–Watson statistic, for which we will include examples below, since regression software often computes this statistic more or less automatically, while the autocorrelations of the residuals require extra user effort.

Recall that we assumed that the phenomenon being modelled is capable of being explained by a linear relationship in the variables. This condition of *linearity* of the problem is one that has, historically, been imposed because techniques for nonlinear estimation were unavailable or considered unreliable. If a nonlinear relationship is required, we should not be using something that can never address this reality. One of us (JCN) has prepared many papers and three books that deal with nonlinear modelling, along with software that runs on computers of any size. See also Chapter 20 and the references therein. If the true model is *not* linear but we still use a linear model, then we have *misspecified* the model. Our results will then generally be incorrect, particularly any extrapolations used as forecasts.

Dummy variables

Dummy variables (or *indicator* variables) are artificial variables, usually made up of zeros and ones, designed to allow specific events or phenomena to be taken into consideration in our model. For example, if we wanted to look at production of armaments in the USA, we might make a dummy variable with zeroes for each year but the WWII period (for the USA this is 1941–45).

> *Question*: What about Korea and Vietnam? Should we use 0.5 for these years perhaps? Or some other value?

In a forecasting situation, dummy variables are useful to allow us to estimate *additive seasonal models*. We choose one season as our 'base' model. Usually this is a simple trend line, that is,

$$y(t) = \text{intercept} + \text{slope} * t$$

The intercept is the coefficient of the constant, usually given the parameter name $b(0)$. The slope will often be given the name $b(1)$.

We then generate dummy variables for each of the *other* seasons. Note that we need one less variable than the number of seasons, since our model now has the conceptual form

$$y(t) = \text{base\_model} + [\text{adjustment for the season corresponding to time } t]$$

The adjustment is zero when the season corresponds to the base season. The adjustment will be whatever amount is appropriate – possibly negative – for each other season. Thus, for quarterly data (four seasons), let us suppose that season 4 (Winter) is our base. The data matrix X then has a structure that starts with the four rows (cases)

)	Constant	Time	Season 1	Season 2	Season 3
	1	1	1	0	0
	1	2	0	1	0
	1	3	0	0	1
	1	4	0	0	0

The effect is that every Winter, our model predicts that the value of the quantity of interest will be at the level of the trend line (namely the base model given by the intercept plus the slope times the time index t). Since all the dummy variables for the other seasons are zero, there is no adjustment of this prediction. In the second season, we again calculate the trend. This will be higher or lower by the magnitude of the slope. That is, we adjust the base model by one time increment times the slope. Then we note that all the dummy variables for this time period are zero *except* that for season 1 (Spring), so we adjust by the coefficient estimated for the dummy variable corresponding to season 1. This is shown graphically in Figure 12.1.

The regression coefficient of the dummy variables for season i gives the amount to be added (possibly negative!) to the base model, that is, to the trend line prediction for the appropriate time point t. The base model with no adjustment gives the season 1 behaviour. Of course, we may also have other variables. If those other variables are seasonal, then we may get some unexpected results in our regression output due to collinearity (see below for more discussion).

Our experience in teaching the multiple regression approach to additive seasonal modelling via dummy variables is that while the ideas are simple, the details of setting up the data and interpreting the results require careful attention. For both set-up and interpretation, the fact that we can choose any of the seasons as the base for modelling is a source of confusion. That is, in a quarterly model, our base linear trend may apply to any of the Spring, Summer, Fall or Winter seasons, with appropriate sets of parameters for the adjustments or offsets for the other seasons. We can have four apparently different additive seasonal models that are equivalent. In fact, given one model, we can easily work out all the other possibilities. That is, all the models must have the same set of *differences* between the seasonal values. Let us start with the format above, where Season 4, or Winter, is our base. Our model, since it is based on Winter, we will choose to write

$y\_model(Winter, t) = b0(Winter) + t * b1(Winter) + Adjustment(Winter, t)$

where

$Adjustment(Winter, t) = b2(Winter, Spring)$ if t corresponds to Spring
$Adjustment(Winter, t) = b3(Winter, Summer)$ if t corresponds to Summer
$Adjustment(Winter, t) = b4(Winter, Fall)$ if t corresponds to Fall

Our notation for the dummy variable coefficients has redundant information, but we want to be clear that Winter is our base season and the others are offset from it in our model. Now suppose instead we chose Summer as the base period. Then

$y\_model(Summer, t) = b0(Summer) + t* b1(Summer) + Adjustment(Summer, t)$

but since the slope of the trend should *not* be changed in the modelling,

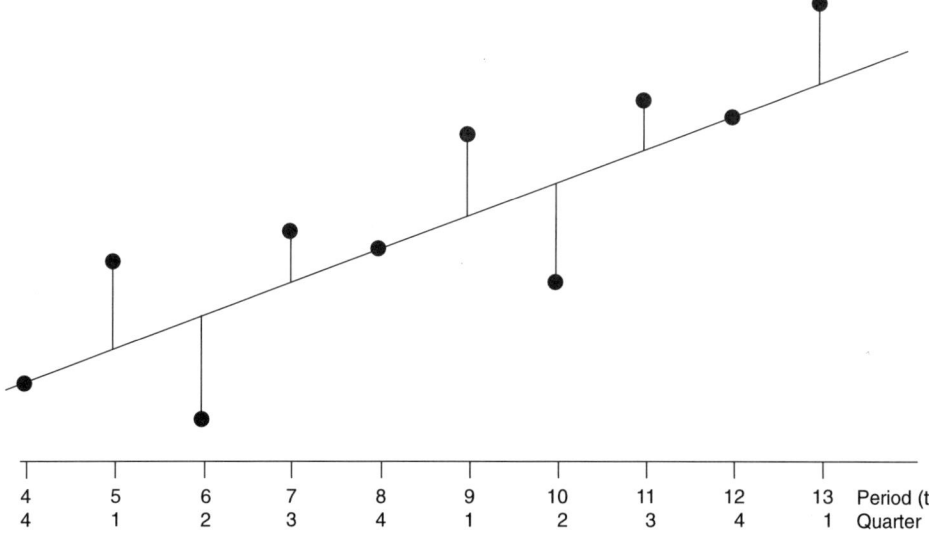

Figure 12.1. Additive shift model via dummy variables.

$b1(\text{Summer}) = b1(\text{Winter})$

The intercept changes by the difference in level of Summer from Winter.

$b0(\text{Summer}) = b0(\text{Winter}) + b3(\text{Winter, Summer})$

The offset adjustment for Winter in the 'Summer' model is the opposite of that for Summer in the 'Winter' model,

$b4(\text{Summer, Winter}) = -\,b3(\text{Winter, Summer})$

[Note that we have chosen to keep the seasons in their natural order in numbering the regression coefficients. We must carefully keep track of the meaning of our regression coefficients.] The other adjustment values are found as

$\text{Adjustment}(\text{Summer}, t) = b2(\text{Summer, Spring})$
$\qquad = b2(\text{Winter, Spring}) - b3(\text{Winter, Summer})$ if t corresponds to Spring

$\text{Adjustment}(\text{Summer}, t) = b3(\text{Summer, Fall})$
$\qquad = b4(\text{Winter, Fall}) - b3(\text{Winter, Summer})$ if t corresponds to Fall

The Appendix has an example of the equivalence of additive models using different base seasons. A related 'detail' that gives trouble is that of keeping track of the season associated with which a particular time period, t. This is, as we repeat, a matter needing constant discipline.

A real example

As before, we load our Quarterly Traffic Fatalities in Canada data into Minitab and delete the observations for 1975–81. Then we set up the quarterly dummy variables for the first 13 years (the Trial model period). In the listing below, we have used the Minitab **set** command to create the dummy variables, but we note that many people find it easier to build these using the 'Data' window, or spreadsheet. We wanted to keep a log file here, so used the script approach. Note that we also set up a series of values for the time period and for the quarterly dummy variables during the validation forecast period, since we will need these later to generate our validation tests.

```
Information on the          MTB > set c21             MTB > set c32
Worksheet                   DATA> 13(0 1 0 0)         DATA> 3(0 0 1 0)
Column  Count  Name         DATA> end                 DATA> end
C1         62  fqmdata       MTB > set c22             MTB > set c33
C2         52  fqtdata       DATA> 13(0 0 1 0)         DATA> 3(0 0 0 1)
C3         10  fqvdata       DATA> end                 DATA> end
C4         62  mperiod       MTB > set c23             MTB > delete 11:12
C5         52  tperiod       DATA> 13(0 0 0 1)         c31-c33
C6         10  vperiod       DATA> end                 MTB > set c30
Constant       Value         MTB > name c21 'q2dummy'  DATA> 81:90
Name                         MTB > name c22 'q3dummy'  DATA> end
K47         1.00000          MTB > name c23 'q4dummy'  MTB > name c31 'q2vdmmy'
K48      1975.00             MTB > set c31             MTB > name c32 'q3vdmmy'
K49         4.00000          DATA> 3(0 1 0 0)          MTB > name c33 'q4vdmmy'
                             DATA> end
```

We now compute the regression model for the trial period. Note that we save a lot of information in case we need it. The instructions to do this can be provided, as here, in a script, or else by selections in a mouse-click dialogue. We omit commenting on all possible features of the output to keep our presentation brief.

```
MTB > Name c50 = 'COEF3' c51 = 'FITS3' c52 = 'RESI3' c53 = 'TRES3' &
CONT>        c54 = 'HI3' c55 = 'COOK3' c56 = 'DFIT3' K52 = 'MSE3' &
CONT>        c57 = 'PFIT3' c58 = 'PLIM3' c59 = 'PLIM4'
MTB > Regress 'fqtdata' 4 'tperiod' 'q2dummy' 'q3dummy' 'q4dummy';
SUBC>    Coefficients 'COEF3';
SUBC>    Fits 'FITS3';
SUBC>    Residuals 'RESI3';
SUBC>    Tresiduals 'TRES3';
SUBC>    Hi 'HI3';
SUBC>    Cookd 'COOK3';
SUBC>    DFits 'DFIT3';
SUBC>    MSE 'MSE3';
SUBC>    Constant;
SUBC>    VIF;
SUBC>    DW;
SUBC>    Predict C30 C31 C32 C33;
SUBC>       PFits 'PFIT3';
SUBC>       PLimits 'PLIM3'-'PLIM4';
SUBC>    Brief 2.
```

```
Regression Analysis
The regression equation is
fqtdata = 939 - 4.71 tperiod + 304 q2dummy + 580 q3dummy + 351 q4dummy
```

Predictor	Coef	StDev	T	P	VIF
Constant	939.17	46.94	20.01	0.000	
tperiod	-4.7129	0.7712	-6.11	0.000	1.0
q2dummy	304.10	32.65	9.31	0.000	1.5
q3dummy	580.27	32.68	17.76	0.000	1.5
q4dummy	350.83	32.73	10.72	0.000	1.5

```
S = 83.23       R-Sq = 88.1%      R-Sq(adj) = 87.0%
Analysis of Variance
```

Source	DF	SS	MS	F	P
Regression	4	2399721	599930	86.60	0.000
Residual Error	47	325583	6927		
Total	51	2725303			

Source	DF	Seq SS
tperiod	1	185127
q2dummy	1	326
q3dummy	1	1418230
q4dummy	1	796038

```
Unusual Observations
```

Obs	tperiod	fqtdata	Fit	StDev Fit	Residual	St Resid
1	29.0	624.0	802.5	29.6	-178.5	-2.29R

```
R denotes an observation with a large standardized residual
Durbin-Watson statistic = 1.32
```

```
Predicted Values
```

Fit	StDev Fit	95.0% CI		95.0% PI	
557.4	31.6	(493.8,	621.0)	(378.3,	736.5)
856.8	31.6	(793.2,	920.4)	(677.7,	1035.9)
1128.3	31.6	(1064.7,	1191.9)	(949.2,	1307.4)
894.1	31.6	(830.5,	957.7)	(715.0,	1073.2)
538.6	33.8	(470.6,	606.6)	(357.9,	719.3)
838.0	33.8	(770.2,	905.9)	(657.2,	1018.7)
1109.4	33.8	(1041.4,	1177.4)	(928.7,	1290.1)
875.3	33.8	(807.3,	943.2)	(694.6,	1056.0)
519.7	36.1	(447.1,	592.4)	(337.2,	702.2)
819.1	36.1	(746.5,	891.7)	(636.6,	1001.6)

The model fits reasonably well from the R_squared statistic. Figure 12.2, a time plot of the residuals (which we saved into column c52), shows the inverted 'U' pattern we saw with the *ad hoc* seasonal models, but no obvious heteroskedasticity. For the moment let us ignore this temporal pattern. The distribution of the residuals, seen in Figure 12.3 (we could have used a histogram) is roughly mound-shaped. If we can assume that the errors are Gaussian – and we should do so with our fingers crossed – then the regression output shows that all the regression coefficients are significantly different from zero. The F-statistic, now redundant, also supports a conclusion that at least one of the coefficients is non-zero.

The Durbin–Watson (DW) statistic for the trial model estimation of our data has been computed as 1.32. This is less than 2, indicating possible positive serial correlation of first order. We have 52 observations and 4 predictor variables. Tables of the Durbin–Watson statistic (e.g. Makridakis *et al.*, 1998, pp. 630–631) give lower and upper critical values (at 5% significance) of 1.42 and 1.67 for 50 observations and 4 predictors. At this level of significance, we reject the null hypothesis ('no first-order serial correlation') and conclude that there *is* first-order serial correlation. We already know this, of course, from the inverted 'U' pattern that we can see. If we look at the 1% significance critical values of 1.24 and 1.49 (again for 50 rather than 52 observations), our computed DW statistic is in the 'inconclusive' zone for the test.

We will not comment on all the information in the regression output here. Our exposition is long enough already.

```
Descriptive Statistics (edited)
Variable  N   Mean  Median  StDev   Minimum   Maximum      Q1       Q3
RESI3     52   0.0   -1.8    79.9    -178.5     144.7     -67.6     75.4
```

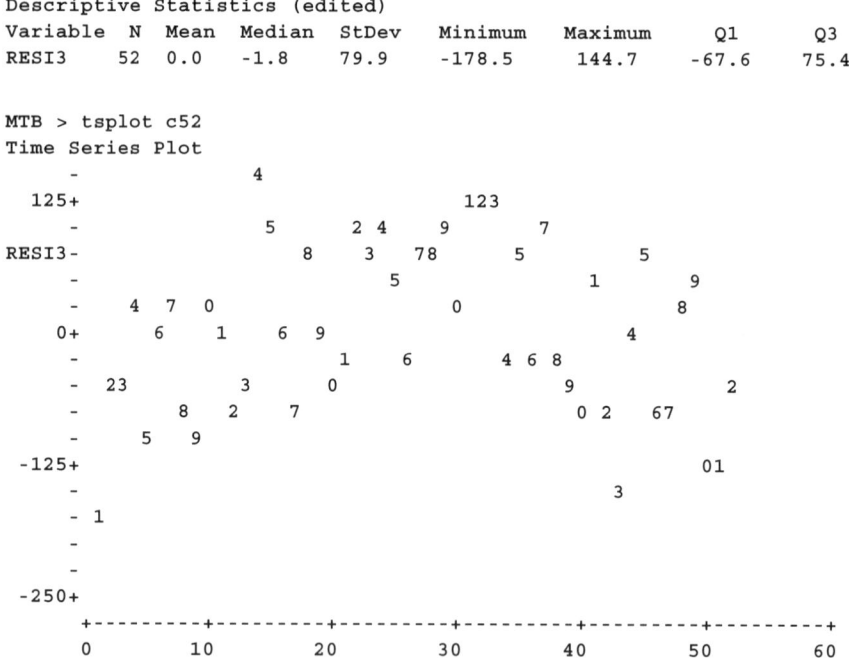

```
MTB > tsplot c52
Time Series Plot
        -                4
    125+                                        123
        -                     5        2 4      9          7
   RESI3-                          8       3    78        5               5
        -                                     5           0           1         9
        -            4   7  0                                                 8
     0+            6      1     6  9                                   4
        -                           1      6          4 6 8
        -    23                3       0                  9                        2
        -            8   2     7                          0 2      67
        -        5   9
   -125+                                                              01
        -                                                        3
        -  1
        -
        -
   -250+
         +---------+---------+---------+---------+---------+---------+
         0        10        20        30        40        50        60
```

Figure 12.2. Trial estimation residuals for an additive seasonal model of Quarterly Traffic Fatalities in Canada.

```
MTB > stem c52
Stem-and-leaf of RESI3
N  = 52    Leaf Unit = 10

      1      -1 7
      2      -1 5
      4      -1 32
      4      -1
      8      -0 9988
     13      -0 77777
     19      -0 555544
     22      -0 322
     (5)     -0 11000
     25       0 011
     22       0 3333
     18       0 445
     15       0 67777
     10       0 89999
      5       1 01
      3       1 22
      1       1 4
```

Figure 12.3. Distributional shape of residuals of the trial model for Quarterly Traffic Fatalities in Canada.

We now compute and graph the validation forecast errors (labelled vres3 below).

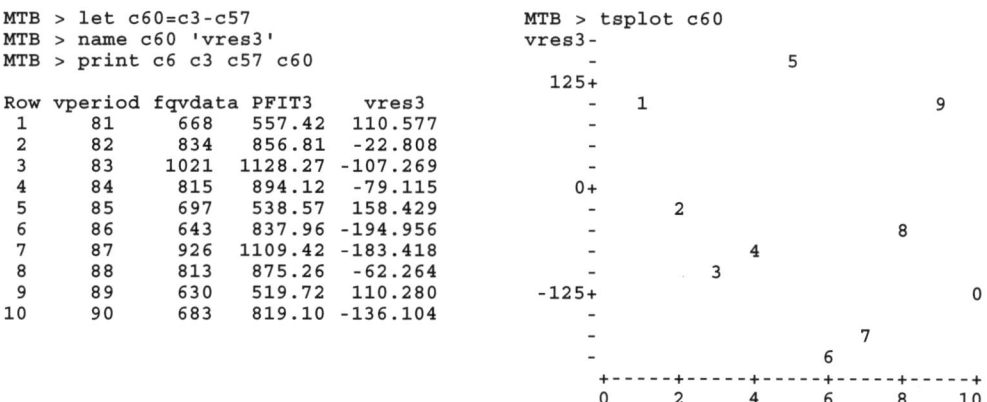

```
MTB > let c60=c3-c57                    MTB > tsplot c60
MTB > name c60 'vres3'                  vres3-
MTB > print c6 c3 c57 c60                    -                    5
                                        125+
Row vperiod fqvdata PFIT3     vres3          -    1                        9
  1     81    668    557.42  110.577        -
  2     82    834    856.81  -22.808        -
  3     83   1021   1128.27 -107.269        -
  4     84    815    894.12  -79.115       0+
  5     85    697    538.57  158.429        -        2
  6     86    643    837.96 -194.956        -                         8
  7     87    926   1109.42 -183.418        -               4
  8     88    813    875.26  -62.264        -           3
  9     89    630    519.72  110.280    -125+                         0
 10     90    683    819.10 -136.104        -
                                             -                   7
                                             -               6
                                             +-----+-----+-----+-----+-----+
                                             0     2     4     6     8    10
```

We combine the results of the computations in Figure 12.4, which shows the data and forecasts over time for the trial model. It is clear that the forecasted seasonality has too much spread during the later time periods. Although the trial model is not fully satisfactory, we will proceed to compute the main model and forecasts.

```
Descriptive Statistics (edited)

Variable  N    Mean  Median   StDev    Minimum  Maximum     Q1      Q3
vres3     10   -40.7  -70.7   127.1     -195.0    158.4   -147.9   110.4
```

To estimate the main model, we must first generate the seasonal dummy variables for the full data. This ends in the middle of a year; we adjust our series accordingly. In this script, we chose to generate full years of data then delete the last two points. We also generate dummy variables for the main forecast period. This starts in mid-year so must be adjusted at its beginning.

```
MTB > set c21                          MTB > set c31
DATA> 16(0 1 0 0)                      DATA> 3(0 1 0 0)
DATA> end                              DATA> end
MTB > set c22                          MTB > set c32
DATA> 16(0 0 1 0)                      DATA> 3(0 0 1 0)
DATA> end                              DATA> end
MTB > set c23                          MTB > set c33
DATA> 16(0 0 0 1)                      DATA> 3(0 0 0 1)
DATA> end                              DATA> end
MTB > delete 63:64 c21-c23             MTB > note Must delete the first 2 rows
MTB > set c30                          (beginning of 1997)
DATA> 91:100                           MTB > delete 1:2 c31-c33
DATA> end
```

Now we can compute our main regression model as before. We omit the instructions and some of the output less relevant to our needs, but include the column names of saved information.

```
The regression equation is
fqmdata = 996 - 5.27 mperiod + 259 q2dummy + 537 q3dummy + 318 q4dummy
```

Predictor	Coef	StDev	T	P	VIF
Constant	996.02	43.70	22.79	0.000	
mperiod	-5.2747	0.6359	-8.29	0.000	1.0
q2dummy	258.84	31.68	8.17	0.000	1.5
q3dummy	537.19	32.19	16.69	0.000	1.5
q4dummy	318.26	32.20	9.88	0.000	1.5

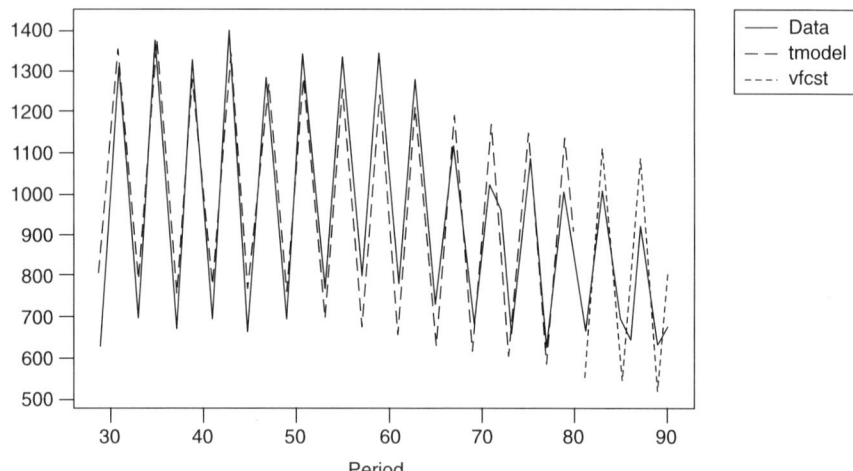

Figure 12.4. Trial model and validation forecasts for Quarterly Traffic Fatalities in Canada.

```
S = 89.58          R-Sq = 86.0%      R-Sq(adj) = 85.1%

Unusual Observations
Obs     mperiod      fqmdata         Fit    StDev Fit     Residual     St Resid
 1        29.0        624.0         843.1        29.4        -219.1       -2.59R
R denotes an observation with a large standardized residual
Durbin-Watson statistic = 1.54

Predicted Values
    Fit   StDev Fit           95.0% CI                  95.0% PI
 1053.2      30.8      (   991.5,   1114.9)   (   863.5,   1242.9)
  829.0      30.8      (   767.3,    890.7)   (   639.3,   1018.7)
  505.5      31.1      (   443.1,    567.8)   (   315.6,    695.4)
  759.0      31.1      (   696.7,    821.4)   (   569.1,    948.9)
 1032.1      32.5      (   966.9,   1097.3)   (   841.3,   1223.0)
  807.9      32.5      (   742.7,    873.1)   (   617.1,    998.8)
  484.4      32.9      (   418.4,    550.3)   (   293.3,    675.5)
  737.9      32.9      (   672.0,    803.9)   (   546.8,    929.1)
 1011.0      34.4      (   942.2,   1079.9)   (   818.9,   1203.1)
  786.8      34.4      (   718.0,    855.7)   (   594.7,    978.9)
```

The fit of this model is not as good as that for the trial period. We could use this to support a conjecture that the seasonal pattern is changing. The Durbin–Watson statistic is slightly 'better', because it is in the inconclusive range for a 5% significance level, and in the 'no first-order serial correlation' zone for the 1% significance level. We do not generally use the DW statistic because we have to rely on tables. Hypothesis tests for the regression coefficients, on the other hand, are easily carried out using the p-values provided in the Minitab output. Actually, the time plot of the residuals (which we will omit to save space) no longer shows the inverted 'U' clearly. The stem and leaf diagram is clearly mound-shaped.

Thus we have the situation that the results of the Main modelling regression calculation show poorer fit than the trial model, but the distribution of the residuals is 'nicer' in the sense that it conforms to a desired shape. In forecasting, however, fit is generally more important. We want good predictions more than being able to provide statistical tests about them. The summary statistics for the residuals follow. Figure 12.5 presents the graph of the data, model and forecasts.

```
Descriptive Statistics (edited)
Variable  N     Mean   Median   StDev   Minimum   Maximum      Q1       Q3
RESI2     62    -0.0    9.9      86.6    -219.1     156.7     -64.6     82.4

Stem-and-leaf of RESI2      N = 62
Leaf Unit = 10

    1    -2 1
    2    -1 5
    8    -1 433200
   17    -0 999887665
   29    -0 444444333000
  (14)    0 00111111223334
   19     0 55788889999
    8     1 0001334
    1     1 5
```

We note that the model still appears to have seasonal swings that are too wide in the latter periods. Note that the forecasts, displayed above, are stored in PFIT2.

Estimating regression models

Estimating the parameters in regression models, that is, fitting them, is a task we delegate to computers. However, it is worthwhile having some idea of the work the computer is doing for us. In particular, multiple linear regression is generally accomplished by minimizing the sum of squared residuals, that is, the sum of the squares of the deviations between the model and the data. Indeed, it is often referred to as least squares regression, but you should be aware of other principles that may be used to estimate the model.

Minimizing a sum of squared residuals or some other objective function by adjusting parameters in a model is a classic viewpoint of estimation. Unfortunately, what actually goes on inside the computer may or may not correspond to this strategic approach. The 19th century statisticians and mathematicians who developed the ideas did not have personal (or even impersonal!) computers, but they did know their calculus. They therefore took the sum of squared residuals as a function of the parameters and the data, $S(b, \text{data})$, and proceeded to try to minimize it by traditional calculus methods. This leads to the *normal equations*, which are a set of n equations in n unknowns. Most students start to squirm in their seats at this point in a lecture on the subject!

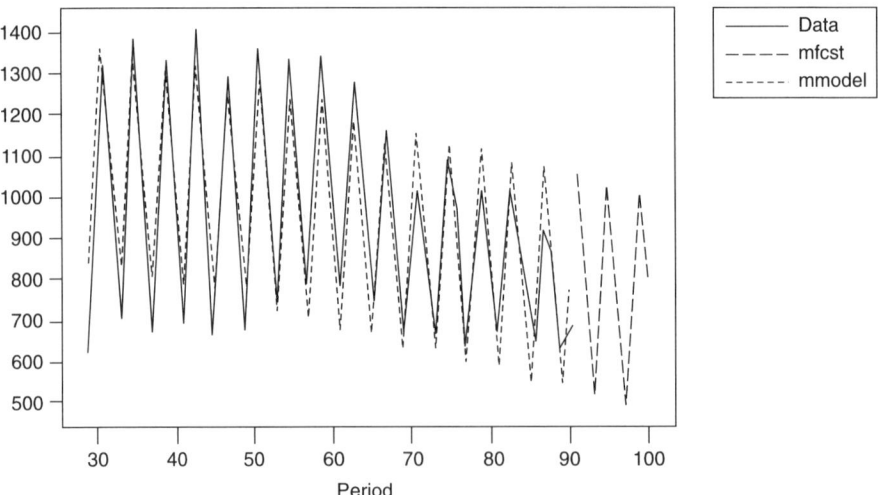

Figure 12.5. Data, main model for regression on dummy variables, and forecasts.

Much of the traditional textbook material on multiple regression concerns the linear algebra of solving the estimation equations. This is both difficult and dry for most business students, so we will look at the underlying objectives of the main methods.

- *Least squares*: we try to minimize the sum of the squares of the deviations

$$y(i) - y\_model(Xs, bs) = r(i, Xs, bs) = i^{th} \text{ residual}$$

by adjusting the bs. Note that the *form* of the function $y\_model(\ )$ does *not* have to be linear, and that the same ideas apply to nonlinear modelling (also called nonlinear regression).

- *Weighted least squares*: we minimize the sum of the quantities

$$r(i, Xs, bs)^2/w(i)$$

where the *weights* $w(i)$ are chosen to fit the specifics of the problem, such as a knowledge that the variability of the data around the model is *not* constant. A typical choice for $w(i)$ is to use an estimate or prediction of the variance $\sigma^2(i)$ of the data about the model for the values of the data (Xs) that correspond to data point i.

- *Maximum likelihood*: if we have a model of the probability a given observation is observed, then the likelihood all the data is observed is the product of these probabilities. We formulate this function, then maximize this function. In practice, it is usual to take the logarithm of the function (which converts the product to a sum), then to minimize ($-2 * \log(\text{likelihood})$). The minus sign converts a maximization to a minimization; the factor 2 is an artefact to remove factors of 0.5 that often arise in practice.

The availability of optimization in spreadsheet software allows us to 'try out' such ideas with both linear and nonlinear models. However, we accompany this suggestion with a warning. At an informal lunch during the Society for Industrial and Applied Mathematics (SIAM) Optimization Meeting of May 1996, a well-known and respected optimization researcher pointed out that the Generalized Reduced Gradient algorithm used by Microsoft Excel 'didn't work when Lasdon published it and still doesn't work now'. This was not intended to mean it never carries out the optimization task. It does, in fact, 'work' most of the time, creating an unwarranted confidence for those occasions where it produces a result that is *not* the optimum. We have used the optimization tools of spreadsheet packages and believe that they can be extremely helpful for quickly estimating forecasting models as long as the user takes the trouble to verify that the results are sensible. The easiest way to do this is to graph the model and data together to check visually that they are in accord. We present an example in Chapter 20.

The use of optimization methods does *not*, unfortunately, help us in obtaining estimates of the variability of the parameters unless we are prepared to compute the loss function for a lot of values of the parameters. Even then, the variances of the parameters are not readily available and awkward to compute. On the other hand, as long as the sizes of validation errors and in-sample errors for the trial and main model estimation periods are reported clearly, we do not see this as a major issue. That is, the important matter is to tell our clients the level of error they can expect in our forecasts, possibly including information on the distribution of these errors. Whether we get the traditional standard errors for model coefficients is, in our opinion, of minor interest.

Collinearity

Collinearity is essentially a situation where two or more of the predicting variables compete to explain the same variation or pattern in the $y()$ data we want to model. Unfortunately, it is not always easy to recognize or interpret properly. Moreover, regression models for forecasting often exhibit collinearity because we must sometimes use several explanatory variables to get a good enough fit. While most statisticians treat collinearity as a 'bad thing', it need not be a reason to throw away collinear regression models for forecasting, as we shall now explain.

A typical situation arises if we wish to model prices of foods that provide us with protein, say red meat, poultry, fish, eggs and cheese. If we have price or price index data over a range of observations on all the prices, we could model one of these, say egg prices, in terms of the others. However, we would expect that these prices would move in similar patterns, and should expect collinearity. This will mean that we cannot determine the precise values of the model parameters since we do not know which of the variables relating to meat, poultry, fish or cheese is 'causing' the change in egg prices. Even worse may be the situation where the variation in the price of eggs is related to a difference between very similar variables, since these differences will clearly have few digits of information. Such digit cancellation is a common concern in numerical computation. In multiple regression, it is easily seen where two variables are very similar. Special methods may be needed to uncover the problem where more than two variables 'add up' to another. We have discussed this problem at length elsewhere (Nash, 1976a and 1979).

Less obvious is that seasonal dummy variables can exhibit collinearity even if we leave one season out, as we should. We have usually seen evidence of collinearity in monthly data, where we need 13 parameters for an additive seasonal model (2 for the trend, 11 for the seasonal dummies). We suggest avoiding linear regression for more than 5 or 6 parameters. If possible, keep the number of parameters less than 10% of the number of observations.

When using regression for forecasting (there are many other applications), we hope to have a good fit to the modelling data. This translates to a high R_squared, typically higher than 0.9. We may or may not care that the model parameters are well-defined if the fit is good. However, we should be careful not to make any interpretation of the parameters, for example, to say that a $1 per pound increase in poultry prices will raise egg prices by 15 cents a dozen. This is because the underlying situation may be such that a $1 increase in poultry can only occur if there is also a 50 cent per pound increase in fish and a $1.25 per pound increase in red meat. That is, the movements of the variables are tied together in special ways – the essence of the collinearity.

Example of collinearity in regression

As an example, we use data from Bowerman and O'Connell (1979, p. 25). This example was the first one in a book on forecasting. Unknowingly, the authors used an example where collinearity is quite severe, but they did not notice it! (In later editions of the book, the example has been changed.) The scenario is that of forecasting the demand for a detergent product called FRESH using its price (variable *price*), the average price of competing products (*avprice*), the time period (*time*) and the expenditure on advertising (*adexp*). The script contains the units for these variables. *Reminder*: Define and document all variables and related features of the data file. When writing this section, we could not remember the background information, but fortunately it was **with** the data. A more complete exposition of this example is provided on the PFM Web site, with only an outline here. The traditional Minitab output yields a model:

```
The regression equation is
demandy = 7.76 - 0.00474 time - 2.39 price + 1.60 avprice + 0.510 adexp

Predictor      Coef      Stdev   t-ratio     p
Constant      7.759      2.457      3.16   0.004
time      -0.004738   0.005024     -0.94   0.355
price       -2.3905     0.6403     -3.73   0.001
avprice      1.6039     0.2961      5.42   0.000
adexp        0.5104     0.1265      4.03   0.000

s = 0.2352    R-sq = 89.7%   R-sq(adj) = 88.1%
```

Note that the output is hardly 'unusual', although from its t-statistic, `time` does not appear to be a significant variable, and we may want to re-calculate the regression without this variable. Given that the Minitab output does not appear unusual, what other methods can we use to detect the collinearity? Some of the most effective are matrix calculations based on eigenvalue or singular value decompositions of the data array for our regression. We used the package MATLAB to do this, but any software that performs similar matrix calculations would suffice. Our results are given in Table 12.1.

The approach we used is to combine the predictor variables (a constant, `time`, `price`, `avprice`, `adexp`) to create a set of 'new' predictors that are orthogonal, since orthogonal variables are certain to be linearly independent and to explain different 'directions'. They are prepared by a method that is equivalent to the method of *principal components*. The 'new' variables that are linear combinations of the original ones are formed in a way that preserves the overall variability in the predictor variables. In particular, the new variables can be ranked according to the proportion of the total variability they carry. Now we can form regression models of the variable (*y*) that we wish to model against one or several of the 'new' predictors.

Since we have formed the new variables as linear combinations of the original ones, we can easily back-transform the models in the 'new' orthogonal variables to obtain models written down in terms of the original predictors. That is, the models in Table 12.1 are given as the coefficients that multiply the original variables.

In Table 12.1, the model Matlab_full uses all the new variables, and involves the five original predictors (this count includes the constant). Suppose we use less than all five of the new variables to generate models. Model Matlab_3 uses the three new variables that 'explain' the largest proportion of the variability in the predictors. Matlab-2 uses just two of the new variables. Note that the models using fewer than five predictors have very different parameter values from the full solution. For example, the constant in the 4-component solution Matlab_4 has a different sign from that in the full solution! While these 'partial' models are not as 'good' in fitting the data as the full model, they still do quite well.

We leave to the PFM Web site and to books on multivariate statistics how the principal components approach (carried out, for example, via the singular value decomposition; see Nash 1979) helps us to detect and deal with collinearity. While we have considerable interest in the subject, we believe a more complete discussion belongs elsewhere.

Table 12.1. Possible models of the FRESH data obtained by principal components regression

Coefficient	Minitab	Matlab_full	Matlab_4	Matlab_3	Matlab_2
price	-2.39	-2.3905	-0.4319	0.1824	0.438
avprice	1.6	1.6039	1.3344	0.3467	0.4644
adexp	0.51	0.5104	0.7607	0.9825	0.7532
constant	7.76	7.7586	-0.1287	0.0646	0.1168
time	-0.00474	-0.0047	-0.0036	-0.0046	-0.0037
R_squared	0.897	0.8973	0.8549	0.8042	0.7675

Other uses of regression in forecasting

In this chapter we have mainly considered the use of multiple linear regression using dummy seasonal variables to obtain additive seasonal models. There are, however, other uses of regression in forecasting.

Sometimes we have a variable X that we believe we can forecast easily, and also find that we can develop a regression model for the quantity of interest Y in terms of this X, along with dummy variables or the time indicator that can be easily calculated for future times. Clearly, we can then forecast X, then use our regression model to find forecasts for Y. While not very common in our experience, it may serve as an alternative method on occasion.

Econometric modelling, which we treat only briefly in this book in Chapter 19, is an important tool for developing causal models used in policy analysis. Individual regression models are linked together in a large system of equations and inequalities to allow simulations of economic systems. Generally, if we look at the forecasts for a single variable, then the forecasts developed from econometric models have larger errors than those found from much simpler techniques. Parallel opinions are found in Makridakis *et al.* (1998, Chapter 6). However, econometric models typically will forecast several hundred variables simultaneously along with their interactions and interdependencies. Moreover, they allow the analyst to try 'what if' experiments with various policy controls such as interest or taxation rates, or 'shocks' such as a sudden rise in the price of oil.

A use of regression we do feel is worthwhile in the toolkit of the practical forecaster is its application to attempts to model residuals from standard methods – to 'tweak' a reasonably good model to make a very good one. We also present this approach in Chapter 19.

Exercises

E12.1. What should we do as good forecasters / statisticians before considering the work with the FRESH data complete?

E12.2. In the analysis of the Quarterly Traffic Fatalities in Canada data earlier in this chapter, the Minitab script includes regression subcommands Tresiduals, Cookd, Dfits, VIF and DW. Use the Minitab Help facility, and/or books on multivariate statistics, to prepare a short paragraph describing the output these subcommands provide. Are these facilities useful for forecasting? *[If your course uses different software, your instructor may wish to adjust this exercise.]*

E12.3. Use regression to develop an additive seasonal model for the Dry Cleaning data.

Forecasting trend and season II: smoothing methods

Why smooth?

Smoothing methods are among the most widely used in business and administrative contexts. They are also among the simplest to use, although their statistical foundation can be quite complicated to state clearly and to formalize. In essence, smoothing methods aim to allow a forecast to be prepared based on past experience that is not 'too sensitive' to local disturbances. Smoothing methods may be considered pattern fitting methods, where the purpose of the smoothing is to 'average out' the noise.

Clearly the main issue is 'How do we distinguish genuine change from random noise?' As a point of departure, let us use a *stationary* time series: one for which the long-term average level is constant. We can forecast a stationary time series by its average. Thus, noise is superposed on the model

$$y\_predicted = constant = mu\_Y = \mu_Y$$

with our full model being

$$y = y\_predicted + error$$

Our difficulty is deciding how to tell when or if the model (i.e. the average) is changing versus a situation where we simply have a few noise disturbances randomly in one direction. Ultimately, no computational or statistical method can decide such matters – they are questions of understanding or belief about the system of interest. It is also of interest to decide if the change is a *trend* that is a gradual up or down movement, or a *level shift*.

Moving averages

One approach to smoothing is a moving average. This is an average over a time 'window' that we slide along the time axis. Moving averages may be simple or compound. We will develop a notation to allow us to express the moving averages succinctly. Thus,

$$\mathrm{MA}(n, t, y) = (\sum_{i=1}^{n} y_{t+1-i}) / n$$

is the average of n data points taken backward from time point t. Note that we include the time point t. We can form weighted moving averages, for example,

$$\mathrm{WMA}(n, t, y) = (\sum_{i=1}^{n} w_i \, y_{t+1-i}) / n$$

The simple moving average MA( ) has weights all equal to $1/n$. If we form the Moving Average *operator* which is like a command acting on a time series, then we can talk of, say, a 3-point moving average, which has a window that is three time points wide. We can apply such operators in sequence to generate compound moving averages. That is, we can apply one moving average operator after another. Like differentiation, these operators act on whatever object is on their right-hand side. Thus,

$$\mathrm{MA}(4) \, (\, \mathrm{MA}(3) \, Y \,) \, = \mathrm{MA}(4) \, \mathrm{MA}(3) \, Y$$

is the 4-point moving average of the series formed by computing the 3-point moving averages of the original series Y. It turns out that we can expand the formulas for these compound moving averages and derive the *weights* of the corresponding weighted moving average.

Example: Table 13.1 is based on data that we introduce later as an example of forecasting with moving averages. Here we simply want to show the results of computing a 3-point moving average of the series Y. We lose the first two elements of the new series because we cannot compute a moving average of three points until we reach the third element of Y. In a time series, this is at the third time point. The arrows show two instances of the contributions of the Y column to the MA3(Y) column, and one instance of contributions of MA3(Y) elements to the MA4(MA3(Y)), a compound moving average. Each element of MA(4,MA(3,Y)) needs 6 raw data elements for its computation, and the first value is placed in the 6th position in the right-most column of Table 13.1.

One aspect of moving averages (and similar functions of time series) that causes both students and practitioners a great deal of trouble is where they should be placed on the time axis. Placement, in our experience, is *the* tricky issue of forecasting, even though it involves no complicated mathematics nor any unusual thinking. It simply requires great care and attention to details. In Table 13.1, we have placed the averages at the first time position where there is enough data to compute the average. We call this *'as soon as data is available'* placement. However, the moving average of an even number of points can also be said to 'belong' at the mid-point of the window, that is, at the 'average' time point. We do not necessarily place it at this *central* placement, since we may wish to put the computed moving average opposite the last (most recent) data point for which we can compute the average. Clearly, it is *very* important to document the convention for placement. In draft work, we recommend drawing braces, arrows, lassoes or some other indicator so the reader knows what information has been derived from which data. In final reports, similar indications, as illustrated in Table 13.1, are useful to the reader, as well as ourselves when we return to review our work. As a check, it is often useful to use a simple, derived, time series such as ($3 + 2 * t$) to check placement.

Table 13.1. Illustration of the relationships between data elements and their moving averages

time index	Y	MA3(Y)	MA4(MA3(Y))
			First item
1	20.71	1st 3 items	needs 4
2	22.22	in Y averaged to	numbers
3	21.02	21.3167 <--3rd MA row	from MA3,
4	20.91	21.3833	6 from
5	22.07	21.3333	original
6	20.87	21.2833	21.2175 <-- data
7	21.13	21.3567	21.2450
8	22.53	21.5100	21.6500

Properties of the moving average MA(*n*)

We summarize some important features of moving averages:

- MA(*n*) X has *n*-1 fewer data elements than X. This is simply a result of averaging. It affects placement, and introduces missing values into time series.

- The longer the window (*n*), the more 'smoothed' the resulting series.

- The length of the MA can only be chosen in a context, and this choice is critical to the usage of the MA as a forecasting tool.

Trended data

Since the purpose of our moving average is to 'average out' noise that has been superposed on a stationary (fixed value) quantity, we need to think of special procedures to deal with data that have trend. First, we will consider simple linear trend. Consider a straight line model

$$X(t) = b\,t + a$$

To make things more concrete, we take the MA(3) of this series and place the result at the time endpoint of the window ('as soon as data is available' placement).

$$S1(t) \quad = \mathrm{MA}(3)\, X(t) = [X(t) + X(t-1) + X(t-2)]/3$$

$$= b\,(t + t - 1 + t - 2)/3 + a\ = b\,t - b + a = b\,(t - 1)\ +\ a$$

If we take the MA(3) of $S1$ and place the result at the time endpoint of the window, we get (summarizing!)

$$S2(t) =\ \text{MA}(3)\ \text{MA}(3)\ X(t) =\ b\,t - 2\,b + a\ = b\,(t - 2) + a$$

Thus

$$S1(t)\ -\ S2(t)\ = b$$

and we can then substitute to find parameter a also. It turns out the general formulas for a window of length n is

$$b(t) = 2\,(S1(t) - S2(t))\,/\,(n - 1) \qquad\qquad a(t) = 2\,S1(t) - S2(t)$$

Many books omit the time subscripts on $a()$ and $b()$, but we want to include them here because, with data that is not perfectly linear, we will get different values for a and b at each time point for which they are available. That is, we get one pair of slope and intercept estimates at each time point for which the first and second smoothers $S1$ and $S2$ can be computed. We can compute forecasts k periods forward from time t using

$$F(t + k,\ t) = b(t)\ k + a(t).$$

Moving averages thus offer a fairly simple technique for forecasting either stationary or linear trended series. While placement of computed elements of the method with respect to the time points is a tedious nuisance, the method is conceptually easy and can be applied by anyone who can do simple arithmetic. A disadvantage that was a concern before personal computers were common, indeed when calculations were performed on hand-cranked adding machines, is that we need to keep and work with the latest n data points for each series we need (X, $S1$, $S2$). When very many series must be observed and forecast simultaneously, as in supermarket inventory systems that may track up to 30,000 items, the need to keep n elements at hand in the main memory (or RAM) may still be a concern.

Learning about smoothing methods

Moving averages are important in the historical and pedagogical aspects of forecasting, but are now less prominent in usage. They do help in the understanding of other methods. We believe that the best way to obtain a true comprehension of smoothing methods is actually to carry out the calculations. Hand calculation is far too tedious to be sensible, but spreadsheet software makes the task relatively straightforward. Once we understand the methods, we use automatic computation. Few packages include moving average methods, however, as they are generally superceded by exponential smoothing approaches.

The approach we suggest is to generate data that are appropriate (or not) to the methods by suitable modification of a series of generated random numbers. We warn, however, that it is important to recognize:

- The spreadsheet random number generators usually produce *uniformly* distributed numbers, that is, (real) numbers that have equal probability of falling between the limits specified for the generator. There are annoying details over the inclusion or exclusion of the end points.

- The random number functions may 'recalculate' the values they generate from time to time, depending on a particular 'starting value' or *seed* that is used in generating the pseudo-random sequence. We generally want to have repeatable exercises – students rightly get confused when the numbers on the screen do not match the printout in their text or handout. Therefore we must find a way to 'fix' the random number series. One way to do this is to learn how to control the seed, but this tumbles us into the technical details of a particular spreadsheet package, or even a particular version thereof. An easier approach is to use a separate spreadsheet to generate some numbers, then save the *values* of the pseudo-random series in a file and read or copy it whenever we need those values. This also allows us to generate and use numbers from different distributions, for example the Gaussian distribution, which can be done in other packages, such as Minitab.

- We also recommend that the numbers be formatted to a known precision *before* they are saved to a file. We generally save them as decimal numbers suitable for use in a plain text file since we like to be able to print the numbers or transmit them as electronic mail messages, even though there may be some computational advantages to using binary formats.

The MA-TEST data (matest.txt) includes both a stationary and simple trended series to illustrate how smoothing methods work on these two main classes of data. We could also include a sudden level shift or seasonality. It is important to note that applying the 'wrong' method generally leads to poor forecasting models and poor forecasts. Even so, smoothing methods often behave 'not too badly' even when mis-applied because of the inherent stability of the models in most situations. Moreover, we believe that it is important to gain experience with the mis-application of methods.

Figure 13.1 shows the application of MA3 and MA7 smoothers to a pseudo-random stationary series. Figure 13.2 shows the use of double moving average forecasts (1-period ahead forecasts only) to a trended series perturbed by pseudo-random fluctuations. We have used a double MA4 and a double MA8 forecast and, for comparison, have used a single MA4 smoother as a forecast.

Interestingly, even though the test data were, by construction, based on a linear trend, for which the double moving average is designed as a forecasting approach, the single moving average has error measures that are slightly smaller than either of the double MA approaches.

This is likely a consequence of the size of the fluctuations relative to the magnitude of the slope of the underlying trend.

Stationary series with MA3 and MA&

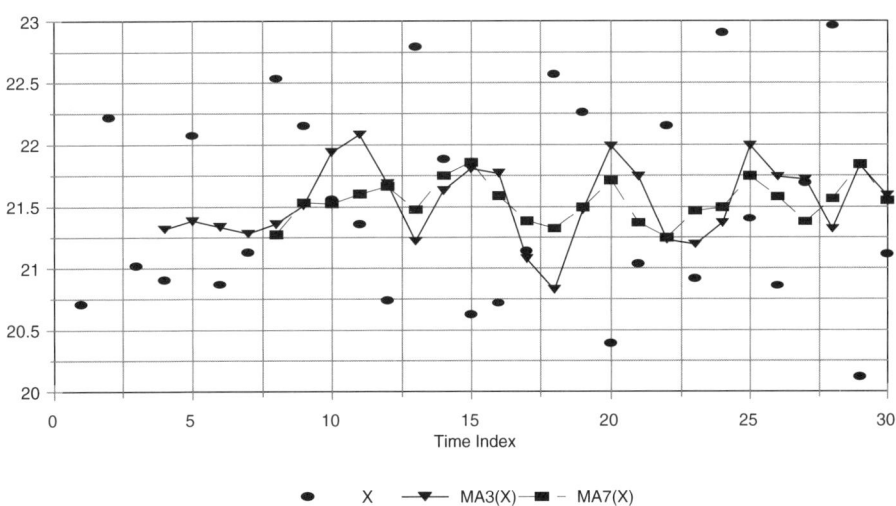

Figure 13.1. MA3 and MA7 operators applied to a pseudo-random stationary series. The shortest series is the MA7 series. We found Quattro version 7 did not make adding the legend for the MA7(X) easy.

Difficulties with moving averages

We have already noted that placement of elements of the moving average calculations at the appropriate time points needs careful attention. This is only a difficulty when we are calculating with a spreadsheet or 'by hand', or else coding a computer program. Another disadvantage is that we need to keep the latest N data points in memory, where N is the length of the moving average, for each series we are working with. This was a problem when computer memory and disk storage was expensive, and is still a concern where very many series must be followed. A typical situation of interest may be forecasting the stock of non-perishable items in a supermarket, where between 5,000 and 30,000 products may be stocked.

Moving averages on trended series

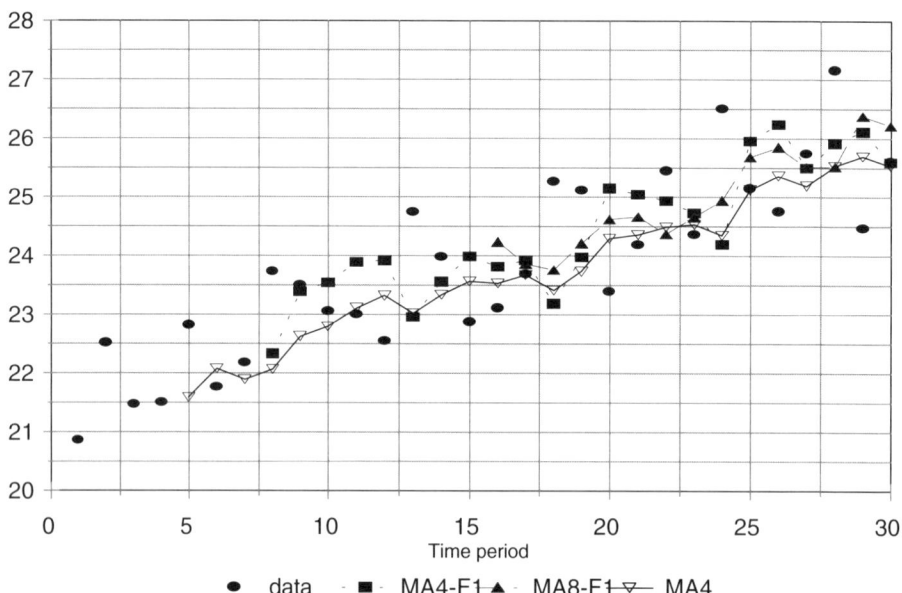

Figure 13.2. Application of moving averages to a trended series generated from pseudo-random numbers. The longest series is the data, the series starting at period 5 is an MA4 of this data positioned as a 1-period ahead forecast. The series starting at time point 8 is a double moving average 1-period ahead forecaster using MA4. The shortest series, starting at time point 16, is a double moving average 1-period ahead forecaster based on MA8.

Exponential smoothing

Exponential smoothing was motivated in part by the concern to avoid holding N points per series. It is similar in philosophy to moving average techniques. That is, it is based on the idea that a stable pattern underlies the data, but noise has been added that obscures this pattern. Consider the naive forecast: $F(t + 1, t) = Y(t)$. That is, we use the last available data value as our forecast for the next period. If we already have a forecast for $Y(t)$ in $F(t, t - 1)$, and we believe this to have some validity, then we could in some way *mix* the new data (naive forecast) and the existing forecast. A linear sum provides

$$F(t + 1, t) = \alpha \ Y(t) + (1 - \alpha) \ F(t, t - 1)$$

This is the general form for SES, *Simple Exponential Smoothing* or *Single Exponential Smoothing*, which can be used to forecast (mainly) stationary series. Another form for the forecast is obtained by algebraic manipulation:

$F(t + 1, t) = F(t, t - 1) + \alpha (Y(t) - F(t, t - 1))$ = old forecast + α * error in old forecast

Note that substitution for $F(t, t - 1)$, $F(t - 1, t - 2)$, etc. gives a power series in $(1 - \alpha)$ which is the origin of the name 'exponential smoothing'.

Some difficulties with SES are:

1. The choice of the smoothing parameter α;

2. The choice of the initial 'forecast' $F(1,0)$.

Most forecasting books talk of the SES method having only 1 parameter, but it can be argued that *both* α and $F(1,0)$ are parameters. Usually we estimate $F(1,0)$ as an average of the first few, say 5, observations, and take a value such as 0.1 for α. We justify these choices below.

The larger the value of α, the more quickly the forecast reacts to changes; the smaller it is (like the longer the window of a moving average) the more 'smooth' the forecast series will be. SES is intended to be used for quantities that are more or less stationary, such as the demand for standard products in a supermarket. In such applications it generally works quite well, although we make no claims for highly accurate forecasts. The use of SES forecasts in such applications is to track long-term movements in the sales or inventories of products. We want a 'quick and dirty' way to avoid transient noise in the data. However, SES does eventually follow the data if there is a trend.

We could use an optimization technique to choose $F(1,0)$ and α, but caution that this may be counter-productive. We do not want a model that reacts 'too quickly' to changes in the data in case these are just noise. Thus, setting the starting 'forecast' to an average of a few early values in the data, and keeping the smoothing parameter relatively small is sensible. If the 'fit' of SES models is sensitive to $F(1,0)$ and α, we probably should not be using them.

What are the forecasts from SES? Recall that the supposed underlying model is stationary. Therefore,

$F(t + k, t) = F(t + 1, t)$

for any $k > 0$. *This means the forecast is the same for all future periods.* Generally, however, SES is used for 1-period ahead forecasts.

We continue our examples introduced for moving averages. Note how poorly SES performs on the trended series. The graph in Figure 13.3 is the Simple Exponential Smoothing applied to the uniformly distributed example data. Note that the 'forecast' has been started with a quite poor value of $F(1,0)$, so takes a number of steps to 'chase' the data to a reasonable value if the smoothing parameter α is small (0.1 in the graph). On the other hand, when α is 'large' (in our case 0.9), we see that the forecast responds too quickly to the data (the jagged line). The spreadsheet listing is omitted to save space and avoid overwhelming our readers.

The graph in Figure 13.4 shows how Simple Exponential Smoothing works with a trended series. When the smoothing parameter is small (a lot of smoothing and little reaction to the data),

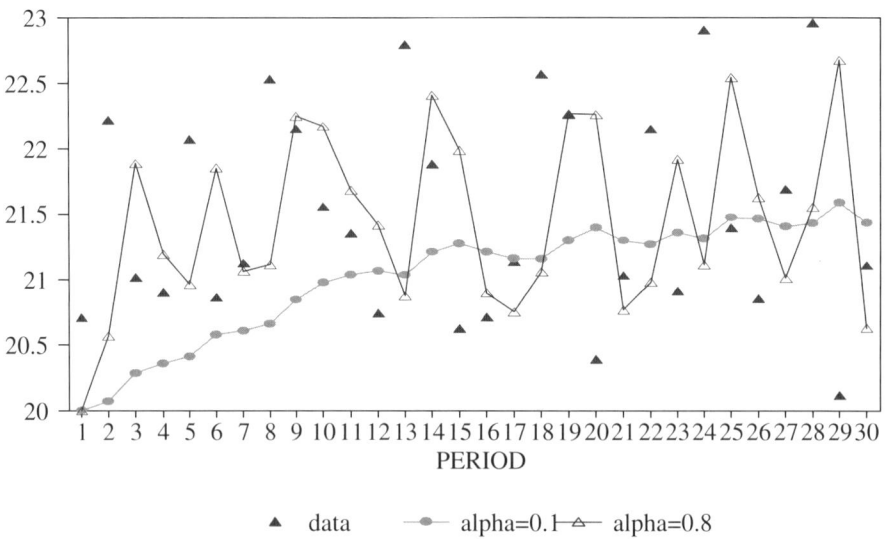

Figure 13.3. Simple exponential smoothing applied to a supposedly stationary series. The starting value of the smoothers is a deliberately poor choice. Note how a small value of the smoothing constant alpha gives a higher degree of smoothing.

the model trails the data. When the smoothing parameter is large, we chase the data, but get a jagged forecast line. Again, we omit the spreadsheet listing used to generate these graphs.

Modern statistical software (e.g. SPSS, Minitab and others) includes exponential smoothing techniques in the repertoire of methods. These are available by mouse-click from menus. Such ease of use can leave students with little facility to check that computations have been correctly carried out. For purposes of *learning* the methods, we again recommend the use of a spreadsheet. This allows the tedium of the calculations to be performed by the computer, while the student is forced to enter the appropriate formulas and position the results in the appropriate cells of the spreadsheet.

Double exponential smoothing

Like the double moving average, double exponential smoothing (DES) exists to cope with linear (or additive) trend. There are several approaches, of which two are Brown's 1 parameter linear method and Holt's 2 parameter method. In these cases, the initialization of the forecast, $F(1,0)$, could be regarded as another parameter.

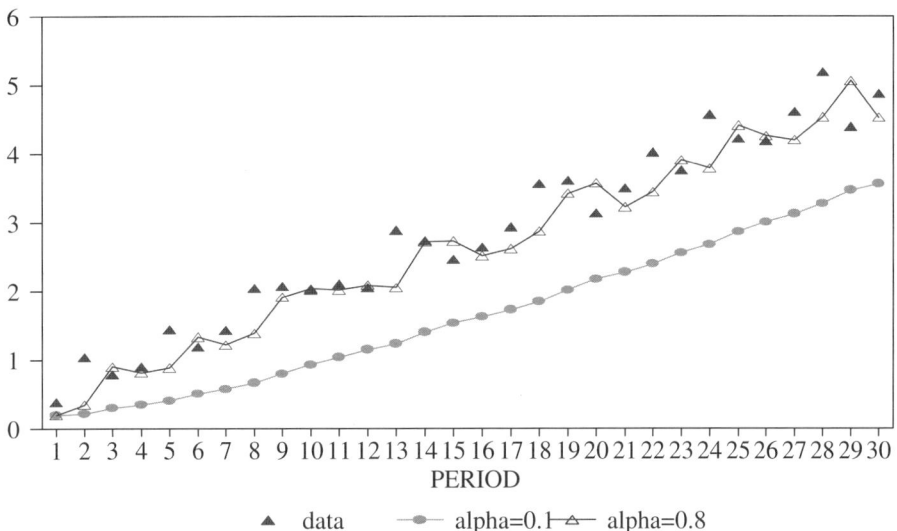

SIMPLE EXPONENTIAL SMOOTHING
TRENDED DATA SERIES

Figure 13.4. Simple exponential smoothing applied to a trended series. Note how the smoother using alpha=0.8 is able to 'chase' the data series.

Brown's 1 parameter ES – a double exponential smoothing

We begin as with single exponential smoothing (SES) and compute the smoother

$$S1(t) = \alpha\ Y(t)\ +\ (1 - \alpha)\ S1(t - 1)$$

where $S1(0)$ has the role of $F(1,0)$ in SES. As with double moving averages, we then form the second smoother

$$S2(t) = \alpha\ S1(t) + (1 - \alpha)\ S2(t - 1)$$

That is, we smooth the smoothed series. Note that we will also need $S2(0)$. We now compute

$$a(t) = 2\ S1(t) - S2(t)\ \text{ and}$$
$$b(t) = \alpha\ (S1(t) - S2(t)) / (1 - \alpha)$$

We forecast using

$$F(t + m, t)\ =\ a(t) + b(t)\ m$$

where m is the number of periods ahead we are forecasting. Note the similarities between Brown's method and the double moving average method.

Holt's 2–parameter exponential smoothing

Holt's 2–parameter ES has a slightly more complicated structure. We are again seeking a model with a linear form

$a + b * \text{time}$

However, we now want to estimate (and smooth) the slope b. If we have a value for the slope at $(t - 1)$, that is $b(t - 1)$ and a value for the level of the series at the same time point, that is $S(t - 1)$, we forecast the level at time t to be

$S(t - 1) + 1 * b(t - 1) = S(t - 1) + b(t - 1)$

The smoothing equation for the data series level is thus

$S(t) = \alpha \, Y(t) + (1 - \alpha) \, (S(t - 1) + b(t - 1))$

This is the first smoother. We need starting values $S(0)$ and $b(0)$ for the level and slope of the series. The Holt method then smooths the slope parameter

$b(t) = \gamma \, (S(t) - S(t - 1)) + (1 - \gamma) \, b(t - 1)$

This is using the difference in the level at adjacent points as the updating value, or 'data', for the slope. Thus, the formula for $b(t)$ is just SES on the slope series. Note that we have two (different) smoothing parameters, α and γ. We forecast with linear form

$F(t + m, t) = S(t) + b(t) \, m$

where m is number of periods ahead we wish to forecast. Some issues in double exponential smoothing should seem familiar. First, we need to choose the smoothing parameters. Note that small values give the most smoothing, while large values cause more rapid adjustment to the data. We also need starting values for the smoothed series. We suggest values that are more or less averages of the numbers we expect to get. One approach is to choose values from the graph of the time series to compute slopes and starting (intercept) values, although once again the time-placement issue arises. As before, our results should not be too sensitive to choices of starting values and smoothing parameters.

Reminders

1. We need to choose the smoothing parameters. Note that small values give maximum smoothing, large values cause more rapid adjustment to the data.

2. We need starting values for the smoothed series. Choose values that are more or less averages of the numbers we expect to get. You can choose values from the graph for slopes and intercept values. Establishing starting values in any form of smoothing method is, in our opinion, a topic

BROWN'S 1 PARAMETER DES
TRENDED RANDOM SERIES (alpha .125)

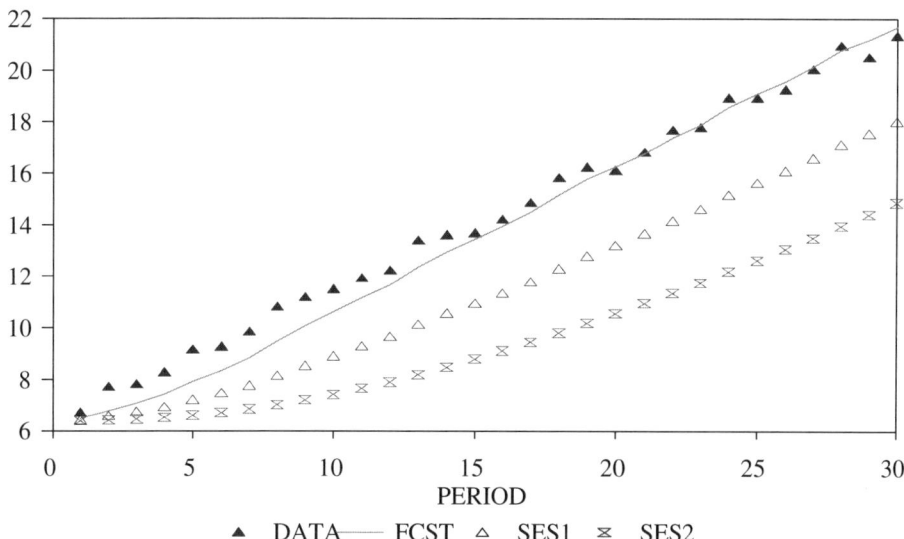

Figure 13.5. Brown 1-parameter double exponential smoothing applied to trended example data.

HOLT 2 PARAMETER DES
(alpha=0.125 gamma=0.125)

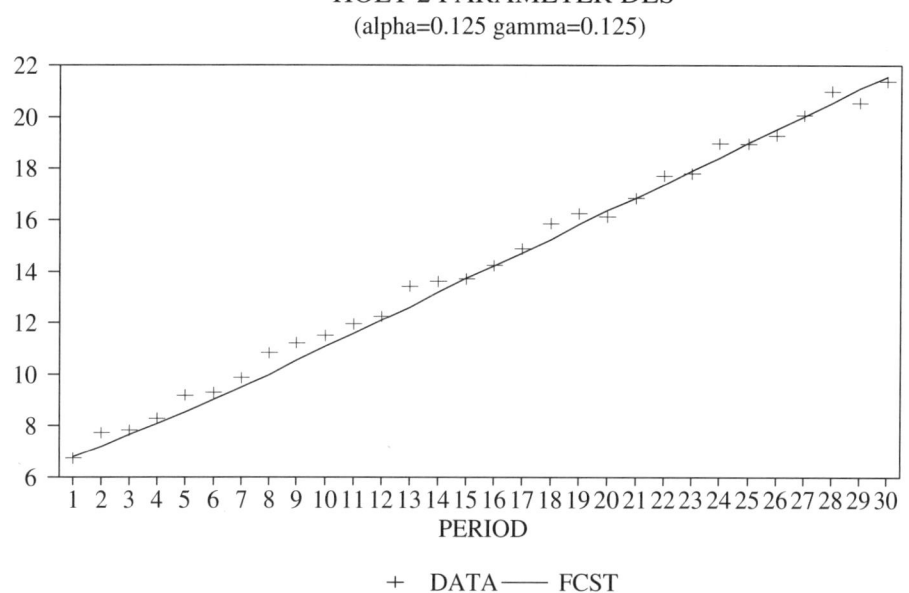

Figure 13.6. Holt's 2-parameter double exponential smoothing applied to the trended example data.

that students and novices to forecasting find difficult to understand. Even automated techniques may sometimes choose poor starting values, as we shall see later in examples that use Minitab to run the computations. This is manifest by early 'fits' being poor. To get 'fair' values for the measures of fit, we should compute them only for the latter part of the estimation period, but this will sometimes mean that we have to intervene manually to correct the values produced in automatic procedures. For an example, see Figure 13.8.

3. It is important to get the time placement of smoothed series and forecasts right. Make sure you practise this before trying it seriously.

4. Our results should not be too sensitive to choices of starting values and smoothing parameters.

Continuing our example, we now apply the Brown 1-parameter method to the trended example data series. Figure 13.5 shows the results. We again omit the spreadsheet listing. We have used $\alpha = 0.125$. Note that the starting smoother value is not a very good choice, but that the method eventually gets on track to model the data quite well. Figure 13.6 shows the Holt 2-parameter double exponential smoothing applied to the trended example data series. We use $\alpha = \gamma = 0.125$. Note that it is instructive to 'play' with the smoothing parameter values in these methods and look at the graphs as a tool for learning.

Seasonal series – Winters' method

We now have smoothing methods for stationary series and series with linear trend. How shall we handle series with seasonal variation? Clearly we must now add seasonality. Winters' method (Winters, 1960) uses a multiplicative seasonal model, which is made clear in the formula below for the m-period ahead forecast, but which can be expressed in words as

forecast(time) = (level + slope * time) * seasonal_factor(time)

One thing we *must* know in advance or else determine by graphs is the length of seasonality L. That is, L is the number of periods in a cycle, so $L=12$ for months in a year. There will be L seasonal_factor values that are used to forecast, but these will be updated by their own smoothing formula. We denote the level by the series $S()$, the slope by $b()$ as in Holt's method, and the seasonal_factor by $I()$. You may think of $I()$ as referring to 'inflation' to have a mnemonic. We smooth the level of the series by the formula:

$S(t) = \alpha * Y(t) / I(t - L) + (1 - \alpha)*(S(t - 1) + b(t - 1))$

with α the smoothing parameter. Note that we have divided $Y(t)$ by the available seasonal_factor. This is an attempt to discount the seasonal variation. We also increment the last available level, $S(t - 1)$ by the last available estimate of the slope, $b(t - 1)$. With these adjustments, the formula is just SES.

The trend smoother we will use is identical to that in Holt's method, that is,

$$b(t) = \gamma*(S(t) - S(t - 1)) + (1 - \gamma) * b(t - 1)$$

where γ is the smoothing parameter.

The seasonal smoother is new. It has the formula

$$I(t) = \beta*Y(t) / S(t) + (1 - \beta) * I(t - L)$$

where β is the smoothing parameter. The data value $Y(t)$ divided by the smoothed level $S(t)$ is an estimate of the appropriate seasonal_factor, and this ratio is thus 'data' for the seasonal SES calculation. Note that we must compute the smoothers in the given order, that is, first level, then slope, then seasonality. The m-period ahead forecast is calculated

$$F(t + m) = (S(t) + b(t)\ m) * I(t - L + m)$$

The biggest difficulty most students experience when trying to learn Winters' method using a spreadsheet or hand calculation is that of initiating the seasonal_factor series $I()$. The initialization of the level smoother and the trend smoother also give students some concern. To initialize the seasonal factors we want some measure that approximates the ratio of the quantity of interest during a particular season to its overall trended level. One suggestion is to take two or three cycles of data and average all the elements to get a quantity we can call Y_bar_start. Let k be the number of 'starting' cycles, i.e. $k=2$ or 3. If the average of the data elements for a single season (over the chosen cycles) is divided by Y_bar_start, we have an estimate of the seasonal factor for that season. We repeat for all seasons. Y_bar_start itself provides a value for $S(0)+b(0)*(k*L+1)/2$, allowing us to estimate $S(0)$ if we have an estimate $b(0)$. If Y_bar_1 and Y_bar_2 are the averages of the data for the first and second cycles, then a starting slope is

$$b(0) = (Y\_bar\_2 - Y\_bar\_1)/L$$

Another choice is $b'(0) = (Y(2*L) - Y(1)) / (2*L - 1)$. Our experience is that the method will tolerate any *reasonable* approximation to the level, slope and seasonal factors, since the smoothing tries to adjust the model components to appropriate values.

Warning: data and result placement is once again the critical problem.

The example in Table 13.2 and Figure 13.7 (see below) shows a spreadsheet development of the Winters' multiplicative exponential smoothing method applied to a seasonal data series from the text by Bowerman and O'Connell (1979). This is a small enough example that it can easily be used to learn the details of Winters' method.

Minitab has a macro to perform Winters' seasonal exponential smoothing. Unfortunately, the default mechanism for selecting starting values for the level ($S(0)$) and slope ($b(0)$) are such that the forecasts do not fit the data very well in the early periods. This is illustrated with the Bowerman and O'Connell data in the left-hand graph in Figure 13.8. Table 13.3 gives the

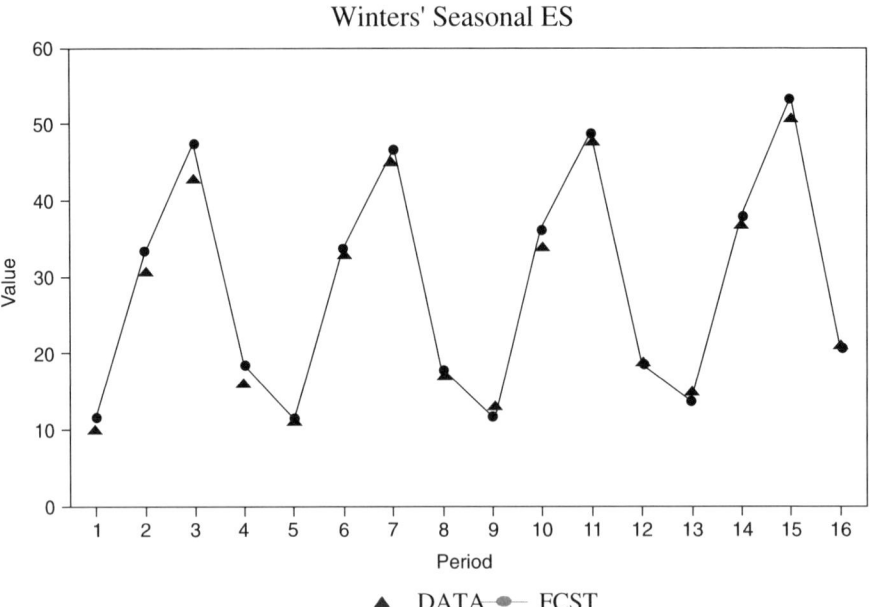

Figure 13.7. Seasonal data example and the Winters' method forecasts. See Table 13.2 for the calculations.

measures of fit for the Minitab default method for the initialization of the level, slope and seasonal factors in comparison with a method based on averaging appropriate sets of raw data. One of our students, Xin Zhou, incorporated this into a Minitab macro which we now have substituted for Minitab's version after a few minor modifications. This new macro allows us to use the Minitab initialization, the averaging method described above (which Xin Zhou kindly calls Nash's initialization, although we do not claim its invention), or values we can input in the command line 'call' of the macro. The averaging method for initializing the smoothed series uses the first two seasonal cycles, i.e. $2*L$ data points, where L is the number of data points in a seasonal cycle. L must be 'known' or discovered by inspection of the time plot. We get the initial slope estimate by averaging the data for the first and second seasonal cycles. In our example, these are the averages of data elements 1 to 4 and 5 to 8 respectively. Call these averages $A1$ and $A2$. Then, if we think a simple straight line model underlies the trend, a good estimate of the initial slope is given by the value

slope(0) = ($A2$ - $A1$) / L

that is, the difference between the averages divided by the length of a cycle, since the 'average' of the time indices for the first cycle is $L/2$, while that for the second is $3*L/2$. We can now estimate the level at time point 0 by taking the average of the data in the first 2 cycles, which is ($A1 + A2$)/2 and decrease this amount by the appropriate multiple of the initial slope. Since the

Table 13.2. Example of use of Winters' seasonal exponential smoothing method from Bowerman and O'Connell (1979, p. 225)

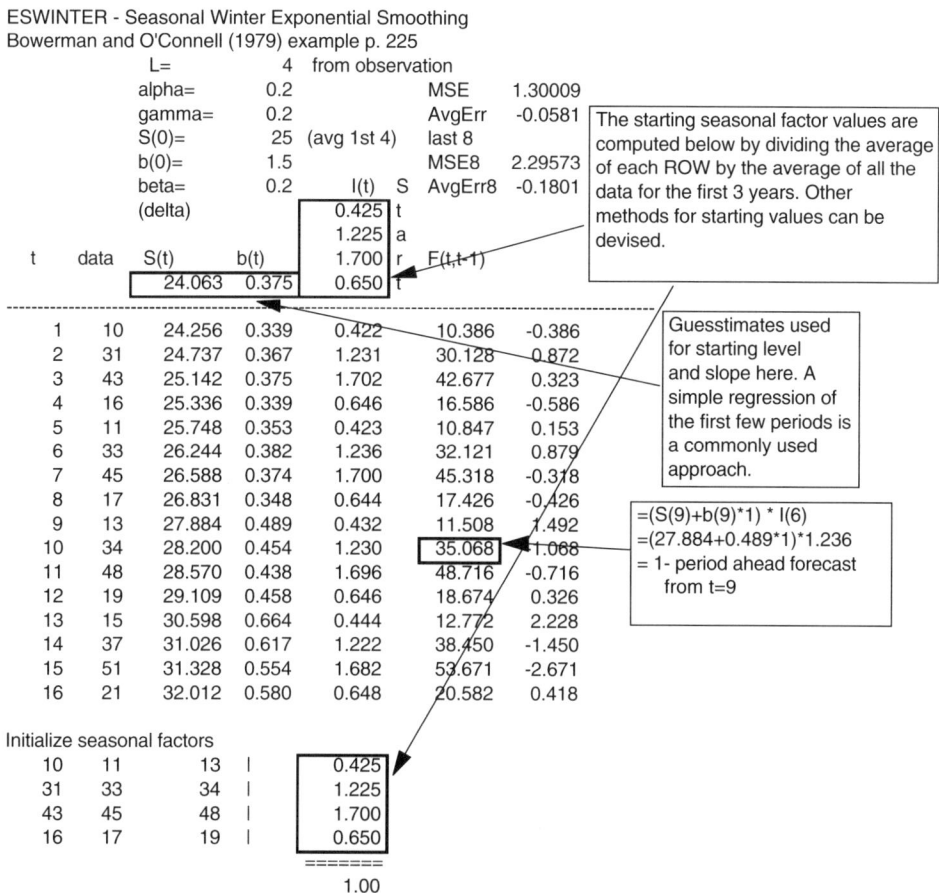

ESWINTER - Seasonal Winter Exponential Smoothing
Bowerman and O'Connell (1979) example p. 225

			L=	4	from observation		
			alpha=	0.2		MSE	1.30009
			gamma=	0.2		AvgErr	-0.0581
			S(0)=	25	(avg 1st 4)	last 8	
			b(0)=	1.5		MSE8	2.29573
			beta=	0.2	I(t) S	AvgErr8	-0.1801
			(delta)		0.425 t		
					1.225 a		
t	data	S(t)	b(t)		1.700 r	F(t,t-1)	
		24.063	0.375		0.650 t		

The starting seasonal factor values are computed below by dividing the average of each ROW by the average of all the data for the first 3 years. Other methods for starting values can be devised.

t	data	S(t)	b(t)	I(t)	F(t,t-1)	error
1	10	24.256	0.339	0.422	10.386	-0.386
2	31	24.737	0.367	1.231	30.128	0.872
3	43	25.142	0.375	1.702	42.677	0.323
4	16	25.336	0.339	0.646	16.586	-0.586
5	11	25.748	0.353	0.423	10.847	0.153
6	33	26.244	0.382	1.236	32.121	0.879
7	45	26.588	0.374	1.700	45.318	-0.318
8	17	26.831	0.348	0.644	17.426	-0.426
9	13	27.884	0.489	0.432	11.508	1.492
10	34	28.200	0.454	1.230	35.068	-1.068
11	48	28.570	0.438	1.696	48.716	-0.716
12	19	29.109	0.458	0.646	18.674	0.326
13	15	30.598	0.664	0.444	12.772	2.228
14	37	31.026	0.617	1.222	38.450	-1.450
15	51	31.328	0.554	1.682	53.671	-2.671
16	21	32.012	0.580	0.648	20.582	0.418

Guesstimates used for starting level and slope here. A simple regression of the first few periods is a commonly used approach.

=(S(9)+b(9)*1) * I(6)
=(27.884+0.489*1)*1.236
= 1- period ahead forecast from t=9

Initialize seasonal factors

10	11	13		0.425
31	33	34		1.225
43	45	48		1.700
16	17	19		0.650
				=======
				1.00

average of the time period counter for the first 2 cycles is $(L + \frac{1}{2})$ (i.e. on the boundary between the cycles), we suggest

$$level(0) = (A1 + A2)/2 - (L + \tfrac{1}{2}) * slope(0)$$

There are some other details of Minitab's approach to Winters' method that are worth mentioning, as their documentation is difficult to find. We have found that most software packages have such 'details' and caution our readers to check results carefully. We discuss more of these details in our presentation of the application of Winters' method to the Quarterly Traffic Fatalities in Canada data.

Table 13.3. Comparison of Minitab and 'Nash' initialization procedures for Winters' method applied to the Bowerman and O'Connell data. Both trials use level, trend and season smoothing constants of 0.2

Accuracy measure	Minitab initialization	Nash initialization
MAPE:	15.6078	4.13541
MAD:	4.3032	1.00125
MSD:	28.4791	1.91691

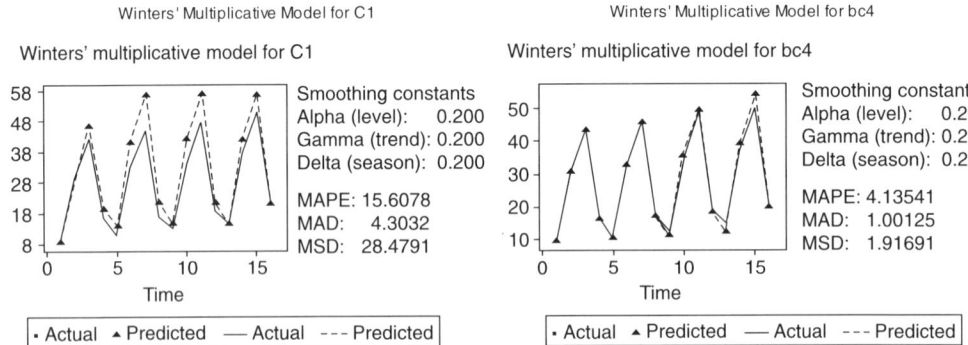

Figure 13.8. Minitab's multiplicative Winters' method. The left-hand graph uses the default Minitab method for computing starting values for the level and slope. The right-hand graph uses the 'Nash' method described above. The measures of fit displayed on these graphs are given in Table 13.3 and are not intended to be readable in the figure.

Other smoothing methods

There are a great many variations on the theme of smoothing. Other disciplines have used the same ideas but with very different jargon and general approach. Some authors prefer to think of smoothers as *filtering* the data to extract the *signal* from the *noise*. We do not propose to present details of such approaches here, but believe that the practising and practical forecaster should be aware that such methods exist. A few, very selected, examples are :

• the Kalman filter (see, for example, Harvey, 1989, and some discussion in Chapter 19), which was developed mainly in the engineering community for signal processing and tracking in situations that generally have a lot of data.

• the Additive Winters' or similar method, where we use seasonal shifts rather than factors. An additive seasonal model on the logarithms of our data back-transforms to a model that multiplies seasonal factors times an exponential trend.

• Gardner's (1985) characterization of ES methods.

- Adaptive Response Rate Single Exponential Smoothing, with the unfortunate acronym ARRSES, is intended for situations where a level shift may occur from time to time. To allow for the shift, the SES equation is modified so that there is a *different* smoothing parameter at each period. Clearly, there has to be a mechanism to allow the smoothing parameter to be altered sensibly. Wilson and Keating (1998, page 115 ff) or Makridakis *et al.* (1998, Chapter 4) for the formulas. The latter reference cites work that suggests this method does not work terribly well, which agrees with our limited experience with it. The difficulty is that it is too easy to get good fits over the estimation period, then have poor validation or actual forecasts.

- Tracking signal methods (Makridakis *et al.*, 1983, p. 114; Makridakis *et al.*, 1998, p. 616, Bowerman and O'Connell, 1979, p. 133) are intended to flag changes in a process by performing calculations similar to those in exponential smoothing. They are akin to methods for the detection of outliers in statistics or to methods for informing operators that a process has gone out of control in statistical quality improvement.

- The extension of the Holt/Winters' smoothing method by Chatfield *et al.* (1999) to situations where the variance of the random disturbances in a process is non-constant. This links smoothing methods to a variety of more formal modelling methods.

- The modifications to Holt's method (and similar methods) by Wright (1985, 1986) to allow such methods to be applied when data are irregularly spaced. This may be because data are missing. However, we feel a more important situation arises when data have been measured at irregularly spaced time points. An example arises with government surveys of smoking behaviour of Canadians, where surveys were conducted at different times of year or were omitted in some years for budgetary or other reasons.

A real example

Minitab makes exponential smoothing very easy to carry out provided we can specify the smoothing parameters. Fortunately, as we have suggested above, a smoothing method should not be terribly sensitive to these values. While we have run many of our computations from scripts because of the level of detail involved, we actually recommend the interactive approach here. We do, however, continue to use a script to load the data, that is, we issue the command

```
exec 'fq98-0'
```

to load the main (C1), trial (C2) and validation (C3) Quarterly Traffic Fatalities in Canada data into the usual set of columns, along with appropriate columns (C4–C6) for the time periods. We delete data for 1975–81 as usual.

For the trial data period, we will use Winters' multiplicative seasonal method. The default smoothing parameters suggested by Minitab is 0.2 for all three values. This actually works quite well, but by writing a simple script, we are able to try a number of values. Our own choice is to try the 27 values made up of all combinations of 0.1, 0.2 and 0.3. The results are summarized in Table 13.4, to which we will return in a moment.

The rationale that underlies our choice is that we truly want a *smoothing* method, so the constants used should not be too large. We do, however, want to try values 'around' the defaults. Values much larger than 0.3 for the smoothing constants suggest that the series that the constant applies to – namely the artificially created level, trend or seasonal factor series – is changing rather rapidly. This may mean that we should not be using smoothing. In some cases we will want to use values larger than 0.3 or smaller than 0.1 for one or more of the smoothing constants, but in such cases we will want to look carefully at our problem, data and results.

From Table 13.4 we see that the parameter set (0.2, 0.2, 0.3) gives the smallest mean square deviation (MSD). Note that Minitab uses MSD rather than MSE (mean squared error). We believe that the distinction is that the MSD is the sum of squared deviations (SSD = SSE) divided by the number of data elements in the model, while MSE is SSD divided by the number of data elements *minus the number of parameters in the model*. These small details cause a lot of trouble for students. We will use the (0.2, 0.2, 0.3) parameters for the rest of our analysis, including the estimation over the main model period (all the data). However, we note that by the Mean Absolute Percentage Error (MAPE) and Mean Absolute Deviation (MAD) criteria, the parameters (0.1, 0.2, 0.3) are slightly better.

Some workers have suggested using a nonlinear optimization technique to find the 'best' set of parameters. At one time, we ourselves suggested using either the Hooke and Jeeves or Nelder-Mead minimization techniques to find the parameters *and* the starting smoothers. However, we now believe this is overkill. Nevertheless, the 'traditional' wisdom that the smoothing parameters lie in the range 0.1 to 0.3 can be challenged. Chatfield (1997) gives some discussion and references.

Minitab lets us generate forecasts automatically, and we therefore generated forecasts for the 10 quarters that make up our validation period, along with the errors, which we DESCribe.

Variable	N	Mean	Median	TrMean	StDev	SE Mean	Minimum	Maximum	Q1	Q3
verror	10	47.0	43.5	48.4	60.7	19.2	-61.4	143.4	2.3	95.0

Row	vperiod	fqvdata	FORE1	verror
1	81	668	583.214	84.786
2	82	834	742.641	91.359
3	83	1021	972.850	48.150
4	84	815	817.301	-2.301
5	85	697	553.582	143.418
6	86	643	704.423	-61.423
7	87	926	922.132	3.868
8	88	813	774.129	38.871
9	89	630	523.949	106.051
10	90	683	666.204	16.796

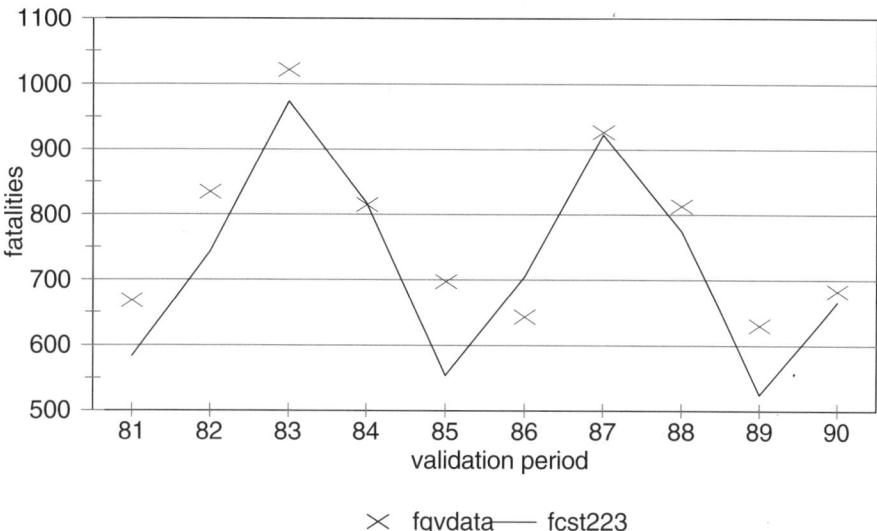

Figure 13.9. Validation of Winters' method using the trial model developed with the smoothing parameters (0.2, 0.2, 0.3).

Although we have *not* established that the parameters (0.2, 0.2, 0.3) are appropriate for the main model estimation of the Quarterly Traffic Fatalities in Canada data, we will proceed to use them for the main model and forecasting so that there is consistency with our trial model and its validation. The measures of fit are shown on the lower graph in Figure 13.11. The trial period modelling is presented in the upper graph of this figure.

The regular Minitab output does not provide the model used for forecasting, which we regard as unfortunate, so we have extracted the numbers from the Minitab data worksheet where they are stored and we have displayed them in Table 13.5. We note that 'cut and paste' between the Minitab data worksheet and common spreadsheet software such as Excel or Quattro is a convenient method for quickly transferring data between the packages. Table 13.5 also gives the forecasting model for the main estimation period, that is, using all the data. We have used the last available level and slope, along with the last 4 (i.e. last full seasonal cycle) seasonal indices to provide our models. Some comments on how Table 13.5 was developed are in order, since they illustrate how important details can be in methods such as Winters'.

- If we have *n* data points, then $S(n)$ and $b(n)$ define the final trend equation for future points, so that the trend line equation with a traditional intercept, y_0, has to be worked out as

$y_0 = S(n) - n * b(n)$

Table 13.4. Trial model measures of fit for the Quarterly Traffic Fatalities in Canada data

alpha	gamma	beta	MAPE	MAD	MSD
0.1	0.1	0.1	6.63	62.86	6811.83
0.1	0.1	0.2	5.94	56.97	5672.91
0.1	0.1	0.3	5.63	54.13	5017.24
0.1	0.2	0.1	6.37	59.78	6053.76
0.1	0.2	0.2	5.73	54.86	5156.55
0.1	*0.2*	*0.3*	*5.39*	*52.12*	4646.94
0.1	0.3	0.1	6.43	60.12	5917.98
0.1	0.3	0.2	5.82	55.57	5091.41
0.1	0.3	0.3	5.41	52.42	4615.95
0.2	0.1	0.1	6.5	60.44	5970.47
0.2	0.1	0.2	5.83	55.21	5074.03
0.2	0.1	0.3	5.45	52.2	4567.53
0.2	0.2	0.1	6.56	60.78	5808.76
0.2	0.2	0.2	5.91	55.82	4990.56
0.2	*0.2*	*0.3*	5.47	52.42	*4530.3*
0.2	0.3	0.1	6.63	61.39	5854.61
0.2	0.3	0.2	5.98	56.48	5058.61
0.2	0.3	0.3	5.61	53.61	4614.4
0.3	0.1	0.1	6.69	61.5	6055.31
0.3	0.1	0.2	5.99	56.09	5152.54
0.3	0.1	0.3	5.6	53.04	4639.52
0.3	0.2	0.1	6.74	61.82	6048.49
0.3	0.2	0.2	6.06	56.64	5178.73
0.3	0.2	0.3	5.66	53.5	4691.79
0.3	0.3	0.1	6.8	62.38	6172.09
0.3	0.3	0.2	6.12	57.21	5296.96
0.3	0.3	0.3	5.75	54.17	4816.69

That is, we have to 'step backwards along the line'. The importance of this transformation is to allow us to compare the trial and main models.

- If n is not divisible by L, the length of the seasonal cycle, then we must be very careful with the alignment of the seasonal factors. This occurs with the main estimation model, since the data ends in the middle of 1997, i.e. half way through a cycle, since $n=90$, $L=4$.

- To compute forecasts, we use the trend line equations to compute a trend, then multiply it by the appropriate seasonal factor. We have done this in Table 13.5.

- Since Minitab draws its time plots based on the current index of time series data, its graphs will display a 'local' time index (as in the graphs in Figure 13.11) which may mislead us and others. Thus, the time 'axes' of Table 13.5 and Figure 13.11 are different.

Note that the trial model has been extrapolated beyond the validation period in Table 13.5 and Figure 13.10 so that both the trial and main models can be compared over the forecast period. We see that both models are similar, but that the trial model gives lower fatality forecasts. This

Table 13.5. Developing forecasting equations for trial and main models of the Quarterly Traffic Fatalities in Canada data. Note that the forecasts are for the same period, i.e. t=91...100

Trial model					**Main Model**				
period	season	Level	Slope	Season	period	season	Level	Slope	Season
77	1	878.52	-7.1877	0.7074	87	3	805.490	-9.2036	1.1898
78	2	859.110	-9.6327	0.9123	88	4	797.480	-8.9658	1.0152
79	3	846.130	-10.3019	1.2107	89	1	797.430	-7.1835	0.7664
80	4	834.970	-10.4730	1.0305	90	2	786.380	-7.9561	0.8807
for forecasting					**for forecasting**				
0	1	1672.810	-10.4730	0.7074	0	1	1502.429	-7.9561	0.7664
	2			0.9123		2			0.8807
	3			1.2107		3			1.1898
	4			1.0305		4			1.0152

tperiod	Fore-casts	sindex	trend	season	tperiod	Fore-casts		trend	season
91	871.408	3	719.767	1.2107	91	926.130		778.424	1.1898
92	730.956	4	709.294	1.0305	92	782.202		770.468	1.0152
93	494.318	1	698.821	0.7074	93	584.374		762.512	0.7664
94	627.987	2	688.348	0.9123	94	664.560		754.556	0.8807
95	820.690	3	677.875	1.2107	95	888.267		746.600	1.1898
96	687.784	4	667.402	1.0305	96	749.893		738.643	1.0152
97	464.685	1	656.929	0.7074	97	559.984		730.687	0.7664
98	589.768	2	646.456	0.9123	98	636.531		722.731	0.8807
99	769.972	3	635.983	1.2107	99	850.404		714.775	1.1898
100	644.613	4	625.510	1.0305	100	717.584		706.819	1.0152

is not surprising, since our trial model under-forecast the validation data, as shown by Figure 13.9.

Note that Minitab attempts to provide a confidence interval for our forecasts on the graphs in Figure 13.11. Provision of such information is not part of the traditional exponential smoothing calculations, although the issue of forecast reliability and accuracy is clearly of interest. As we indicate at several points throughout the book, we believe most of our readers will prefer to use the sizes of residuals (during the modelling periods) and validation errors as a guide to the likely errors in their final forecasts. These are also displayed in the following output. However, the issue of measures of variability for forecasts is of continuing theoretical and practical interest (see Chatfield *et al.*, 1999, for example).

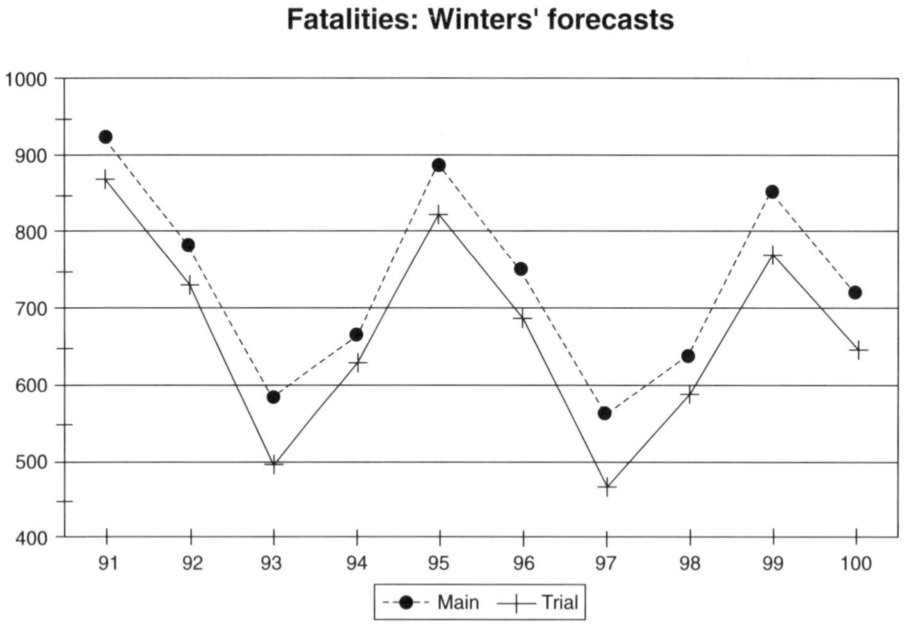

Fatalities: Winters' forecasts

Figure 13.10. A comparison of the trial and main model forecasts for the forecast period. The horizontal axis gives the time index.

Alternative seasonal forecasting with smoothing

We conclude this chapter by noting (see Wilson and Keating, 1998, p. 118 ff) that it may be preferable *not* to apply the Winters' approach to a seasonal series, but first to *deseasonalize* the data by the methods for generating seasonal factors or seasonal shifts discussed in Chapter 10. That is, we compute a set of seasonal factors $I(t)$, of which there will be L different values, one for each season, or seasonal shifts, shift(t), similarly. Then we compute the deseasonalized series

$$z_1(t) = y(t)/I(t) \qquad \text{or} \qquad z_2(t) = y(t) - \text{shift}(t)$$

by application of these factors or shifts to our data $y(t)$. We can now use non-seasonal smoothing methods (or indeed other non-seasonal forecasting methods, see Chapter 16) on the deseasonalized series. When we have the forecasts for the deseasonalized series, we must reseasonalize by reversing the process above. We leave as an exercise the application of these ideas.

Fatalities: trial Winters' model

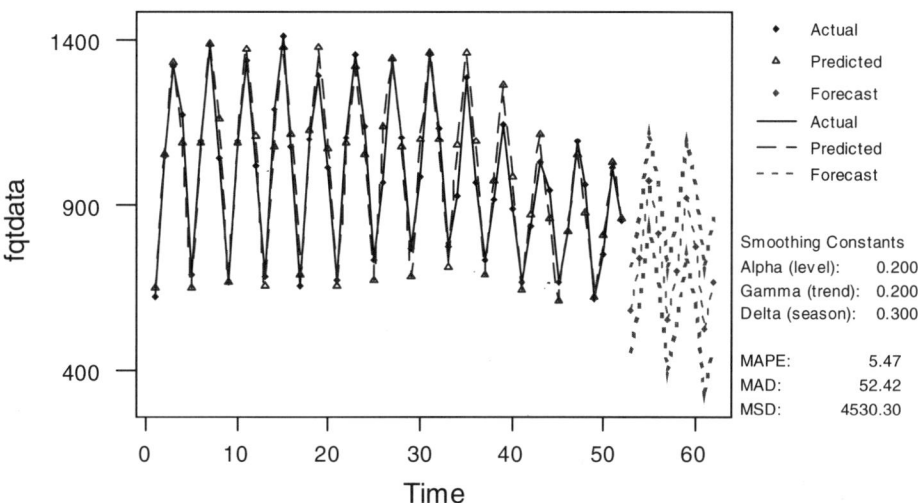

Fatalities: main Winters' model

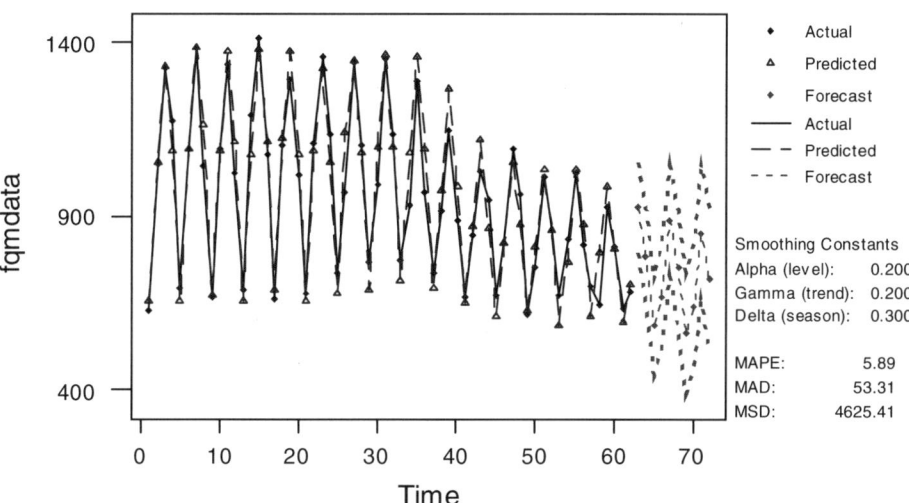

Figure 13.11. Trial and main models for Winters' method as drawn by Minitab. Note that because the first 7 years (28 periods) of data have been deleted, the time axis values are diminished by 28.

Exercises

E13.1. What weighed average is equivalent to the MA4(MA3()) compound moving average.

E13.2. Smoothing methods are often helpful for situations where we have sudden changes of level or slope or both. Generate several series with different properties (e.g. script lvlshift.mtj on the PFM Web site). Try different smoothing methods with different parameters to examine their forecasting performance for a 6 period forecasting horizon. How would you recommend that a forecaster should proceed. This exercise can be structured as a class debate or team competition.

E13.3. Generate the series $(3 + 2 * t)$ for $t=1...10$ and compute the MA3 and MA4(MA3()) series for this. Try both *central* and *'as soon as data available'* placement of the results.

E13.4. a) Generate a set of 30 pseudo-random numbers in the interval $(0, 1)$ and save them to a file. *[A saved set of numbers provides for verifiable results in the exercise.]* Call them rnd(t).
b) Load these numbers into a spreadsheet and from them generate a series $X = 30 + 3*rnd(t)$
c) Generate the series $Y = X + 0.3 * t$.
d) Use moving averages of length 3, 6 and 9 on X and Y as 1-period ahead forecasts. Compute errors and their sums of squares for these six sets of forecasts.
e) Try double moving average forecasts for the Y series using MA3 and MA6 operations. Compute errors and their sums of squares for the two sets of forecasts.

E13.5. Repeat exercise E13.4 using SES for series X with α values of 0.1, 0.3, 0.5, 0.7 and 0.9. Apply SES, the Brown 1-parameter DES and Holt 2-parameter DES to the trended series using α and γ values 0.1, 0.3 and 0.5. You may choose only the cases $\alpha=\gamma$ for Holt to reduce the work.

E13.6. How would you cope with multiplicative trend via exponential smoothing? [This is a thinking exercise. We do not expect you to be able to provide a good answer at this point.]

E13.7. Apply Winters' method to data in Table 13.2. Interpret your results.

E13.8. Re-run a selection of the calculations leading to Table 13.2, saving the residuals. Compute the MAPE, MAD and MSD using the last 4 years of the trial period only. Would this alter our choice of a 'good' set of parameters for the trial modelling period?

E13.9. Try using Holt's or Brown's method on a deseasonalized version of the Quarterly Traffic Fatalities in Canada data. Remember to reseasonalize before computing errors etc.

E13.10. Apply Winters' method to the Dry Cleaning data.

Forecasting trend and season III: Time series decomposition

How does this differ from Winters' method?

In our courses we require students to understand the main details of the classical 'ratio-to-moving averages' time series decomposition. Given the many variants on the time-series decomposition theme, some with high levels of development, there is a rich and sometimes confusing literature on the subject. We do not feel it is worthwhile for students to become too immersed in the details. The foundation of the approach is, however, a cornerstone of modern forecasting methods. We provide some further exposition in Chapter 18.

We have already encountered the essence of time series decomposition in the model(s) used for Winters' exponential smoothing (Chapter 13), in the regression models involving seasonal dummy variables (Chapter 12), and of course in the Pegels' classification (Chapter 3). The idea is to break our time series down into parts, primarily involving trend and season, although the concept can be extended to other 'structure' in the data, in particular, long term patterns to model the so-called business cycle, which we will abbreviate to 'cycle'. The two most common forms of decompositions of the trend/season/cycle type are:

- Additive: data = trend + season + cycle + error

- Multiplicative: data = trend * season * cycle * error

There are possible mixed forms, such as an error that is added to a multiplicative model trend*season*cycle.

Historical notes

The practical uses of time series decomposition appear to have originated in government statistical bureaux in the 1920s as a means to 'seasonally adjust', that is, deseasonalize, various statistics. Quantities such as unemployment rates have natural cycles because much work in agriculture, construction, fisheries, etc. is seasonal. To be able to compare month to month or quarter to quarter, the 'adjusted' figures are preferred, even if they are really based on a forecasting model.

One of the main products of this activity was the ratio-to-moving averages time series decomposition. Today we have spreadsheet and statistical software on computers that can carry out the calculations for us in a variety of ways. In the era of mechanical calculators – essentially adding machines – it was important to be able to structure a calculation so that several machine operators could carry out the 'number crunching' cooperatively. Moving averages are well-suited to the operations of mechanical calculators. Moreover, the decomposition can be set up so some operators carry out the moving averages, others compute the ratios to the moving averages, yet others average the appropriate ratios to get the seasonal factors, and so on. When automatic computation arrived, the work was simply transferred. Eventually the programs were revised to take advantage of the greater computational capability.

For example, the US Bureau of the Census went through a number of versions of time series decomposition, culminating in the Census II (X-11) method. Statistics Canada took this a step further with the X-11 ARIMA method (associated with the name of Estelle Dagum). These have been revisited in the more recent X-12 (Chapter 18). These variants of time series decomposition all have supporters and detractors. We expect our students to comprehend the fundamentals of classical ratio-to-moving-average time series decomposition, as well as to understand the difference between an additive and multiplicative forecasting model. Details beyond the basics we feel should be considered when and if forecasting tasks require them.

Classical time series decomposition

We will present the classical time series decomposition step by step.

1. *Seasonality L*: Our first step is to decide the *length L of seasonality*, e.g. 12 for monthly, 4 for quarters, 5 for business days (no weekend trade). Deciding L may be the most important step in the whole process. If we get it wrong, the entire exercise is wasted, and there is no obvious way to automate this decision. We generally rely on plots of the data to see if there is a seasonal variation.

2. *CMA(L)*: Now form *centred moving averages of length L* of the data. Recall that we defined the moving average of length L of the data as the series labelled MA(L,data). The issue now is to position the moving average of L points at the centre – in the time domain – of the L points that span the moving average. This moving average represents (trend * cycle) since season and error should be 'averaged out'. For L even, we clearly cannot position the moving average at an integer valued time point. For example, for annual series, the 'middle' position is at the time point 6.5. Students often jump to say '6'. This is enough of a difficulty that we will take the time to illustrate. Let us write down time points for an annual series:

data:	1	2	3	4	5	6	7	8	9	10	11	12	13
MA(L):						6.5	7.5						

The first two moving averages will be positioned at 6.5 (for data from times 1–12) and 7.5 (data from times 2–13). The time points are like fence posts, and the moving averages like the fence wire. We need $(n + 1)$ posts for n lengths of fence. Alternatively, the middle position of n posts is $(n + 1)/2$, which is not a whole number if n is even.

For L odd, we will be able to use MA(L, data) to get the centred moving averages. We will position each moving average at the centre position of the data from which it has been computed. For L even, however, we will form the centred moving average by taking the two-element moving average of the moving average of length L. The MA(L,data) is positioned at 'half' time points, so MA(2, MA(L, data)) is back on integer time positions. Thus:

data:	1	2	3	4	5	6	7	8	9	10	11	12	13	14
MA(L):						6.5	7.5	8.5						
CMA(L):						7	8							

The formulas for CMA(L) when L is even can be quite tricky to write down; we direct your attention to the time indices in the summations. Operationally it is easier to use

CMA(L,data) = MA(2, MA(L,data))

$$CMA(L, data, t) = \tfrac{1}{2}(data(t + L/2) + data(t - L/2)) + \sum_{k\,=\,t-L/2+1,}^{t+L/2-1} data(k)$$

We will write CMA(L,data) for the centrally positioned MA(L,data) when L is ODD to save having to distinguish further even and odd cases.

3. *Ratio to moving average*: Divide the data elements by CMA(L,data) for all points where these moving averages are available. This yields data which represent the (season * error) for various time periods.

4. *Preliminary seasonal factors*: Average the (season * error) series just obtained for time periods corresponding to the same season. This should 'average out' the error and leave us with season factors. Typically, the seasonal factors are obtained by using a medial average (trimmed mean) in which the two most extreme values are dropped before the average is computed. Many minor variations in treatment exist. One popular variant expresses the seasonal factors in percentage form, that is, so that an index of 100 becomes a factor of 1.

5. *Normalized seasonal factors*: Normalize the seasonal factors so that their mean is 1 (or 100 if we are using indices scaled to 100). This is to make sure we have no trend component in the seasonal factors. Although the forecasting model is multiplicative, it is traditional that we use an *arithmetic* average to normalize the seasonal factors.

6. *Deseasonalized data*: Divide each data element by the appropriate seasonal factors, giving a 'deseasonalized' data series, that is,

deseasonalized = data / season

In doing this, we caution that it is important to get the right seasonal factor with each data element. When we compute the seasonal factors above, we will *not* get them in the natural order because of the elements 'lost' in computing the centred moving averages.

7. **Trend line**: Compute the trend line using as data the deseasonalized series

$$\text{trend}(t) = a + b\,t$$

using simple linear regression. We used to show students formulas for computing trend lines, but simple regression is now available widely in spreadsheet and other software.

The irregular variation remains. We can still consider further decompositions of this component, but in our courses we have never made such analysis a requirement on assignments or examinations, although we would encourage such analysis in project work.

8. **Irregular component**: Divide the deseasonalized series by the trend model to get the preliminary irregular or (error * cycle) series

$$\text{preliminary irregular} = \text{error} * \text{cycle} = \text{deseasonalized} / \text{trend}$$
$$= \text{data} / (\text{trend} * \text{season})$$

9. **Cycle analysis**: If there appears to be a pattern in the (error*cycle) series, then take long period moving averages (i.e. quite a bit longer than L), or use plots and a spline or cut-out outline of 'cycle' if the pattern is stable and can be assumed to continue into the future. The cycle can be treated in much the same way as the seasonality. Note that in many situations it may not be worthwhile trying to forecast the cycle.

10. **Forecasting model**: Our forecasting model is

$$\text{forecast}(t) = \text{trend}(t) * \text{season}(t) * \text{cycle}(t)$$

with $\text{trend}(t)$ $= a + b * t$

              $\text{season}(t)$ = seasonal factor for time period t

              $\text{cycle}(t)$ = an appropriate index for time t,

                                 = 1 if cycle is ignored.

Generally, we recommend that students only try to model and forecast the cycle if there is clear visual and other evidence that a *predictable* cycle exists. The so-called 'business cycle' is a pattern of increase or decrease in business activity that does *not*, in our view, have sufficient regularity to permit useful forecasting without great care. The issue is that, while there are visible ups and downs in graphs of business activity, these do not have a stable period length. That is, we cannot easily predict when the highs and lows will occur. We consider this further below.

An example

Time series decomposition (TSD) calculations are built into Minitab and other statistical software. Unfortunately, it seems that almost every programmer interprets 'classical' time series decomposition differently. Minitab's TSD is not the method described above. However, it produces a model of the same form, that is, trend line times seasonal factors, although the algorithm to compute this model is not the ratio to moving averages we have presented above.

Table 14.1. Spreadsheet computation of the Time Series Decomposition

Classical time series decomposition TSDECOMP.xls
Bowerman and OConnell (1979, p. 225)   Resid SS=   19.6185

period	data	MA(4) placement!!	CMA(4)	ratio =data/cma	Seasonal Factors	Deseason. =data/St	trend = a+b*t	Predicted =trend*sf	Residual	Irreg
1	10				0.463	21.583	23.162	10.732	-0.732	0.932
2	31	25.000			1.218	25.453	23.765	28.944	2.056	1.071
3	43	25.250	25.125	1.711	1.686	25.507	24.367	41.078	1.922	1.047
4	16	25.750	25.500	0.627	0.633	25.279	24.969	15.804	0.196	1.012
5	11	26.250	26.000	0.423	0.463	23.741	25.571	11.848	-0.848	0.928
6	33	26.500	26.375	1.251	1.218	27.095	26.173	31.877	1.123	1.035
7	45	27.000	26.750	1.682	1.686	26.693	26.776	45.139	-0.139	0.997
8	17	27.250	27.125	0.627	0.633	26.859	27.378	17.328	-0.328	0.981
9	13	28.000	27.625	0.471	0.463	28.058	27.980	12.964	0.036	1.003
10	34	28.500	28.250	1.204	1.218	27.916	28.582	34.811	-0.811	0.977
11	48	29.000	28.750	1.670	1.686	28.473	29.184	49.199	-1.199	0.976
12	19	29.750	29.375	0.647	0.633	30.019	29.787	18.853	0.147	1.008
13	15	30.500	30.125	0.498	0.463	32.374	30.389	14.080	0.920	1.065
14	37	31.000	30.750	1.203	1.218	30.380	30.991	37.745	-0.745	0.98
15	51				1.686	30.253	31.593	53.260	-2.260	0.958
16	21				0.633	33.179	32.195	20.377	0.623	1.031

Trend line from regression of Deseason vs. period (separate sheet)
a = Intercept =   22.56031   Final Seasonal
b = slope =   0.602185   Factors =

	Year 1	Year 2	Year 3	Year 4	Avg.=Raw Seasonal Factors	Raw SFacts/ Normalizing Constant
Qtr 1		0.423	0.471	0.498	0.4639	0.463
Qtr 2		1.251	1.204	1.203	1.2193	1.218
Qtr 3	1.711	1.682	1.670		1.6878	1.686
Qtr 4	0.627	0.627	0.647		0.6337	0.633

Average=   1.00115 = Normalizing Constant

Data above is copied from the 'ratio to moving avg' column and is CAREFULLY positioned. The Paste is performed saving VALUES, not formulas.

Notes: 1) The final seasonal factors have been formatted to 3 decimals
        2) The final seasonal factors are Pasted as VALUES.

Our experience is that the Minitab TSD generally works well, and this was also the conclusion of an MBA student (Xin Zhou) who examined the Minitab TSD macro and tried some modifications. However, a colleague observed a case where it did not perform very well. As always, we must look at our results.

As with the smoothing methods, we will use Minitab's macro to do the calculations for our example, but recommend learning the details and ideas via more or less manual calculations with spreadsheet software. Our first (simple) example in Table 14.1 is the same one used to illustrate Winters' method in Chapter 13.

Figure 14.1 shows the intermediate calculations in graphical form. Note that we use the deseasonalized series to get the trend line. Figure 14.2 shows the result of multiplying the trend by the seasonal factors. We see clearly how well classical ratio-to-moving averages time series decomposition succeeds in this case. Indeed, this method for forecasting seasonal series is generally very reliable unless there are sudden level shifts or changes in the underlying situation. We cannot expect the simple decomposition to model such shifts. We note that while Table 14.1 is based on a Microsoft Excel 97 spreadsheet, we were forced to rebuild the spreadsheet in Corel Quattro 7 to draw Figure 14.2. Such details are a continuing nuisance for forecasters.

In the time series decomposition treatment, we are presuming a *linear* trend and single level of *seasonal variation*. Clearly more complex patterns require more complex decomposition models.

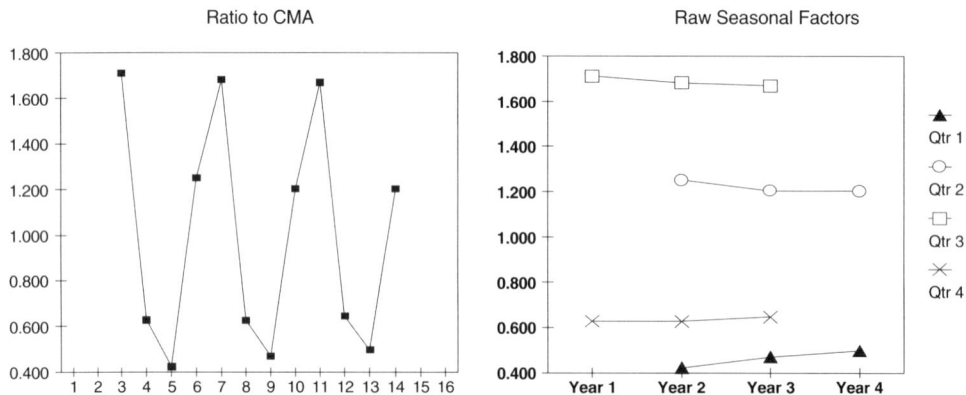

Figure 14.1. Time series decomposition example. The left-hand graph shows the ratio of data to centred moving averages. The right-hand graph shows these data split by year as a way to check for the stability of the seasonal factors.

Traditionally, time series decomposition uses the ratio of the data to the model (predicted) to provide a description of the lack of fit. This series is called the irregular component and it is a *relative error*. We measure fit via residuals, so should be computing deviations from the model

$$(a + b*t) * \text{season}(t)$$

Once we have the true residuals, we can prepare all the usual measures and indicators of model

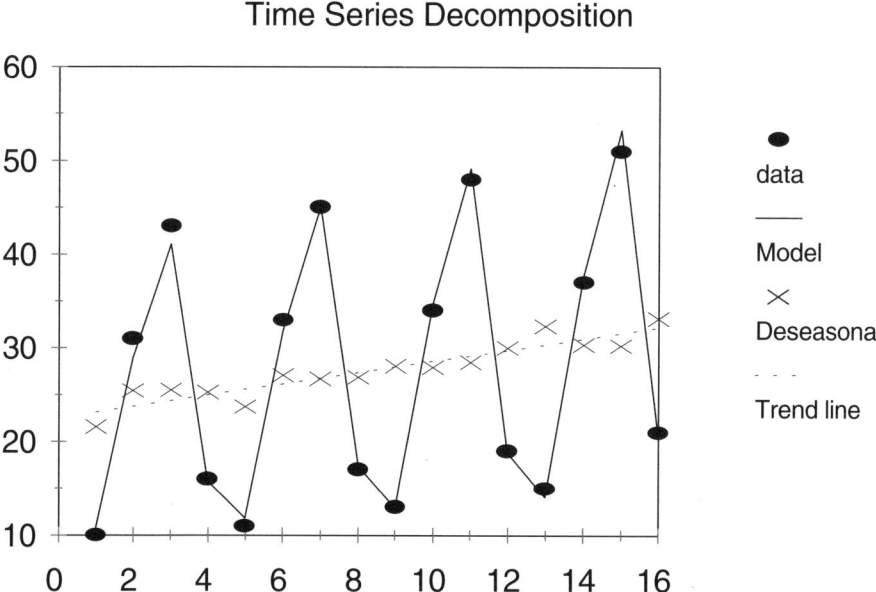

Figure 14.2. Time series decomposition: data and model for the simple example with deseasonalized data and trend line. (Graph drawn with Corel Quattro 7.)

quality presented in Chapter 3, most particularly MSE and R-squared and the distribution and fit graphs. Note that the sum of squared residuals is displayed for the example in Table 14.1. To complete our treatment, we display the irregular component in Figure 14.3.

Further analysis of the irregular component

The history and folklore of time series (and the related topic of *seasonal adjustment* of data) includes a considerable literature on the further refinement of the model by including long term cycles. Much of this discussion is influenced by considerations of the so-called 'business cycle'. We will not enter into this minefield of controversy. Business cycles are not of fixed period, and might better be titled 'long term variation in business activity'. Because the period length is not fixed, it is difficult to predict them. Moreover, the shape and amplitude of the variations is not regular, which adds to our woes.

For situations where there is a visible cyclic pattern in the irregular component of our decomposition, and this is of a regular shape and frequency, we should certainly attempt to provide a forecast. In our view, situations where such cycles are likely to be seen do *not* relate to traditional sales or administrative data based on the calendar, but to phenomena such as traffic

TSD: Irregular component

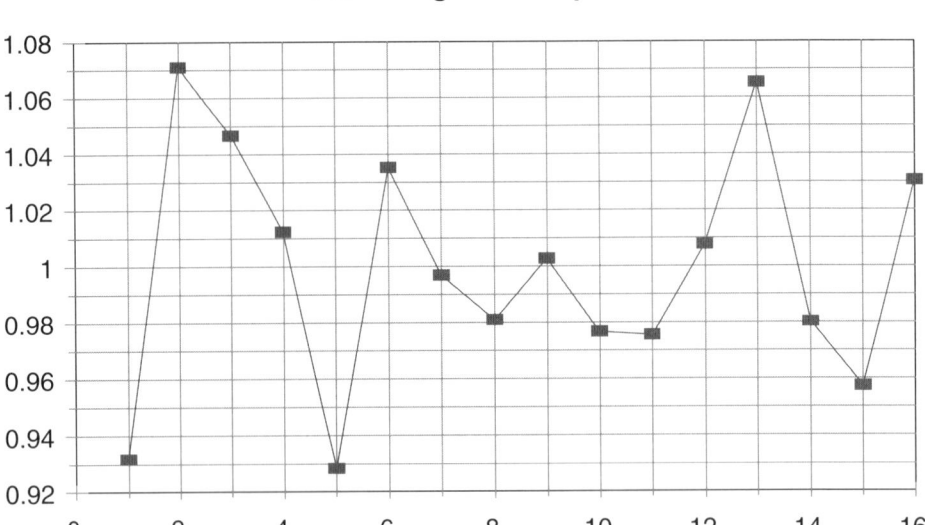

Figure 14.3. Irregular component for a simple time series decomposition example.

looked at on an hourly basis for roads or telephones in a large city. The cycle will be imposed by the weekly and possibly monthly variations that are superposed on the more obvious daily pattern.

Forecasting such cycles may be attempted in various ways:

- We may repeat the processes of the classic decomposition, but this time working with the irregular component (or possibly the true residuals). This means deciding on suitable long term centred moving averages, taking ratios to these moving averages, finding cyclic factors and normalizing them, then producing a new 'irregular' component.

- If the irregular component or the residuals show very clear and reproducible pattern(s), we could use tracing paper, a deformable 'ruler' or acetate sheet to trace the pattern, then advance the pattern one or more cycles forward via a simple mechanical shift of our trace along the time axis. We have already discussed such an approach in Chapter 10 in presenting ruler forecasts.

A real example

Minitab allows us to perform the classical decomposition very easily. An edited output is given below. Figure 14.4 shows the data, main model and forecasts for time series decomposition applied to the Quarterly Traffic Fatalities in Canada data. Minitab also presents other information in subsidiary graphs, which are also shown. These are useful for discovering weaknesses in the approach for a particular set of data. Note that Time Series Decomposition (TSD) does not usually provide dispersion estimates for the forecasts. In our treatment here, we have also omitted the validation step.

```
Trial Model                                   Main Model
           Data          fqtdata                       Data          fqmdata
Length       52.0000                           Length       62.0000
NMissing    0                                  NMissing    0

Trend Line Equation                           Trend Line Equation
Yt = 1096.47 - 3.97558*t                      Yt = 1121.41 - 5.25310*t

Seasonal Indices                              Seasonal Indices
Period    Index                               Period    Index
   1      0.715780                                1      0.730082
   2      0.978892                                2      0.978176
   3      1.27429                                 3      1.26088
   4      1.03104                                 4      1.03086

Accuracy of Model                             Accuracy of Model
MAPE:         6.68                             MAPE:         7.01
MAD:         62.99                             MAD:         62.66
MSD:       5815.80                             MSD:       6058.76

Forecasts                                     Forecasts
  Row   Period      FORE1                       Row   Period      FORE2
    1       53      634.01                         1       63      996.678
    2       54      863.17                         2       64      809.443
    3       55     1118.58                         3       65      569.432
    4       56      900.96                         4       66      757.796
    5       57      622.63                         5       67      970.184
    6       58      847.61                         6       68      787.782
    7       59     1098.32                         7       69      554.092
    8       60      884.56                         8       70      737.242
    9       61      611.25                         9       71      943.690
   10       62      832.04                        10       72      766.121
```

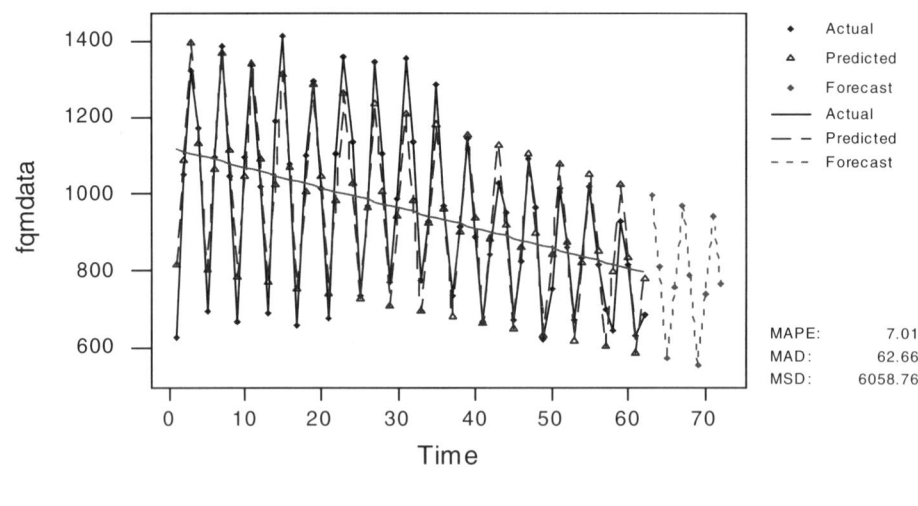

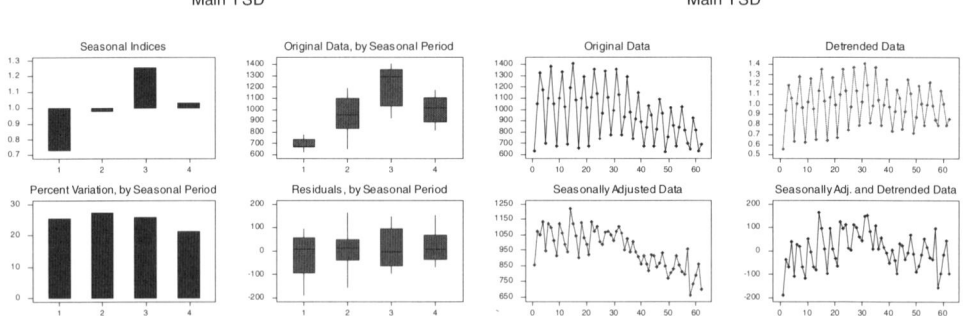

Figure 14.4. Graphical output of multiplicative time series decomposition for the main modelling data of the Quarterly Traffic Fatalities in Canada data.

Exercises

E14.1. Using the forecasts from the trial modelling period (these are shown above, but you could also re-create them from the original data by applying TSD), carry out the validation step for TSD forecasting.

E14.2. Apply time series decomposition to the Dry Cleaning data.

15

ARIMA and related models for forecasting

What are ARIMA models?

ARIMA models are a class of time series models that involve previous data values and previous values of errors. The Box–Jenkins methodology is employed to estimate and forecast with these models, although other methods could be used to carry out these tasks. We will begin with the *models*, and ignore how to use them for the time being. Some of these models are very difficult to visualize, so, as usual, we will rely on graphs as much as possible. For many of our readers, the principal reason for getting to know something about ARIMA models is that they are often suggested in discussions of forecasting. However, it is our experience that many people who suggest or ask about them are familiar only with the name, rather than how to use them. We have also noted questions about ARIMA models in job interviews for government and public utility positions. Rarely have we found them to be very much better than choices that are simpler to explain, such as Winters' method. In many cases, however, the simpler methods can be expressed in the ARIMA framework, permitting a unification and comparison of the ideas in mathematical terms.

The major difference between ARIMA models and the (mostly regression type) models we have considered so far is the fact that the error terms are an important part of the structure of the new models. That is, we forecast future values of our time series as a function of previous data values and previous error values.

One difficulty with ARIMA models is that their handling needs quite a lot of algebra, albeit quite simple algebra. In our courses, we require students to be able to deal with this algebra to the extent of being able to write down the forecasting form of a specified ARIMA model. The rationale for this otherwise onerous condition is that we have encountered highly paid consultants who were unable to perform a 1-period ahead forecast on a simple ARIMA model without returning to their office to 'run the computer'. Garbage in, garbage out!

ARIMA stands for Auto-regressive Integrated Moving Average. The 'auto-regressive' means that the time series of interest is influenced by its own earlier values. The 'integration' implies actually that we model *differences* of the series of interest, and must work back to the actual series. Such differences are used to remove trend, so that ideas based on stationary series can be used. The 'moving average' is a misleading expression which, in fact, refers to terms in the model that are similar to those occurring in the expansion of exponential smoothing formulae. These involve previous error terms.

We will concentrate on the notation used by Minitab and other software to work with ARIMA models. It is a very efficient, but rather cryptic, way to express them. We spend considerable class time with students to train them to be able to decode the ARIMA expressions into algebraic forms that can be used to forecast, but these latter equations are often very complicated. In general, an ARIMA model is specified by means of an expression of the form:

ARIMA $p\ d\ q\ P\ D\ Q\ L$

where the parameters p, d, q define the *non-seasonal* part of the ARIMA model, and the P, D, Q, and L define the *seasonal* part of the model. In fact, L is the seasonality, that is, the number of time periods that make up one season. WARNING: do not confuse the lower and upper case parameters!

Some useful notation

The *back shift* or *lag* operator B is a clever 'trick' of mathematicians that allows us to work out usable formulas for ARIMA models from the shorthand '$p\ d\ q$' notation just introduced. Like differentiation operators or the MA(K) operators we have already encountered, B acts on any quantity to its right. In fact, B, applied to any item, object or quantity containing a time index t, replaces t with $(t - 1)$ everywhere in the object. Thus,

$B\ y_t = y_{t-1}$
$B\ x(t - 4) = x(t - 1 - 4) = x(t - 5)$
$B\ (y_t - y_{t-1}) = y_{t-1} - y_{t-2}$

We can apply B many times, thus 'squaring' or 'cubing' it.

$B^2\ y_t = y_{t-2}$ $\qquad\qquad$ $B^3\ y_t = y_{t-3}$

The *difference* operator, DIFF, is simply given as

DIFF = $1 - B$

Thus,

DIFF(y_t) $= y_t - y_{t-1}$

DIFF^2(y_t) $=$ DIFF (DIFF(y_t)) $= y_t - 2\ y_{t-1} + y_{t-2}$
$\qquad\qquad\quad = y_t - y_{t-1} - (y_{t-1} - y_{t-2})$

DIFF is a simple polynomial in *B*. We can create other, more complicated polynomial expressions in *B*, e.g.

$$1 - parameter1 * B - parameter2 * B^2$$

where *parameter1* and *parameter2* are simply numbers. Such expressions are operators and act on objects on their right, with *t* being replaced by *(t - 1)*, *(t - 2)* etc.

We can also define differences that span more than 1 time period. Thus,

$$(1 - B^4) \, y_t = y_t - y_{t-4}$$

We can call this operator DIFF_4. The Minitab command DIFF 4 (with a space rather than an underscore) will actually carry out the operation. Such differences are useful in modelling seasonal data, as we shall see.

Note that the polynomials in B may be multiplied together in any order – in mathematical jargon, multiplication is *commutative* – since *t* never appears anywhere inside these operators. We now have a convenient framework for expressing and working with ARIMA models.

Non-seasonal ARIMA models

For non-seasonal data, we need only the notation

ARIMA *p d q*

i.e. the lower-case or non-seasonal control values. The simplest of all these models is an ARIMA 0, 0, 0 model, which is really too simple for most textbooks to include, and generally cannot be 'estimated' by software. (This is likely a programming fault, as the model is very simple to estimate – we need only take the mean of the series.) It has the form

$$y_t = \mu + error_t = constant + error_t$$

That is, a stationary model plus noise, which is the same model as for SES. Many authors use μ (Greek mu) as the constant in ARIMA models, even though it is not truly the mean in many ARIMA models, although it clearly will minimize the sum of squared deviations between model and data for this simple case.

An AR(1) model (the simplest autoregressive model) is denoted ARIMA 1 0 0, and is written down

$$[1 - AR1 * B] \, y_t = \mu + error_t$$

where mu is a constant (the mean in this case) and AR1 is a parameter of the model. To get the forecasting form, we want y_t on its own on the left-hand side of the equations, with only earlier time periods or constants appearing on the right-hand side. Expanding,

$$y_t - AR1 * y_{t-1} = \mu + error_t$$

Rearranging gives the forecasting form

$$y_t = \mu + AR1 * y_{t-1} + error_t$$

An MA(1) model, denoted ARIMA 0 0 1, puts the expression in B on the right-hand side where it acts on the errors. We change the parameter name to MA1.

$$y_t = \mu + [1 - MA1 * B] \, error_t \; = \mu + error_t - MA1 * error_{t-1}$$

The simplest model with a difference ($d = 1$) is ARIMA 0 1 0 (a non-stationary random model)

$$DIFF \; y_t \; = \; \mu + error_t = \; [1 - B] \, y_t$$

or

$$y_t \; = \; \mu + y_{t-1} + error_t$$

Often the constant is taken to be zero in this model. We can usually instruct software to include or exclude the constant. In our experience, the criterion used by software that causes the default inclusion or exclusion of a constant is not obvious. We recommend trying both options in practice.

We can have combined forms, such as ARIMA 1 0 1

$$[1 - AR1 \, B] \, y_t = \mu + [1 - MA1 \, B] \, error_t$$

$$y_t = AR1 \, y_{t-1} + \mu + error_t - MA1 \, error_{t-1}$$

Non-seasonal models are built up as follows from the general expression ARIMA (p, d, q)

$$(1 - \sum_{j=1}^{p} ARj \, B^k) \, (1 - B)^d \, y_t = \; \mu + (1 - \sum_{k=1}^{q} MAk \, B^k) \, error_t$$

The parameters of these models are the values of the constant, μ, the ARj's and the MAk's. In our naming of the parameters, we have followed the convention used by Minitab, which we find quite clear and unambiguous on output listings. Traditionally, the parameter ARj is written ϕ_j, that is, lower case Greek 'phi' with subscript j. Similarly, the MAk is written as a lower case Greek 'theta' with subscript k, θ_k.

Seasonal models

Seasonal models are constructed in a similar fashion, but now we add in some seasonal parts involving backshifts and differences of order L. For example, the model

ARIMA 0 0 0 1 0 0 12

(a Seasonal AR1 model) is written

$[1 - SAR12 * B^{12}] \, y_t = \mu + error_t$

where SAR12 and the constant μ are the parameters, which simplifies to

$y_t = SAR12 * Y(t - 12) + \mu + error_t$

Note: we use SAR12 as the *first* Seasonal AR parameter, because it multiplies $Y(t - 12)$. We must warn readers that some workers call this SAR1, since it is the first seasonal parameter of the autoregressive type. Traditional nomenclature, following the use of upper case P for the number of terms, uses an upper case Greek phi, usually with the subscript 1 rather than 12, that is, Φ_1.

The ARIMA 0 0 0 0 1 0 12 model uses a seasonal difference

DIFF_12 $= [\, 1 - B^{12} \,]$

which implies that this January's data value will be similar to that of last January. The forecasting form is

$y_t = y_{t-12} + \mu + error_t$

The Seasonal MA model ARIMA 0 0 0 0 0 1 4 has the forecasting form ($L = 4$, or quarterly model):

$y_t = \mu + error_t - SMA4 * error_{t-4}$

The parameter SMA4 is the first MA parameter for this quarterly data, having $L=4$. The traditional notation is an upper case Greek theta with subscript 1 for the 'first' MA parameter of this type, that is, Θ_1. And once again, some authors use MA1.

We must be a little careful with a two-parameter Seasonal model of either the AR or MA type. This is because the backshifts 'jump' by increments of the seasonality L. Thus ARIMA 0 0 0 2 0 0 4 is decoded as

$(1 - SAR4 * B^4 - SAR8 * B^8) \, y_t = \mu + error_t$

The general model form looks complicated, but each piece is simple to apply.

ARIMA $p\,d\,q\,P\,D\,Q\,L$

is initially written down

$$(1 - \sum_{i=1}^{P} \text{SAR}iL\ B^{i*L}\,)\,(1 - B^{L})^{D}\,(1 - \sum_{j=1}^{p} \text{AR}j\ B^{j}\,)\,(1 - B)^{d}\ y_{t}$$

$$= \mu + (1 - \sum_{m=1}^{Q} \text{SMA}mL\ B^{m*L}\,)\ (1 - \sum_{k=1}^{q} \text{MA}k\ B^{k}\,)\ \text{error}_{t}$$

In this equation, the parameter SMAmL will be SMA8 for m=2 and L=4. That is, we must evaluate the index in order to write down the SAR and SMA parameters.

The simplification of the expression above to a forecasting form involves a lot of algebraic manipulation, but all the steps are simple. Indeed, Minitab and other packages are able to carry out the simplifications with quite elementary computer code. The object is the same – to isolate y_{t} alone on the left-hand side so that we can use the equation for forecasting. Note that it is usual to have no constant for models where there is either a regular or seasonal difference. Unless Minitab or other software provides a value for the constant, you may generally take it to be zero. It is possible to force either a constant or no constant using special commands.

Estimation and use of ARIMA models: the Box–Jenkins methodology

The ARIMA models exist whether or not we have tools to estimate the parameters in them. Indeed, with some sort of general function minimization tool, such as that we shall encounter in Chapter 20 for nonlinear modelling, we can compute residuals for an ARIMA model if we provide a set of parameters, generate the sum of squares, then try to minimize the sum of squares by varying the parameter values. This is, in fact, what underlies the tools in Minitab and other software for estimating the parameters. However, the Box–Jenkins methodology has some additional features and shortcuts to make the process more efficient. The steps in the estimation process are as follows:

Step 0. Get the data into the analysis software system. This is always necessary, but we want to reinforce the message. We need to ensure that we have no *missing values*, since otherwise the software will not work. With appropriately generalized techniques, we believe this shortcoming could be avoided, but not without some assumptions about the behaviour of the phenomenon at hand. The issue is that we need data at every time point in our estimation time period in order that we can compute the errors $e(t)$ that go into the ARIMA models. When there are missing values, we are forced to impute values, which can be quite dangerous to our final conclusions unless we are *very* careful. Also, we should perform any preliminary analysis of the data necessary to check for special values that could be 'outside' the structure of the ARIMA models.

For example, we do not want to model situations where our business was closed because of a storm or the death of an owner. Such situations should be treated as missing values and the 'data' imputed, even though this may be difficult.

Step 1. Conduct a preliminary analysis of the data. We need to decide if there is trend and/or season in the data. Several approaches are useful. First, we consider trend.

- Trend can be visualized directly in a graph of the data versus time. Recall that PLOT is sometimes better than TSPLOT in Minitab since it can show the trend more clearly.

- The ACF of data with trend will not 'die away' quickly.

In the next chapter, we consider some methods for removing trend. For the moment, we assume that the time series we are working with has at most a linear trend. We can try to remove such a trend by taking the first differences of the series. If the first difference does remove the trend from non-seasonal data, then the parameter d is set to 1 in the ARIMA model. For a stationary series, we need no differencing and $d = 0$.

If we have data that are believed to be seasonal, we must decide what is the length of the seasonality L. This is more or less external to the Box–Jenkins methodology, although the ACF and PACF functions may give some guidance. In the case of a seasonal series, we could try a *seasonal* difference rather than a regular one to remove trend. The choice here can be difficult. Are we dealing, in a monthly data series with seasonality, with trend year-to-year or month-to-month? In many cases it will be nearly impossible to decide, and the decision will be deferred to the trial model stage below (Step 3). Generally we will *not* see both d and D non-zero.

Step 2. We suppose that we now have a series Z that has no trend. To provide for future discussion, we will name the series that can be rendered stationary by differencing as W. That is, we presume that a transformation function $T()$ exists so that our original series y can be transformed to

$$W = T(y)$$

that has just linear trend. ($T()$ can be the identity if y does not need transformation.) To remove any linear trend, we apply some difference function $D()$, either of lag 1 (non-seasonal) or lag L (seasonal) so that we arrive at a stationary series

$$Z = D(W)$$

Thus

$$Z = D(T(y))$$

We should now look at the ACF and PACF functions of Z in order to get some suggestion of possible values for the ARIMA parameters p, q, P and Q. Note that the d and D parameters are already more or less determined in the preliminary analysis, though some trial and error refinements may be necessary, particularly if there are autoregressive or seasonal autoregressive terms in our model.

Many textbooks on Box–Jenkins ARIMA modelling spend a good deal of page space on 'rules' for choosing the p and q parameters from ACF and PACF patterns. Over the years we have become disenchanted with such treatments. We do not find the 'rules' work well enough to make them worthwhile, and the power of modern personal computers is such that we can try lots of models quite easily, especially if we use a script to run them automatically. However, the ACF and PACF do have a role to play as diagnostics of the remaining pattern and as signposts of directions for modelling. That is, we use them with residuals to check for a pattern our model has not addressed and as indicators for change in the set of p and q parameters. Remember that we are looking at the ACF and PACF of a supposedly stationary series, that is, either the Z series or a set of residuals from modelling. The modelling itself, however, uses the series W.

Here are some pointers (we revisit these ideas in the next chapter):

- Spikes in the PACF in low-order lags (1 or 2) suggest AR1 or AR2 terms in the non-seasonal part of the model. Usually the ACF will show exponential decay, possibly with oscillations.

- Spikes in PACF at suspected seasonality length L or its multiples suggest SAR terms (1 or 2 maximum in any usual situation) in the seasonal part of the model. The ACF may oscillate and/or decay.

- Similar spikes in the ACF suggest MA or SMA terms in the non-seasonal and seasonal parts of the model respectively. The PACF will decay exponentially, possibly with oscillations, for purely MA models.

- Mixed models, despite anything we have read in books, often give quite messy ACF and PACF structures, especially when seasonality is present.

Step 3. We now examine some trial models using our *undifferenced* series W. Note that we do not work with the Z series here. As usual, we will subset the data series if possible, and set aside some observations in order to allow validation tests. We try a selection of models using forms for which the ACF and PACF have provided hints. As we have said, textbooks make much of the ACF/PACF analysis, but it is no more than a signpost. We find it helpful to plan the list of models to try and to put these into an executable script, which Minitab and most major packages allow. This is very important if you have a slow computer or a large data set. It is important to make sure that you save the output, that is, use the OUTFILE command in Minitab or similar record of activity in other systems.

Using the output, we build a summary table of the results of modelling.

- The SSE (sum of squared errors) or MSE (mean squared error) for the models to give us the fit over the estimation period

- We note models for which there was a convergence failure or cases where backforecasts do not die away rapidly. The backforecasts use the ARIMA model and reorganize it so that we can compute early values of the model from later ones, that is, run backwards in time. The ARIMA models are an example of a general mathematical structure called difference equations or recurrence relations. They can give rapidly growing values of the model at each step, suggesting the model is not stable. There may be ways to employ such models, but they are not for general usage and we suggest dropping such models from consideration.

- We will want to flag models where there are small t-ratios for some of the parameters indicating that these parameters are not needed in the model and could be dropped. Of course, t statistics depend on the assumption of Gaussian errors $e(t)$, which may not be true. Nevertheless, we will use the t-ratios as an indicator of the utility of the parameters.

- Models with 'large' chi-square values in the Ljung–Box statistics for different lags suggest that there is a pattern remaining in the residuals. We do not generally save the residual series at this point. Some software actually will give the p-value for a test of the Ljung–Box statistics, but a simple rule of thumb is that values roughly equal to the number of degrees of freedom imply little remaining pattern.

Step 4. We choose the 'best' model(s) from the trial set and re-run them to compute validation forecasts and to save residuals and predicted values. With the residuals, we can run PACF as well as ACF graphs. We can look at the closeness of data and predicted values over the trial estimation and validation periods, and also get prediction intervals for the forecasts. Moreover, we can check the distributional properties of the residuals to see if the assumption of a Gaussian distribution of errors can be supported.

Step 5. At any step above, we may return to a previous step to try more or different models or transformations. Note that we may also have to back-transform

$$y\_model = T^{-1}(W\_model)$$

to compute 'true' measures of fit and to check the assumptions via distributional plots and other graphs and tests.

Step 6. When we are satisfied that we have chosen a suitable model, we apply it to the full data series to compute forecasts. In so doing, we should still check the fit and that the parameters are more or less similar to those found in the trial estimation period.

Step 7. If W is a transformed version of our data Y, we need to back-transform to compute predictions and measures of fit, particularly if there are other models and forecasts that are candidates for use.

A real example

We will assume that we have already carried out a preliminary analysis of our Quarterly Traffic Fatalities in Canada data, and that we believe that ARIMA models offer a sensible possibility for our data. We already know that it is quarterly data, so our season length is fixed at 4. We are fairly sure that there is a slight downward trend in the data. This should imply that we need to difference our series before examining the autocorrelation function. The ACF and PACF of the data (here we have used the Main data, but one could also use the Trial data) are shown at the right. We note that the ACF declines rather slowly, supporting the contention that there is trend present in our data. The pattern of spikes every 2 lags likely reflects the quarterly seasonality, on top of which there is some semi-annual pattern

```
ACF C1                                                    PACF C1
ACF of fqmdata                                            PACF of fqmdata
 -1.0 -0.8 -0.6 -0.4 -0.2  0.0  0.2  0.4  0.6  0.8          -1.0 -0.8 -0.6 -0.4 -0.2  0.0  0.2  0.4  0.6  0.8
  +----+----+----+----+----+----+----+----+----+            +----+----+----+----+----+----+----+----+----+
  1                    XXXXXX                                1                    XXXXXX
  2       XXXXXXXXXXXXXX                                     2       XXXXXXXXXXXXXX
  3                    XXXXX                                 3                    XXXXXXXXXXXXXXXXXXX
  4                    XXXXXXXXXXXXXXXXXXXXXXXXX             4                    XXXXXXXXXXXXXXXXX
  5                    XXXXX                                 5              XXXXXXXXX
  6       XXXXXXXXXXXXXX                                     6                    XXXX
  7                    XXXX                                  7                    XXX
  8                    XXXXXXXXXXXXXXXXXXXXX                 8                    XXX
  9                    XXXXX                                 9                    XXXX
 10       XXXXXXXXXXXXX                                     10                    XXX
 11                    XX                                   11                    XX
 12                    XXXXXXXXXXXXXXXXXX                   12                    XX
 13                    XXXX                                 13                    XXXX
 14       XXXXXXXXXXXXX                                     14                    XX
 15                    XX                                   15                    XXX
```

To get rid of the trend, which we take to be linear for the present, we will use a difference of lag 1, that is, a non-seasonal first difference. This is accomplished with the Minitab command

        diff 1 c1 c11

Now we look at the ACF and PACF of the first difference. Note that we could also try a seasonal first difference using the Minitab command

        DIFF 4 c1 c12

then consider the ACF and PACF of column c12.

The ACF of the differenced series is slowly declining, while there are just a few peaks in the PACF. This suggests autoregressive models. We again get the pattern every 2 lags in the ACF and there are hints of this in the PACF also. We will try an autoregressive model with just one term in the non-seasonal part and one in the seasonal part. This gives the output below.

```
MTB > acf c11 <-- the 1st difference of the data
    -1.0 -0.8 -0.6 -0.4 -0.2  0.0  0.2  0.4  0.6  0.8
     +----+----+----+----+----+----+----+----+----+----+-
  1                                X
  2    XXXXXXXXXXXXXXXXXXXXXXXX
  3                                XX
  4                                XXXXXXXXXXXXXXXXXXXXXXXXXX
  5                                X
  6    XXXXXXXXXXXXXXXXXXXXXXXX
  7                                XX
  8                                XXXXXXXXXXXXXXXXXXXXXX
  9                                XX
 10      XXXXXXXXXXXXXXXXXXXX
 11                               XXX
 12                                XXXXXXXXXXXXXXXXXXXX
 13                               XXX
 14      XXXXXXXXXXXXXXXXXX
 15                              XXXX
```

```
PACF of c11 <-- 1st difference of the data
    -1.0 -0.8 -0.6 -0.4 -0.2  0.0  0.2  0.4  0.6  0.8
     +----+----+----+----+----+----+----+----+----+----+-
  1                                X
  2    XXXXXXXXXXXXXXXXXXXXXXXX
  3                        XXXXXXXXXXX
  4                                XXXXXXXXXXXX
  5                          XXXXX
  6                                XX
  7                               XXX
  8                                XX
  9                                XX
 10                                XX
 11                                X
 12                                X
 13                               XXX
 14                                XX
 15                                XX
```

```
MTB > arima 1 0 0 1 0 0 4 c2
ARIMA model for fqtdata
Estimates at each iteration
Iteration        SSE      Parameters
     0        2244958    0.100   0.100  802.884
     1        1657550    0.115   0.250  655.874
     2        1169171    0.136   0.400  510.812
...
     7         190389    0.459   0.990    4.641
     8         189954    0.466   0.994    2.366
     9         189928    0.466   0.994    1.821
Unable to reduce sum of squares any further
* ERROR * Model cannot be estimated with these data
```

We have chosen not to also include a difference here because AR terms can often approximate a difference. Indeed, we now see that a difference may be a better choice for this data. Note that we included both a non-seasonal and seasonal AR term because the PACFs of order 2 and 3 as well as 4 are non-zero. Minitab thinks it cannot proceed with the computation of the estimates. We want to point out something quite important for analysing ARIMA outputs. This is that the current approximations to the parameters in the model are displayed for each iteration of the estimation algorithm. In particular, we note that the second parameter, that for the Seasonal Autoregressive (AR) term of the model is approaching 1 in value, and ends at 0.994. If it were exactly 1, we would have a seasonal difference. This suggests we try the following model (we have abbreviated the output).

```
MTB > ARIMA 1 0 0 0 1 0 4 'fqtdata';
SUBC>    Constant;
SUBC>    Forecast 52 10 c12;
SUBC>    Brief 2.
ARIMA model for fqtdata

Estimates at each iteration
Iteration     SSE       Parameters
    0        207486    0.100  -17.310
...
    5        185766    0.423  -11.159
Relative change in each estimate less than  0.0010
Final Estimates of Parameters
Type       Coef     StDev      T       P
AR    1    0.4230  0.1362    3.11   0.003
Constant -11.159  9.152     -1.22   0.229
Differencing: 0 regular, 1 seasonal of order 4
Number of observations:  Original series 52, after differencing 48
```

```
Residuals:      SS =    184671  (backforecasts excluded)
                MS =     4015  DF = 46

Modified Box-Pierce (Ljung-Box) Chi-Square statistic
Lag                12            24          36          48
Chi-Square       13.7          22.4        39.4           *
DF                 10            22          34           *
P-Value         0.187         0.439       0.242           *
```

We now compute the validation residuals into column 13

```
MTB > name c12 'tf100010'
MTB > let c13=c3-c12
MTB > name c13 'tr100010'
```

There may be some structure still left in the data that we can extract. We attempt another model of the same sort where there is an MA parameter (involvement of the error terms).

```
MTB > ARIMA 1 0 0 0 1 1 4 'fqtdata';
SUBC>   Constant;
SUBC>   Forecast 52 10 c14;
SUBC>   Brief 2.

ARIMA model for fqtdata
Estimates at each iteration
Iteration        SSE      Parameters
     0        202341     0.100      0.100    -17.310
 . . .
     8        182768     0.403      0.148    -11.342
Relative change in each estimate less than   0.0010

Final Estimates of Parameters
Type            Coef        StDev         T          P
AR    1        0.4028      0.1384       2.91      0.006
SMA   4        0.1484      0.1503       0.99      0.329
Constant     -11.342       7.807      -1.45      0.153

Differencing: 0 regular, 1 seasonal of order 4
Number of observations:  Original series 52, after differencing 48
Residuals:      SS =    180239  (backforecasts excluded)
                MS =     4005  DF = 45

Modified Box-Pierce (Ljung-Box) Chi-Square statistic
Lag                12            24          36          48
Chi-Square       11.7          18.6        36.0           *
DF                  9            21          33           *
P-Value         0.229         0.608       0.328           *
```

Note that the iterations list the parameters in the order AR1, SMA1, Constant. (If in doubt, look at the 'Final Estimates of Parameters' below the iterations.) We remember to name our trial forecasts for this model, a never-ending chore. Both models have similar values of the Mean Square Error (MS). The constant in both models is apparently not statistically significant. The SMA4 parameter in the second model is also not significant. For both models, the residuals appear to have little pattern from the evidence of the Ljung–Box statistics. These statistical tests are, of course, subject to an assumption that the errors are Gaussian distributed.

```
MTB > name c14 'tf100011'
MTB > let c15=c3-c14
MTB > name c15 'tr100011'
MTB > print c6 c3 c12-c15
```

The choice of a seasonal MA term in this model is partly motivated by the desire to try a term with different possibilities, and partly by 'gut feeling'. Trial and error is sometimes the way. Let us look at the Validation data and residuals for the two models. The latter model has a slightly better fit over the Trial modelling period, but is a more complicated model. Note that we display fits and residuals for each model.

Row	vperiod	fqvdata (Actual)	tf100010 (Fits)	tr100010 (Resids)	tf100011 (Fits)	tr100011 (Resids)
1	81	668	561.998	106.002	571.148	96.852
2	82	834	717.149	116.851	727.826	106.174
3	83	1021	987.098	33.902	996.376	24.624
4	84	815	835.883	-20.883	848.828	-33.828
5	85	697	541.482	155.518	556.112	140.888
6	86	643	697.311	-54.311	710.426	-67.426
7	87	926	967.546	-41.546	978.025	-52.025
8	88	813	816.453	-3.453	830.094	-17.094
9	89	630	522.103	107.897	537.223	92.777
10	90	683	677.954	5.046	691.476	-8.476

We also want to look at the pattern of the errors. Note below that the y axes are shifted! We may have to be careful to get similar *scales* for the axes, then arrange them appropriately, as here, so we can make comparisons. Another approach is to draw them on the same graph in a multiple plot. The 1 0 0 0 1 1 model is slightly better, and we choose it for generating our Main model and forecasts, despite some reservations about the significance of the parameters. Let us now generate the main estimation model and the forecasts.

```
Descriptive Statistics
Variable   N   Mean   Median   StDev   Minimum   Maximum     Q1      Q3
tr100010  10   40.5    19.5     75.0    -54.3     155.5    -26.0   110.1
tr100011  10   28.2     8.1     74.9    -67.4     140.9    -38.4    99.2
```

ARIMA model for fqmdata   <-- This is for the MAIN data
```
Estimates at each iteration
Iteration         SSE      Parameters
    0          256554    0.100    0.100   -17.243
    1          239751    0.250    0.213   -14.196
    2          238002    0.307    0.263   -13.090
    3          237948    0.310    0.274   -13.045
    4          237942    0.310    0.278   -13.070
    5          237941    0.310    0.279   -13.083
    6          237941    0.310    0.280   -13.088
    7          237941    0.310    0.280   -13.090
Relative change in each estimate less than  0.0010
```

```
Final Estimates of Parameters
Type         Coef        StDev        T         P
AR    1     0.3095      0.1281      2.42     0.019
SMA   4     0.2798      0.1293      2.16     0.035
Constant  -13.090       6.170     -2.12     0.038
```

```
Differencing: 0 regular, 1 seasonal of order 4
Number of observations:  Original series 62, after differencing 58
Residuals:     SS =   233786  (backforecasts excluded)
               MS =     4251  DF = 55
```

```
Modified Box-Pierce (Ljung-Box) Chi-Square statistic
Lag              12        24        36        48
Chi-Square     12.9      19.9      35.3      43.6
DF                9        21        33        45
P-Value       0.169     0.529     0.360     0.531
```

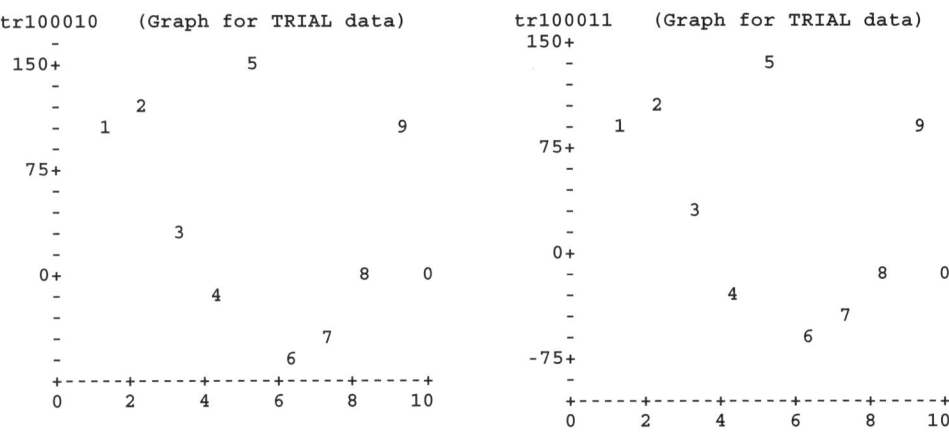

Note: the period number on below the horizontal axis is Minitab's internal index, not our quarter number! The vertical axes of these graphs have the same scale, but different limits.

```
Forecasts from period 62
                           95 Percent Limits
Period        Forecast       Lower        Upper
  63 (qtr91)   932.69       804.88      1060.50
  64           794.46       660.67       928.26
  65           619.93       485.57       754.28
  66           659.31       524.90       793.72
  67           912.27       748.69      1075.84
  68           775.05       608.95       941.15
  69           600.83       434.49       767.17
  70           640.31       473.94       806.67
  71           893.29       702.74      1083.84
  72           756.09       563.38       948.80
```

We note that Minitab offers to generate many graphs that may be useful in interpreting the output. We will not show them here, but we do discuss the models further in the next chapter.

We included a warning above that Minitab has its own internal time index that does *not* correspond to our situation. We need to be *very* careful in interpreting our output. Here we have manually added an indication of the quarter number to the output.

Exercises

E15.1. Generate the forecasting form for the following non-seasonal ARIMA models.

ARIMA 1 1 1

ARIMA 2 0 0

ARIMA 0 1 2

E15.2. Generate the forecasting form for the following seasonal ARIMA models.

ARIMA 0 0 0 0 0 1 4

ARIMA 0 0 0 1 0 0 12

ARIMA 1 1 1 0 0 1 6

ARIMA 1 0 0 1 1 1 4

E15.3. Use the log(toys) data from the toysdata.mtb script on the PFM Web site and generate the (seasonal) forecasting form for the model

ARIMA 0 1 1 0 1 1 4

then apply this to the data and residuals given in the data set to get a forecast for period 61. For period 69. How would you get the forecasts for "toys" (toy sales)?

E15.4. Generate the validation forecasts for the Quarterly Traffic Fatalities in Canada using the models suggested in this chapter.

Using ARIMA models: other issues and examples

Transforming a series to stationarity

There are a number of ways to consider ARIMA models. We find it helpful to view them as a way to model cyclic patterns in time series that are otherwise stationary. That is, there should be no long term movement in the level of the series if we take a long enough term. Of course, many series that are of interest really do have trend. Very few companies are satisfied with stable revenues; growth is almost always a goal. Thus, we need to deal with the trend in some way.

In our view, there are three major approaches to dealing with the trend, although most books that deal with ARIMA modelling present only the first of these.

1. We can take *differences* between data at current and previous time periods, since differences (like differentiation) turn a linear trend into a constant. Typically, we will use a lag of 1,

$$Z_t = y_t - y_{t-1}$$

or a *seasonal difference* between corresponding seasons, that is, for seasonality of length L,

$$Z1_t = y_t - y_{t-L}$$

This approach is the most popular choice for removal of trend, as it is supported within the structure of ARIMA modelling as we have already seen. Note that second-order differences, that is, double differences, will render a quadratic (parabolic) trend stationary.

2. When we think the trend is nonlinear, or more correctly non-polynomial, differences alone will not remove the trend. An exponential trend (the model for compound interest or any growth involving constant percentage change), is not rendered stationary by differencing, although differencing may seem to work over a limited time interval. In this case we can transform the series so that it can be made stationary with differences. For example, if y_t has or is believed to have exponential underlying trend, then we can take logarithms to get

$$W_t = \log(y_t)$$

which is a particular case of a functional transformation

$W_t = transform(y_t)$

For such transformations to be useful, they must be *invertible*, that is,

$y_t = transform^{-1} (transform (y_t))$

Then W_t can be differenced to obtain a stationary series. Once again the differences can be accommodated within the ARIMA modelling framework.

3. Alternatively, we can simply *subtract* or *divide* out the offending trend. If we call the trend function *trend(t)*, and we have it estimated correctly, then

$V_t = y_t - trend(t)$ or $U_t = y_t / trend(t)$

will be stationary. This approach is, in effect, decomposing our series. We choose subtraction for a series we think has additive trend, and division for multiplicative trend.

These ideas are straightforward 'in theory'. Their execution, in particular the back-transformation to get forecasts for the original series, frequently gives both novice and experienced users difficulties. We return to this issue later in this chapter.

The theory behind many statistical methods, including the Box–Jenkins methodology for ARIMA modelling, also calls for the variance to be stationary. We usually term this *homoskedasticity*, with its absence being *heteroskedasticity*. To render a series homoskedastic, we generally resort to transformations of the data. A common choice is the Box–Cox transformation (Atkinson, 1985, p. 85), which is usually applied to non-time-series data:

$W_{BC}(Y, \lambda) = (Y^\lambda - 1) / \lambda$ if $\lambda \neq 0$, or $\log_e(Y)$ if $\lambda = 0$

The transformation parameter λ can be varied, so that by looking at graphs, we can attempt to find a particular λ, hence a particular transformation, that gives a more or less uniform variation across the data. The same approach, using a normal probability plot, can be used to try to render the transformed data 'normal', i.e. Gaussian.

To provide some structure for practitioners, our suggested procedure is as follows.

Step 1. Prepare a time plot of the data and possibly also ACF and PACF plots in order to discover trend and seasonality.

Step 2. If the data are seasonal and have trend, try a seasonal difference (lag L). If the resulting series still has trend (see step 1), try a regular, lag 1, difference. Although the order of differencing for a series is immaterial, the seasonal difference may be sufficient to remove both seasonality and trend, while the lag 1 difference certainly will not. Note that lags other than 1 or L (the length of the seasonality) will result in series that are generally difficult to interpret.

Step 3. If the data are non-seasonal and have trend, try a difference of lag 1.

Step 4. If there is heteroskedasticity but no trend, try a Box–Cox transformation.

Step 5. If there is trend and heteroskedasticity, removing the trend by differencing may also stabilize the variance, but this must be checked. We may still need to transform the de-trended series.

Step 6. For series with a trend that is more than linear, we avoid double (second-order) differences. With each difference, we cancel some digits in our data and lose information. Instead, we recommend trying a transformation. Taking logarithms is common and often effective for exponential growth or decay, since a linear trend results and can be removed with a difference.

Step 7. Where we have a more complicated trend such as a sigmoid, or S-shape, then the decomposition approach may be preferred.

Let us now assume that Z_t is our stationary series computed from the original y_t.

If we have used a log transformation, we must exponentiate our 'forecasts' when we want to compute measures of fit that are in the same scale as the original data. A source of mistakes is the potential confusion between natural (base e) and common (base 10) logarithms. Make sure you know which of these functions you are using. There are several conventions. For example, $log()$ generally refers to the natural or Napierian (base e) logarithms. However, you must be very careful in reading the work of others. We will use the conventions

$$z = \log\_e(y) = \ln(y) = \log(y) \qquad \text{implying} \qquad y = e^z = \exp(z) \quad \text{and}$$
$$w = \log\_10(y) = \log 10(y) = \log_{10}(y) \qquad \text{implying} \qquad y = 10^w$$

The z and w in the above equations have no special significance. We just wanted different symbols for the two transformed values. Keep in mind that $\log 10(10) = 1$, while $\ln(10) \cong 2.303$.

Developing candidate ARIMA models

It would be nice to have a way to know what combination of p, d and q to use for non-seasonal data, and what set of p, d, q, P, D, Q to use for seasonal ones. We will assume that L, the length of the seasonality, is established, although for some data this can be a serious difficulty. For example, if we are working with daily data, a weekly seasonality may be assumed ($L=7$, or $L=5$ for 'weekday only' series). Nevertheless, we could be dealing with a situation that actually has a bi-weekly (fortnightly) pattern with a strong weekly sub-pattern. Some schools, for instance, run on two-week cycles, as do some shift-work institutions.

As we mentioned in Chapter 15, there is a long textbook tradition of looking at ACF and PACF patterns to suggest the choices of ARIMA models, at least for non-seasonal forms. Our experience has mostly been with seasonal data, where the 'rules' do not seem to be followed very closely by our data and models. However, the ACF and PACF patterns can sometimes be helpful. We liken them to a road signpost that teenagers have been swinging on. When we arrive it is clearly loose in the ground and may have been turned so that we are forced to treat with suspicion the directions indicated by the arms of the signpost. The directions may be useful, but we will want to be careful not to trust them too much.

Let us now consider what patterns in the ACF and PACF plots we will use as signposts. We distinguish particularly, what we shall call:

- *spikes*, that is, individual values of ACF or PACF that are relatively large and stand out, although there may be several in a row, with a clear 'break' afterwards.

- *decay*, exponential decrease in magnitude of the ACF or PACF as the lag index k increases. The form of the decay may involve all positive or all negative autocorrelations, or may exhibit a decaying sine-wave pattern. No matter, we will apply the term 'decay' to all these.

Sometimes, especially with seasonality, we will see a superposition of spikes on decay patterns.

For a non-seasonal ARIMA model, we will choose d to be the level of differencing used to transform our series to stationarity. That is, assuming differencing d times was sufficient to make

$$Z_t = \text{DIFF}^d \, y_t$$

stationary, d is the value we use. Some caution is needed if we first transform y, then difference. For example, if

$$Z_t = \text{DIFF}^d \, W_t = \text{DIFF}^d \, (\, \text{transformation}(y_t)$$

then the d is the number of levels of differencing of W. We will need to back-transform later to get appropriate measures of fit. Usually d is 0 or 1, rarely 2, and extremely rarely larger than 2.

If we now look at the ACF and PACF of the $Z(t)$ series, the following table will provide some suggestion of possible values of p and q. When we model, however, the ARIMA process is applied to W or Y, i.e. the series *before* differencing.

Type of process	ACF pattern	PACF pattern
pure AR	decay	spikes up to order p
pure MA	spikes up to order q	decay
mixed ARMA	variety of patterns, sometimes showing decay in both or spikes in both	

Seasonal AR or MA, if not confused by non-seasonal components, may reveal the P or Q by appropriate spikes in the PACF and ACF, respectively. That is, a pure seasonal AR(2) for L=4 may show clear spikes at lags 4 and 8 in the PACF. A pure seasonal MA(1) for monthly data may have a single clear spike at lag 12. In our experience, clarity is rare, but indications often appear.

Trial estimation of ARIMA models

When modelling, first try the forms suggested by the indicators above, then modify these choices. For example, the ACF/PACF patterns may suggest AR(2). However, we would also try AR(1), AR(3) and even an ARMA(1,1), that is, $p=1$ and $q=1$. The d will be determined as in the previous section. While we could try to model the Z series, we recommend going back to the W or Y series to keep our focus on data as close as possible to the original situation. It is much harder to think of forecasting rates of change than levels, yet differences are rates of change.

With seasonal models, we like to try purely seasonal models first, and only when we are not satisfied with the fit or other properties, to add non-seasonal components. We always want to seek the simplest model that is 'good enough' for our purposes. Some technical points:

- When modelling, the software may suggest that the solution cannot be obtained. This usually means that certain models cannot be estimated from the data because of mathematical constraints. From a practical perspective, it is important that the software can detect such impossible situations, but we need not be burdened by the technical details.

- The software may tell us it can make no more progress. Sometimes, the default starting values for the parameters in the iterative method used to estimate ARIMA models may be poor choices (Minitab uses 0.1 for all parameters except the constant). You can look at the progress of the iterations in most ARIMA software – Minitab displays the sum of squared deviations and the parameter values at each iteration. Sometimes progress could be made with more steps. Save the parameter values and specify their location (column) as starting values for ARIMA estimation and run it again. In our experience, it is commonly necessary to give starting values in order to get a good set of parameter estimates for a model.

- It is not difficult to think of many possible sets of $p\ d\ q\ P\ D\ Q$ values. By preparing a *script* of commands to our software we can automate the process of estimating a variety of trial models. In preparing these, we recommend keeping the amount of stored information to a minimum (particularly for users of student editions of software), since we will in any event revisit models that look promising. To help keep track of the models in the script, we suggest using a *lexicographic ordering* of the numbers that make up the $p\ d\ q\ P\ D\ Q$ set. For example, for a non-seasonal model where we will allow p to be any number between 1 and 3, d to be fixed at 1 and q to be 0 to 2, we get a command sequence for models of the data in data column C10 starting with `ARIMA 1 1 0 c10` then using the following $p\ d\ q$ sets:

      (1 1 1), (1 1 2), (2 1 0), (2 1 1), (2 1 2), (3 1 0), (3 1 1), (3 1 2)

Selecting working models

The processes of the previous section will generate a lot of output for the different possible models. At this stage, these will generally be trial models, that is, estimated from the trial data subset. Our first look at the output will be to consider the fit of the candidate models over this period. For models with approximately the same number of parameters, fit can be judged by the sum of squared residuals. If our models have quite different numbers of parameters, so that the number of degrees of freedom (number of observations in the data set minus the number of parameters) changes markedly between the models, then the mean square error, that is, the sum of squared residuals divided by the degrees of freedom, is a better measure. While fit will be our main criterion for selection, we will – where the software provides the statistics – want to look at the significance of the estimated parameters as indicated by the t-statistics or p-values. We should also note if there is significant autocorrelation in the residuals via the Box–Pierce or Ljung–Box statistics. Remember, however, that test statistics for parameters and autocorrelation assume a Gaussian distribution of errors. We consider checking this assumption below.

We suggest at this stage keeping only the best two to four candidate models. With these, we generate validation forecasts to compare to the actual data we have set aside. For most software, this will involve re-running the estimation with options set to compute the forecasts and storing them appropriately. At the same time, it is a good idea to store the residuals from the estimation period. They can be graphed in a histogram or stem and leaf diagram, or subjected to a normal probability plot, all of which allow us to check the normality assumption mentioned above.

In our experience, examination of the validation data and the parameter and autocorrelation significance tests, along with checking the distributional properties of the estimation residuals generally puts all the candidate models in some jeopardy of being discarded. That is, they all have deficiencies in one or other of the desired properties. What are we then to do?

To provide a pathway to completing our forecasting task, we suggest the following plan of action.

- Where the highest order MA, SMA, AR or SAR parameter is far from significant, we may try dropping it. That is, if the t-statistic of such a parameter is less than 1 in magnitude or the p-value is greater than 0.15 (these are merely guidelines), we can estimate a model without the offending parameter. This idea could be applied to any parameter, but some software (Minitab in particular) only allows us to choose the $p\ d\ q\ P\ D\ Q$ numbers, so that only the highest order in each of the $p\ q\ P\ Q$ categories can be dropped. If we have set up an array of choices as in the previous section, the statistics for this model will already be available, and the suggested model may have already been abandoned.

- We have seen several cases where it is worth forcing a constant to be included or excluded from our model, rather than accept the decision that the software makes for us.

- Having settled the issues of significance (particularly if the residuals are reasonably distributed), we rank our remaining candidate models according to how well the validation forecasts match the actual data. Of course, we may find that one candidate model has several of the observations for one model fit beautifully, with one extremely bad forecast. Another may never be terribly close but never terribly far off the mark. Choices, choices! Here you need to wear the managerial hat.

- From the above steps, the number of candidate models will likely have been reduced. With the remaining candidates, we then run the estimation over the full data set and generate forecasts, storing these and residuals as we do so. We then make our judgement on which model and forecasts to use.
  - Full-period estimation should give similar measures of fit as in the trial period estimation.
  - Parameters of the full-period estimation should be 'similar' to those for the trial period.
  - If the residuals from estimation are reasonably distributed (i.e. mound shaped), then the parameters should be significant and residual autocorrelations 'not too bad'.
  - The properties of the forecasts (graph them!) should be 'reasonable', that is, their behaviour should correspond more or less to those of the data and to the validation forecasts.

On the basis of the above, we should be able to choose a model and a set of forecasts. It may arise, however, that two or more models work more or less equally well. If the forecasts are wildly different, then we may suspect we are in a situation where ARIMA modelling is not going to provide a satisfactory set of forecasts. Back to the drawing board to try to gather more information! If the models may give remarkably similar forecasts, we may want to simply average them or choose the model that gives the 'middle' range of forecasts.

Computing the right measures of fit and error

Suppose that we have transformed our original series y_t into a working series $W(t)$. The transformation will affect the magnitudes of residuals, and may affect different residuals differently, so that measures of fit and error, along with distributional properties, are not comparable with other forecasting methods. (It is the comparison that we seek.) Strictly, we want all our measures of fit to be in the *original* scale – that of y_t. This implies that every time we have a model and forecasts, whether trial or final, we should back transform. How do we do this?

- For forecasts, we compute (using brackets rather than subscripts for the time index)
  $$y\_forecast(t) = transform^{-1}(W\_forecast(t))$$

  We can then graph these and, in the case of validation forecasts, compare them to actual data and compute measures of fit. We would usually do this anyway, so it is simply a matter of making sure we are using the properly scaled quantities.

- To get residuals, we need to use the *fits* from the estimation process, which may only be stored by specifically choosing to do so. Then,

  $y\_fit(t) = transform^{-1}\ (W\_fit(t))$

  $y\_residual(t) = y_t - y\_fit(t)$

- The parameter significance statistics can be judged using the t-statistics or p-values generated by our estimation software. These are based on the W_residuals in the transformed space, so it is the distributional properties of *those* residuals that should be checked.

- The autocorrelations should really be checked using the Y_residuals. This requires a check of the distributional properties of these residuals; and a computation of the appropriate Ljung–Box statistic. This is a lot of extra work, especially with several candidate models.

Practically speaking, we may choose to back-transform our forecasts only for the 'best' ARIMA model. As software improves, we may see the processes automated, as there are no conceptual obstacles to doing this. A management decision must be made as to the relative importance of different aspects of the forecasting task. Clearly the final forecasts should be in the original y_t scale but, at intermediate stages, we may be satisfied with working with the $W(t)$ series.

Computing R_squared for ARIMA models presents another issue. Because of the lags that may refer back in time many periods, the residuals and their sum of squares are only available for a reduced set of time periods. Thus, to be strictly correct, we not compute the R_squared using the sum of squared residuals (SSE on Minitab or similar outputs) divided by the total sum of squares. If the SSE involves m observations after differencing and lags (Minitab prints this information) and there are n observations in the full series used to estimate a model, then we could consider inflating the SSE by (n/m). Alternatively, we could approximate R_squared by

  R_squared_alternate $= 1 -$ MSE / SD

where MSE is the mean squared error from our output, which is computed taking account of the number of available observations as well as the number of model parameters, and SD is the standard deviation of the series we are trying to model. R_squared is not widely used in comparing ARIMA and other models, perhaps because of these definitional issues. In the Appendix we present some formulas for computing different measures of fit.

Alternative seasonal ARIMA modelling

We note that some authors (Wilson and Keating, 1998; the program Pro-cast described in Nash and Walker-Smith, 1989b) have decided to avoid entirely the difficulties of selecting appropriate values for the seasonal ARIMA elements $P\ D\ Q$. Instead, they first find a set of seasonal factors for their data, much as we did in Chapter 10. Then non-seasonal ARIMA modelling is applied to the deseasonalized series. Forecasts are prepared by re-seasonalizing. We also suggested this in conjunction with exponential smoothing methods in Chapter 13. This is similar to the *sequential modelling* approach to be discussed in Chapter 19.

Exercises

E16.1. (A large exercise.) Apply ARIMA modelling to the Dry Cleaning data.

E16.2. Generate a series with exponential growth, add a small 'error' using some random numbers, then apply both differencing and logarithmic transformation and differencing. Try forecasting the series using both approaches and back-transforming results appropriately.

E16.3. Try using non-seasonal ARIMA on a deseasonalized version of the Quarterly Traffic Fatalities in Canada data. Remember to re-seasonalize before computing errors etc.

Comparing and combining forecasts

Descriptive comparison

In almost all applications of forecasting, we will need to compare different forecasts. These may arise from our own work by applying different methods, assumptions or starting conditions, but also may be provided by other forecasters, either published or privately communicated. Whatever the origins of the diverse forecasts, we will need to provide a critical comparison, particularly if the forecasts make very different predictions. We agree with Pierce (1980), whose comment on seasonal adjustment applies to most forecasting models:

> It should therefore be borne in mind that seasonal adjustment models are never more than approximations and that theoretically incompatible models can produce results uncomfortably close to each other and uncomfortably far from the 'truth'.

In this chapter, we will mostly be comparing quantitative forecasts. However, the approaches to comparison that we discuss are also applicable to qualitative forecasts.

Volume of data and information

A particular obstacle to comparing forecasts is that we do not deal with individual pieces of information. A forecast is rarely in the form 'we will sell X units of our product next March'. Instead, as we have seen before, we will usually forecast sales for a set of months. Moreover, from even a single method, we will make such forecasts each month as new data become available. Figure 17.1 is a reminder of the situation that can arise.

Beyond the forecasts themselves, we will also have other information:

- Many methods will produce estimates of the reliability of the forecasts, for instance in the form of standard errors.

- Even when standard errors are not available, we can use the stream of errors from the validation test of our methods to provide a guide to the reliability of forecasts.

	Month for which forecast is made				
Month forecast generated					
November, Year 1	Dec. Year1	Jan. Year 2	Feb. Year 2	Mar. Year 2	Apr. Year 2
December, Year 1		Jan. Year 2	Feb. Year 2	Mar. Year 2	Apr. Year 2
January, Year 2			Feb. Year 2	Mar. Year 2	Apr. Year 2
February, Year 2				Mar. Year 2	Apr. Year 2
March, Year 2					Apr. Year 2

Figure 17.1. Illustration of the timing of forecast generation and target periods.

In most real-world situations, we will have extra information about the events and environment that surround the forecast phenomenon. This can and should be recorded as metadata. We admit, however, that few software packages for statistics or forecasting offer convenient ways to handle metadata sensibly, especially as it must often be tied to observational cells, that is, to a particular variable and time-point location in our data array.

Type of model

Besides the volume of data, we also must deal with many modelling and forecasting methods. The information about models and methods is a form of metadata. Each method or model, however, generates one or more time-series, i.e. variables. The name of a method is far from enough to define the results:

- Methods have parameters (e.g. smoothing constants in exponential smoothing methods).

- Similarly, many methods require us to provide starting conditions. Again, exponential smoothing methods need to be 'started' with initial smoothed series values.

- The selection of the forecasting horizon, the sets of data used for estimating test models and validating them, and for final estimation, may all be important to the forecast outcomes.

Our comparisons must therefore ensure that the method used to generate each forecast is fully documented. This requires a firm resolve and almost iron discipline. We cannot say we always succeed ourselves. Nevertheless, we believe that it is important for fair comparisons to provide this level of description *before* we commence the task of comparing the numbers.

Quantitative measures

By now, you should be familiar with the different quantitative measures of fit and also be able to understand the hypothesis tests for significance of parameters or autocorrelations that statistical packages provide. These assume, of course, that the random fluctuations around our model follow a known distributional pattern (usually a Gaussian distribution). We have also seen how to check this assumption by a stem and leaf diagram, histogram or normal probability plot.

The forecaster has, unfortunately, to make many choices of which tests and checks to make and to report, and we counsel that there is no perfect set of choices. Some issues to consider are:

- Whatever measures of fit are selected, they should be computed for all the candidate forecasting methods to be compared, as far as possible.

- To compare fit, R_squared or adjusted R_squared are particularly useful, though not all methods produce them automatically. They are relatively simple to understand and also force us to compare our models to the simple 1-parameter model found by using the mean of the data as a forecast. Topic 15 of the Appendix discusses calculation of R_squared.

- Care must be taken to ensure that the time ranges over which measures are computed are the same for different methods as far as possible. Note that some methods, for example, ARIMA forecasting, may necessarily eliminate certain periods. (See our real-world example below.)

- Methods may have very good behaviour for forecasting, but show poor 'fit' in parts of the modelling period, either for trial or main model estimation or both. The Winters' method shows such behaviour for the Quarterly Traffic Fatalities in Canada data. Therefore, reliance on the numerical measures alone is foolhardy – it pays to draw the graphs!

- Where residuals can reasonably be taken to come from a Gaussian distribution (or in special cases, some other distribution for which statistical support exists), we should examine the statistical significance of model parameters and the autocorrelations of the residuals. We may also be able to obtain prediction intervals for our forecasts.

- For the validation period, note whether errors have essentially zero mean, otherwise our forecasts can be noted to be 'high' or 'low'. A negative mean error for traditional 'Data - Model' residuals implies a 'high' forecast. Another useful viewpoint is the relative numbers of negative to positive errors, which can be seen simply from the signs of the errors.

- Also for the validation period, we feel that it is useful to report maximum and minimum errors. This gives the 'worst case' for either high or low forecasts, unless all the errors are on one side of zero. Converting these to a percentage is also helpful to give readers an understanding of the scale of the deviations.

There are three sets of measures that we will want to report:

1. for the trial modelling period, where we seek candidate methods to test;
2. for the validation period, where tests are carried out;
3. for the full modelling period, where we recheck the methods that we believe are suitable.

It is important to check if trial and main models are similar. We do not want to have wildly different models in the two periods, since this would suggest that the underlying phenomenon was not regular enough to forecast. That is, our assumption of constancy would not be valid. We can check if models are similar by drawing time plots, preferably as multiple graphs on one panel. Alternatively, we can compare the numerical values of the model parameters, although this requires some experience and judgement. The comparison of model parameters is aided if we can rely on standard errors for the parameters, but this requires that such statistics be available *and* that distributional assumptions, usually of Gaussian errors, be supportable.

Graphical comparison

We prefer graphical comparisons of results to numerical summaries. The numbers help us sort out what we 'see' in the graphs. And what we see falls into three categories:

- *Similarities*, that is, common patterns between different forecast methods and (hopefully!) the modelling data;

- *Differences* between the various methods and, where available, the observed data, particularly for certain observations or regions of the time domain, or for particular seasons; and

- *Patterns*, especially those of error distributions or time plots.

The graphs we have used have quite often had several quantities displayed together. The preparation of such multiple plots on a single display can demand quite detailed knowledge of particular software systems. Over many years, this has been a continuing source of frustrations and annoyances. See, for example, Nash (1994). Some tricks we have used from time to time to overcome the deficiencies of either the software or our knowledge of how to control it are:

- Where other software can conveniently prepare the graphs we desire, we can output the plotting data to a file. For example, the Minitab `print` or `write` commands will let us output columns of data. We prepare a file with a column for *x* information and multiple columns with *y* information. This is then read into the graphing program. We find that Stata offers very good two-dimensional statistical graphing capabilities. Moreover, the Stata graphs offer a vector-graphic output (see Chapter 11). A graph editor (Stage) exists for adding special annotations, although we find this very awkward to use.

- Where we like a graph but not its annotation, we can transfer it via file or copy-and-paste into a graphics program such as Corel Draw. There we add titles, legend, and arrows to features on which we wish to comment. Sometimes we even cover aspects of the imported graph that we do not wish displayed with a box or other shape having a white outline and white fill. New material can be added over the white area.

- Sometimes software will prepare a graph without legend. We put the legend information into a caption when we prepare a report with word processing software.

- For work not needing a machine-readable form, we can use hand-written annotations on printed graphs. For example, with student scripts, 'getting the computer to do it' can waste valuable learning time. Our courses are about forecasting, not particular software packages.

- Graphs of single series, carefully plotted to the same scale, can be transferred to acetate sheets for projection. This can be very effective and we can remove some graphs to allow viewers to focus on comparisons of just those graphs that are deemed important.

Reporting honestly

Forecasters do not set out to be dishonest. The pressures of time and limited resources do, however, often funnel our thinking so that we accept conclusions that we would challenge vigorously if we had more time. Therefore, we make the following recommendations.

- Where possible, tables should be complete enough to enable a reader to verify components. If commentary is needed to understand a table, it should be provided. On the other hand, numerical data is often given to a precision that goes beyond what is needed or reasonable. This is likely a consequence of 'cut and paste' from output files, a practice of which we are occasionally guilty. The extra digits bore or misdirect the reader.

- Graphs should be at a scale useful to see the features upon which we comment. If comparisons are made, then use common axis scales. Time periods should be identified so the user can place the features of the graph in the context of the problem at hand.

- The write-up of the forecasting activity should have enough detail to allow the reader to understand how it has been done, but not be overpoweringly long. Achieving a balance is difficult. We had management students who, asked for a report of 12 pages, handed in 61 pages! They received none of the 25% of the mark allocated to 'presentation' – bad management! Skill in matching the report to the requirements is acquired through practice, practice, and yet more practice.

Combining forecasts

Few forecasting efforts rely on a single method. After we have examined several methods, there will likely be two or three that appear to 'work', as judged by the criteria we have discussed. Should we choose one method, or try to combine the results of two or more? There is some empirical evidence (Chatfield, 1996, p. 84) that the combined forecasts 'work' better than their individual components. How might we proceed to build the combined forecasts?

We could simply average the forecasts for a given time period. We prefer to form the average to get the composite forecast for time point t_pred using only forecasts that are all generated with data up to and including time t_last. While we could combine forecasts generated using different supporting data, we feel that it is too easy to become confused about the origins of the forecasts. We want to know how far ahead our forecasting method is acting.

Simple arithmetic averages can, of course, be replaced by weighted averages. For example, we could weight our forecasts by the reciprocal of the RMSE over the estimation period, or over the validation period. In so doing, we must remember to normalize our average by the sum of the weights. The awkward choice, of course, is which measure to use as our weight. Moreover, there may be subjective elements that suggest we use weights that we, or a subject specialist, provide. Expert opinion can be brought to bear in ways we have discussed in Chapter 5. Clearly, any such choices need to be documented, justified, and monitored for their success as events unfold. It is, of course, possible to use the combination approach in a trial, validation, main estimation and forecasting exercise, although we believe this is more work than can be justified.

Diebold (1997) gives a quite detailed set of examples in his forecasting textbook for economics/finance students of an approach that tries to regress the data on the forecasts to be combined. That is, if we have forecasts f_A and f_B for quantity y at different time points, we perform the linear regression

$$y_t = b_0 + b_1 f_A + b_2 f_B$$

The intercept in this 'model' allows for some correction of bias in the forecasts. Diebold's discussion includes consideration of the error structure and other topics that are more mathematical than we wish to include here, while he is less interested than we are in the possibility of special events that compromise the assumptions about the data.

If we do decide to combine forecasts, prediction intervals will be even more awkward to compute. However, since there is ample evidence that such intervals are nearly always too narrow relative to the true errors that are eventually observed, we prefer to use the general magnitude of the validation errors as our guide to likely future success of our forecasts.

The ideas in this section are similar to those in the more detailed discussion by Bunn (1988). We feel more sophisticated approaches, for example Duong (1989), are beyond general usage.

A real example

Consider trial models for the Quarterly Traffic Fatalities in Canada data. Our hope is that these show very similar properties for all models. Figure 17.2 presents the information graphically.

Variable	N	Mean	Median	StDev	SE Mean	Minimum	Maximum	Q1	Q3
tdata	52	991.1	1014.5	231.2	32.1	618.0	1411.0	769.5	1137.2
tmsfa	52	990.4	1004.9	212.3	29.4	623.5	1386.1	795.6	1127.8
tmmreg	52	991.1	1014.5	231.2	32.1	618.0	1411.0	769.5	1137.2
tmes333	52	1035.5	1056.1	273.2	37.9	524.5	1755.4	831.9	1220.1
tmtsd	52	990.5	993.5	205.2	28.5	645.4	1382.0	806.2	1132.7
tm010011	52	991.1	1014.5	231.2	32.1	618.0	1411.0	769.5	1137.2

Both the descriptive statistics and the graphs suggest, at first glance, that the Winters' method has larger errors than the other methods. However, we see that the excursions away from the data are in the early time periods when the smoothing method is adjusting itself to the data.

More important, in our view, is that all the methods give very similar patterns, and that they do, more or less, conform to the data. In fact, it is very difficult to distinguish the methods, even in the residual graph. Note that the residuals (descriptive statistics given below) are roughly centred on zero, and that there is no pattern that obviously commands attention. When we consider the descriptive statistics for the residuals, we find great similarity, although it is again important to discount the results for the Winters' smoothing technique due to the poor early fit.

Variable	N	N*	Mean	Median	StDev	Minimum	Maximum	Q1	Q3
trsfa	52	0	0.7	-4.8	74.4	-141.9	148.7	-59.5	74.9
trmreg	52	0	-0.0	-1.8	79.9	-178.5	144.7	-67.6	75.4
tres333	52	0	-44.3	-7.3	127.8	-369.4	156.7	-86.1	48.8
trtsd	52	0	0.6	-4.2	77.0	-158.0	171.2	-60.1	64.4
tr010011	48	4	-0.79	1.64	61.92	-152.52	114.75	-34.81	46.05

Turning to the validation period, the models produce the following predictions, which we display so we can check them against the graphs. First we look at the validation forecasts:

Row	vperiod	vdata	vmsfa	vmmreg	Winters' vmes333	vmtsd	ARIMA vm010011
1	81	668	611.74	557.42	581.971	634.01	571.15
2	82	834	873.21	856.81	740.890	863.17	727.83
3	83	1021	1107.59	1128.27	969.992	1118.58	996.38
4	84	815	897.43	894.12	810.582	900.96	848.83
5	85	697	600.02	538.57	547.875	622.63	556.11
6	86	643	856.40	837.96	696.839	847.61	710.43
7	87	926	1086.16	1109.42	911.449	1098.32	978.02
8	88	813	879.99	875.26	760.910	884.56	830.09
9	89	630	588.30	519.72	513.779	611.25	537.22
10	90	683	839.59	819.10	652.787	832.04	691.48

Variable	N	Mean	Median	StDev	Minimum	Maximum	Q1	Q3
vdata	10	773.0	755.0	130.9	630.0	1021.0	661.7	857.0
vmsfa	10	834.0	864.8	185.9	588.3	1107.6	608.8	944.6
vmmreg	10	813.7	847.4	217.8	519.7	1128.3	552.7	947.9
vmes333	10	718.7	718.9	151.2	513.8	970.0	573.4	835.8
vmtsd	10	841.3	855.4	180.3	611.2	1118.6	631.2	950.3
vm010011	10	744.8	719.1	166.3	537.2	996.4	567.4	881.1

Note that all the forecasts show similar values, although the Winters' and ARIMA predictions are below the data in mean value, while the rest of the forecasts are above. The data have a smaller standard deviation than any of the predictions, suggesting that the seasonal pattern is diminishing faster than our methods anticipate. Graphs of the validation period predictions shows the same story, but we need a much larger scale graph to be able to disentangle the different methods.

Now let us look at the errors in the validation predictions, in summary and raw numbers.

Variable	N	Mean	Median	StDev	Minimum	Maximum	Q1	Q3
vesfa	10	-61.0	-74.7	101.7	-213.4	97.0	-157.5	45.3
vemreg	10	-40.7	-70.7	127.1	-195.0	158.4	-147.9	110.4
vees333	10	54.3	51.5	59.3	-53.8	149.1	12.0	98.9
vetsd	10	-68.3	-78.8	92.5	-204.6	74.4	-154.9	22.6
ve010011	10	28.2	8.1	74.9	-67.4	140.9	-38.4	99.2

Row	vperiod	vesfa	vemreg	vees333	vetsd	ve010011
1	81	56.257	110.577	86.029	33.99	96.85
2	82	-39.208	-22.808	93.110	-29.17	106.17
3	83	-86.587	-107.269	51.008	-97.58	24.62
4	84	-82.434	-79.115	4.418	-85.96	-33.83
5	85	96.976	158.429	149.125	74.37	140.89
6	86	-213.399	-194.956	-53.839	-204.61	-67.43
7	87	-160.163	-183.418	14.551	-172.32	-52.02
8	88	-66.991	-62.264	52.090	-71.56	-17.09
9	89	41.695	110.280	116.221	18.75	92.78
10	90	-156.590	-136.104	30.213	-149.04	-8.48

The error means for methods SFA, MREG and TSD are negative, implying that the validation forecasts are, on average, 'high'. Those for the Winters' smoothing and the ARIMA model are low, but these methods have the lowest standard errors. To get the error sum of squares, we can actually do the summation or we can form

$$\text{error sum of squares} = n*(\text{mean error})^2 + (n - 1)*(\text{standard deviation})^2$$

where n is the number of elements in the series. Either approach to the error sum of squares calculation gives us the following values:

vesfa	130337
vemreg	161826
vees333	61173.9
vetsd	123693
ve010011	58477.6

These results suggest that the Winters' method (vees333) and ARIMA (ve010011) are our best methods over the validation period. Note how our 'first glance' at the Winters' Trial modelling fit would have discarded this approach, since the smoothing method requires several periods to 'adjust' to the data.

We now look at the main modelling results. We omit the display of the data and descriptive statistics of these data, but look at summary statistics of residuals and graphs (Figure 17.4).

Variable	N	N*	Mean	Median	StDev	Minimum	Maximum	Q1	Q3
mrsfa	62	0	0.03	1.83	77.94	-168.58	167.90	-59.47	60.79
mrmreg	62	0	0.0	9.9	86.6	-219.1	156.7	-64.6	82.4
mres333	62	0	-37.6	-7.1	122.6	-388.7	169.3	-78.9	50.1
mrtsd	62	0	4.8	14.7	87.3	-215.1	162.9	-56.8	77.5
mr010011	58	4	-0.13	0.41	64.04	-168.27	115.59	-36.64	47.94

Trial models vs time

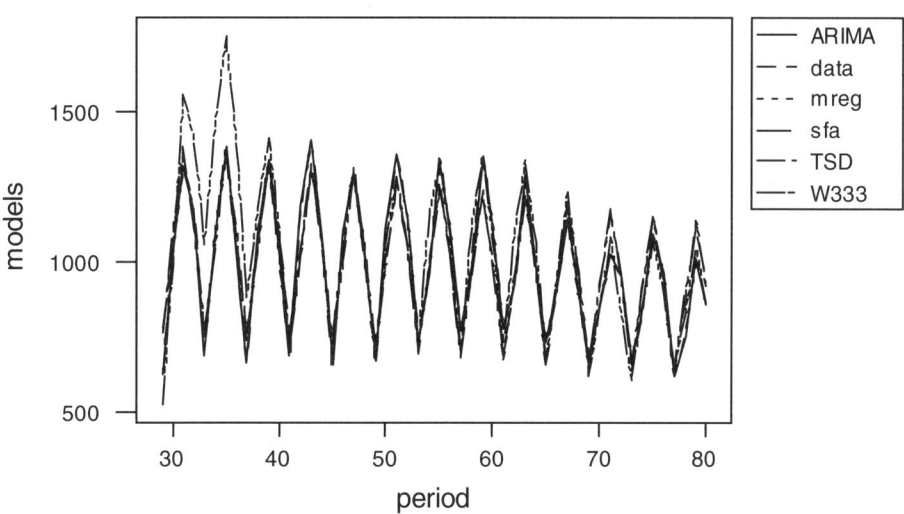

Trial residuals vs time

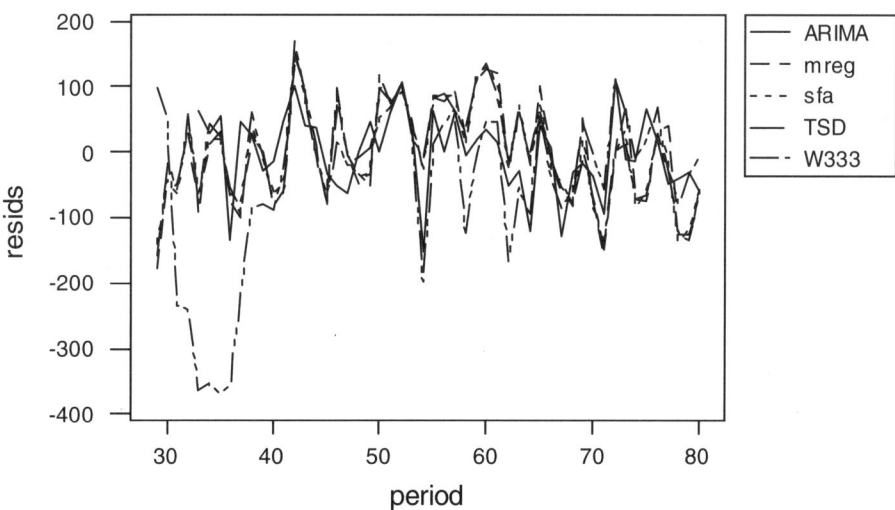

Figure 17.2. Trial models and residuals for the Quarterly Traffic Fatalities in Canada data.

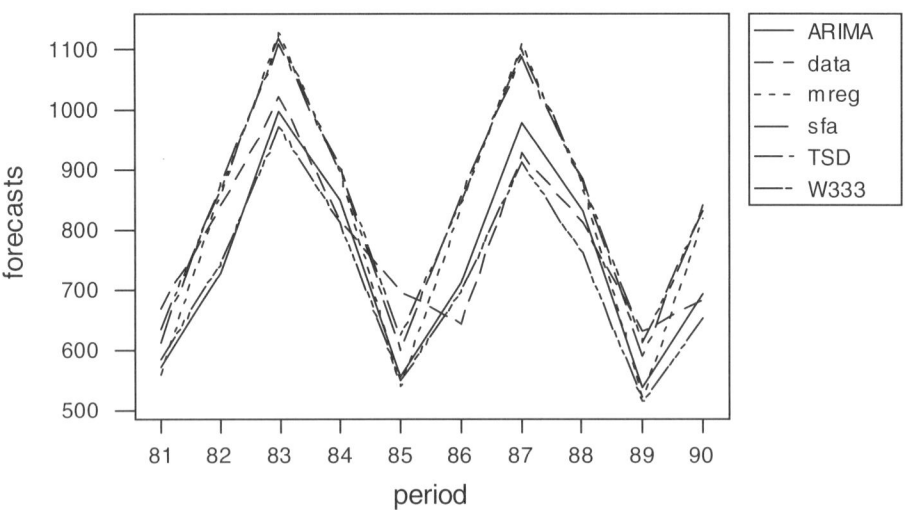

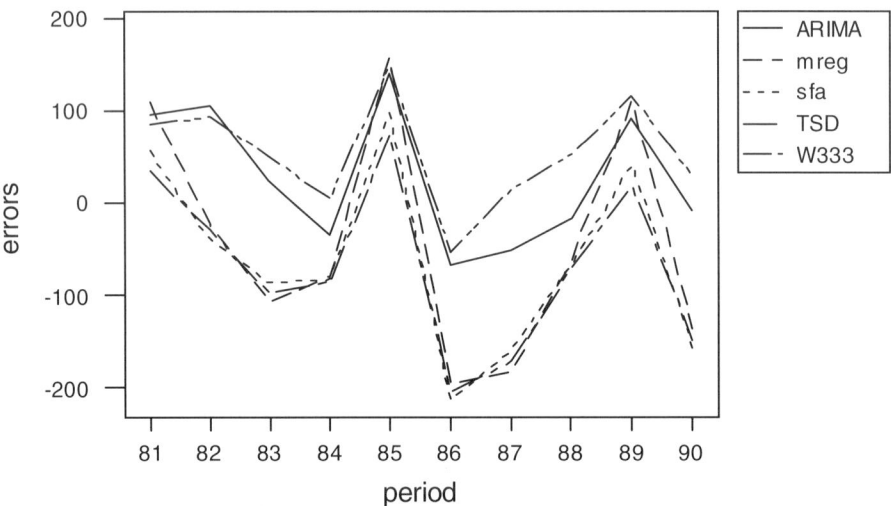

Figure 17.3. Validation period predictions and errors.

Once again the Winters' model (`mres333`) takes some periods to adapt to the data, so that its numerical measures of fit are apparently poor. For the most part, our comments regarding the trial modelling period still hold. While we are looking at the residuals, we should check their distributional properties. The stem and leaf diagrams for the residuals are given as Figure 17.5. Any conclusions for Winters' method will be coloured by the large deviations in early time periods. Indeed, the pattern for the Winters' method shows departures from the mound shape we would like to see. Regression (`mreg`) and time series decomposition (`mrtsd`) are quite acceptable, but the *ad hoc* method (`mrsfa`) and, particularly, ARIMA (`mr010011`) show some excursions from the shape we expect from Gaussian errors. We will want to be a little careful with our use of interpretations of the ARIMA model that are based on distributional assumptions.

We now turn to the forecasts. In tabular form these are:

Row	tperiod	fmsfa	fmmreg	fmes333	fmtsd	fm010011
1	91	1025.73	1053.21	920.837	1064.96	932.69
2	92	832.46	829.01	777.104	808.58	794.46
3	93	567.84	505.47	575.815	502.89	619.93
4	94	773.31	759.04	656.058	753.89	659.31
5	95	1000.81	1032.11	877.025	1043.95	912.27
6	96	812.11	807.91	739.686	787.57	775.05
7	97	553.87	484.37	547.751	481.88	600.83
8	98	754.17	737.94	623.689	732.88	640.31
9	99	975.89	1011.01	833.213	1022.94	893.29
10	100	791.76	786.81	702.268	766.56	756.09

Variable	N	Mean	Median	StDev	Minimum	Maximum	Q1	Q3
fmsfa	10	808.8	801.9	163.3	553.9	1025.7	707.6	982.1
fmmreg	10	800.7	797.4	198.7	484.4	1053.2	679.8	1016.3
fmes333	10	725.3	721.0	127.1	547.8	920.8	611.7	844.2
fmtsd	10	796.6	777.1	204.6	481.9	1065.0	675.4	1028.2
fm010011	10	758.4	765.6	125.4	600.8	932.7	635.2	898.0

The forecasts are presented in graphical form in Figure 17.6. Which of these should we use? Given the experience of the validation period – which we warn may *not* be indicative of the unknown future – we may prefer the forecasts of the Winters' or ARIMA methods. At this point we are forced to consider issues outside those of simply 'fit' of the models to the data.

The attraction of the ARIMA model is that it uses so few parameters. Indeed, we have only three! These are the constant, the non-seasonal AR parameter and the seasonal MA parameter. The stem and leaf plot of the residuals (we are considering the Main modelling period here), as we indicated above, has some 'gaps' from the mound shape that we like. However, if we look at a Normal probability plot, almost all of the points are close to the reference line. (We do not display this graph here, nor would we include it in a report unless Gaussian residuals were a matter of concern. We would, however, report its appearance, as we have done here.) Thus, we have some security in using the standard errors and t-statistics for the parameters, which imply that all three of the model parameters are non-zero. That is, we reject each of the null hypotheses that a parameter is zero.

Accepting that the residuals are Gaussian lets us also use the autocorrelations of the residuals, and, in particular, the Ljung–Box test (Chapter 3). The null hypothesis that the population autocorrelations are zero is accepted for all four choices of lag length presented by

Main models vs time

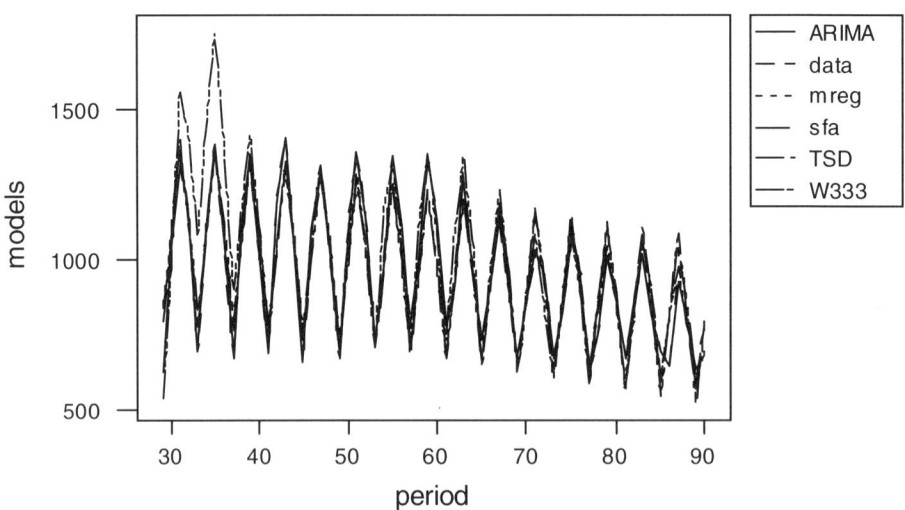

Main residuals vs time

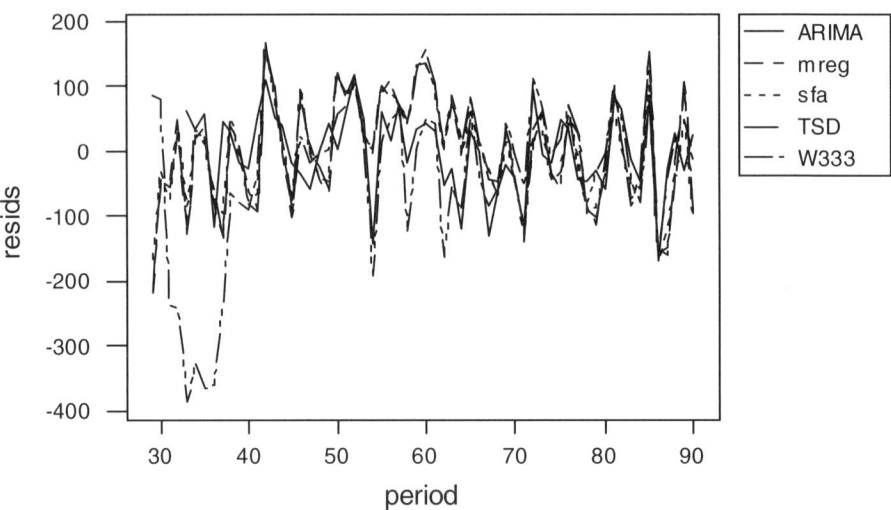

Figure 17.4. Main models and their residuals.

```
mrsfa                       mrmreg                      mres333
Leaf Unit = 10              Leaf Unit = 10              Leaf Unit = 10

    2    -1 66                  1    -2 1                   3    -3 866
    2    -1                     2    -1 5                   4    -3 2
    4    -1 22                  8    -1 433200             4    -2
    5    -1 0                  17    -0 999887665           7    -2 433
    9    -0 9988              29    -0 444444333000       10    -1 966
   15    -0 777766           (14)    0 00111111223334     11    -1 2
   21    -0 554444            19     0 55788889999        22    -0 99877776555
   26    -0 33332              8     1 0001334           (13)    -0 4433321100000
   30    -0 1000              1     1 5                  27     0 00112224444
   (8)    0 00000111                                     16     0 5555666778888
   24     0 2333                                          3     1 01
   20     0 44555             mr010011                    1     1 6
   15     0 67      ›         Leaf Unit = 10
   13     0 8889999          N* =   4
    6     1 00
    4     1 223                  1    -1 6
    1     1                      2    -1 4
    1     1 6                     3    -1 3
                                  6    -1 111
                                  6    -0
mrtsd                             7    -0 6
Leaf Unit = 10                   14    -0 5554444
                                 21    -0 3332222
    1    -2 1                    29    -0 11111000
    3    -1 55                   29     0 01
    7    -1 4321                 27     0 222333333
   17    -0 9998866555          18     0 444445555
   28    -0 44433321100          9     0 66677
  (15)    0 001111222333444      4     0 89
   19     0 677788999            2     1 11
   10     1 0000122
    3     1 556
```

Figure 17.5. Stem and leaf diagrams for the distribution of residuals of the main period models for the Quarterly Traffic Fatalities in Canada data.

Minitab. This is unusually good fortune. Rarely have we seen data that allows us to reduce the residuals to 'noise' in this way. This data set is, after all, derived from real data. Indeed, one of our former students unfortunately became a part of this data in a car accident.

Most people – ourselves included – have difficulty getting a handle on the ARIMA models. They are not intuitive in their structure. All our other models have seasonal factors or seasonal shifts. The regression, time series decomposition and *ad hoc* approaches seem, unfortunately, to over-estimate the seasonal effect in recent time periods. The Winters' seasonal smoothing, true to its intended nature, has adapted to the changes while retaining the trend * season model. There are, however, essentially six parameters in this model – two for the intercept and slope of the trend, and four seasonal indices or factors.

Having narrowed our choice to Winters' method or ARIMA, we could average these forecasts. Alternatively, we could highlight the Winters' and ARIMA traces in the graph above and allow our eye to do the 'averaging' for us.

We have not yet estimated how reliable our forecasts are. Clearly, the standard deviation of the residuals for the main modelling period gives a likely lower bound. This is, in the best case (ARIMA), roughly 65 fatalities per quarter. To put this in context, the average number of

Main forecasts vs time

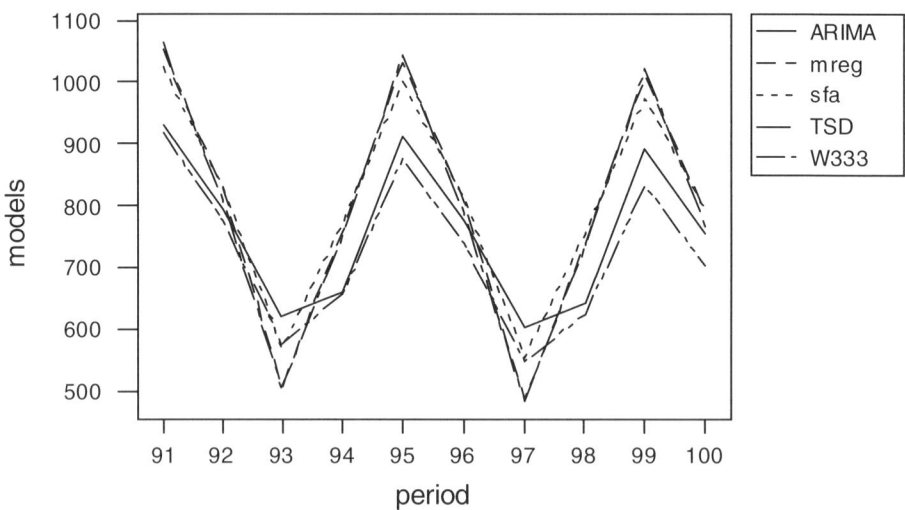

Figure 17.6. Forecasts for the Quarterly Traffic Fatalities in Canada data.

fatalities is about 750 per quarter, so errors of 10% in our forecasts should not surprise us, as the standard deviation in the estimation period is almost certain to be a severe underestimate of the error eventually realized in our forecasts.

ARIMA, under the assumption of Gaussian errors, provides prediction intervals for our forecasts (displayed in Chapter 15). These corroborate what we have just suggested. In fact, the separation between the lower and upper prediction bounds is several hundred fatalities. This may suggest that our efforts to forecast fatalities are in vain. The reality is less pessimistic; we simply must recognize that people do not decide that they are going to kill themselves on the roads this quarter! Events such as the 1997 bus crash in Quebec can seriously perturb single quarter data. That is, there is a natural variation in fatalities which no forecast can predict.

More optimistically, our graphs show that all the methods capture the data reasonably well, especially in its seasonal pattern. The range of the seasonal variation is declining, and the overall rate of fatalities is slowly declining. The downward trend is not necessarily linear, of course, and we have already removed the apparent 1981/82 level shift, but the linear trend model seems to be working reasonably well for the time being.

Review of quantitative forecasting methods

This completes what we believe is a useful collection of quantitative forecasting methods, so we will provide a short review of the techniques and ideas that we have covered.

First, there are methods for forecasting based on *decomposition*, that is, we discover a way to compute the quantity of interest as a composite of some other quantities, and direct our forecasting efforts to the components. Usually, but not always, we will apply such methods to seasonal series. In our discussion, the methods that exploit such ideas are:

- Simple graphical methods based on subset series (the independent seasons) or on the use of traced patterns;

- *Ad hoc* seasonal factor or seasonal shift methods;

- Time series decomposition, in various forms;

- Winters' seasonal exponential smoothing method;

- Dummy variable multiple regression methods.

Secondly, much of our effort to judge these methods has been focused on the *deviations* between the models estimated by the methods and the data they are supposed to fit. These residuals or errors (depending on the time interval of interest) have been used to calculate a variety of measures and to draw graphs showing patterns, both distributional and temporal.

Thirdly, the above methods can all be viewed as using prior data to predict future observations. The errors themselves find a way into models in the ARIMA methods. As we have pointed out, these ARIMA methods can be used for both seasonal and non-seasonal data. They are often extremely parsimonious in the number of parameters needed, but humans do not generally find the models easy to understand.

What methods should you use? We usually find that for seasonal data it is almost always worth trying Winters' method. It is easy to understand and has been proven in many empirical studies to work well. Other practitioners' experience (e.g. Chatfield, 1996) supports this opinion. For non-seasonal data with a linear trend, we would be tempted to use a simple regression or else Holt's double exponential smoothing. If the trend is nonlinear, we would likely want to try methods and models to be described in Chapter 20.

Exercises

E17.1. Compare and combine as appropriate your forecasts for the Dry Cleaning data.

E17.2. (Or class discussion.) Prepare a one-page critique of the overall set of forecasts of the Quarterly Traffic Fatalities in Canada data.

18

Variations on the theme of seasonal adjustment

Origins and motivations

The modelling of data with seasonal and trend components, either in additive or multiplicative combinations, is a well-trodden path for forecasters. Here, we pursue the ideas further to obtain models and/or analyses that help the forecaster – usually us! – cope with real-world forecasting and analytic problems of interest. Trend/season decomposition is taken as the basis of this chapter, but we consider improvements in the approach. Examples of our concerns are:

- The possibility of occasional 'outliers' in the data, for example, due to extreme natural or man-made events, that would distort the forecasting model and hence bias the forecasts;

- Possible missing values in the series;

- Subtle (or not so subtle) interactions of calendar cycles with our model;

- Drift in the trend and/or seasonal pattern;

- End-effects, that is, the recognition that it may be awkward to get estimates for trend or seasonal parameters near the beginning or end of the data.

Unfortunately, much of this work is directed, not to forecasting, but to *seasonal adjustment*. This is the preoccupation of government statisticians in wishing to present data so that month-to-month or quarter-to-quarter figures may be compared on an 'equal' basis. That is, if the unemployment rate has increased, figures should show if it has increased in some 'underlying' way, or if the increase is part of a natural or expected cycle. This practice is common, although we prefer to see the raw data alongside any adjusted information. Indeed, Fischer (1995), in a report to the European Economic Community statistical bureau, Eurostat, recommends against the publication of only seasonally adjusted data, and favours the provision of trend and seasonal components. Some of the justifications for seasonal adjustment are discussed by Dagum *et al.* (1988) with concerns expressed by Abraham and Ledolter (1983, p. 173).

Even though their emphasis is on seasonal adjustment, many workers have made freely available a large part of their software, and forecasters may profit from this. Some of this software, given the technical complexity of the subject, is difficult to set up and use. Recognizing this, workers such as those reviewed by Fischer (1995) have also developed interfaces to make it easier to use. However, it is meant for the seasonal adjustment of large amounts of data with a minimum of intervention, rather than the forecasting of a few series. As a leader in this work, Eurostat is propelled by the need to readjust all its data series to include the newly admitted members of the European Community, a move from 'EUR12' to 'EUR15' data series.

These software tools are very interesting, but are a work-in-progress, so we will address just some of the issues we feel are of ongoing importance.

Goals of decomposition modelling

The most important facet of decomposition methods in this book is how well they forecast. We compute trial and full period estimation residuals, calculate MSE and R_squared, and examine the validation period forecasts to see how well we have forecasted the known results. There are, however, situations where we may want simply to use the decomposition to understand or to approximate the data. In such a case, we would not be concerned with the trial or validation exercise, and only the main modelling effort would be worthwhile. Furthermore, we might prefer to adjust the decomposition to minimize the largest absolute error, so that we could state the worst case behaviour of our approximation model.

The underlying model – additive or multiplicative?

From our initial presentation of Pegels' classification, we have considered that the changes in the trend may be additive or multiplicative, and further that the combination of the trend with the season may be made either a sum or a product. Together with the possibilities of no trend and no season, we have the nine cells of the Pegels' array, although in our experience the 'no trend' situation is generally straightforward to detect if one has enough data. Most real-world situations have external information (or metadata) that inform our judgement about seasonality.

Exceptions – air passenger arrivals in Israel (Exercise E9.5) is a good example – arise if the seasonality follows a different calendar or time scale than the one we use for measurement. One must be careful if the data represent a phenomenon that has no reason to follow our imposed time scheme, which is a human and cultural invention.

Whether our trend model is additive or multiplicative, and whether its combination with seasonality is a sum or a product, are, we feel, among the most difficult decisions for the forecaster. There is always some natural variation or 'error' about whatever model truly underlies the data at hand, and rarely do we have long enough data sets under a believably constant process to see the ultimate dominance of multiplicative over additive change. Moreover, we are never given data for the trend and season separately, so must disentangle the two concerns. Much of the effort in the methods mentioned below is directed to addressing this issue, but they use algorithmic processes to decide which form is appropriate, since one goal is to be able to handle a large number of data series without human intervention. Most forecasters, we feel, will want to be able to decide what model is appropriate using more than just the numerical data. Moreover, we must be prepared for the inevitable changes in process or economic patterns by having a 'system' that is sufficiently adaptable to the ongoing changes in the data.

One approach often used by institutional statisticians is to apply an additive model to the log transform of a data series, thereby effectively applying a multiplicative model to the original series. This makes a lot of sense when we are carrying out our preliminary analysis, as it is easier to find straight-line patterns than to recognize the appropriate shape for an exponential curve. Moreover, estimation of additive models is usually done with regression methods that carry with them measures of reliability (i.e. standard errors), unlike the classical ratio to moving averages (multiplicative) TSD. When fitting models, however, we like to make sure we are minimizing or maximizing exactly the function we intend, and log transforms can get in the way and lead to mistakes. These mistakes are often in the description of the estimation process and may hinder efforts to check our work. For example, if we simply say that we estimated a model by minimizing the sum of squared residuals, we should also state whether these are residuals from the original or log-transformed data.

A related issue affecting combinations of time series such as aggregates or indices is whether to seasonally adjust the combined series or its components. The results, as Dagum *et al.* (1988) attest, are different, yet there may be no philosophical reason to choose between different approaches. This may be a good reason to publish the raw data only!

Massaging the data and metadata

An important improvement in classical time series decomposition is founded on the known, if tedious and annoying, existence of calendar effects in business and administrative data. February is a short month, with 28 days, or sometimes 29. We know the leap-years, but often it is a nuisance to adjust our data. For businesses that are focused on weekends – church collections are an example – the number of weekends (or Sundays or other Sabbath days) in a given calendar month will vary from year to year.

Both the X-11/X-12 family of software and the TRAMO/SEATS packages, and their unification under Demetra include tools for handling trading day and holiday effects, particularly Easter (see http://forum.europa.eu.int/Public/irc/dsis/eurosam/info/data/demetra.htm). Our own interest in this problem is long-standing (Nash and Nash, 1986). As far as we can see from our own attempts to use these software packages, the developers have, as may be expected, taken a Eurocentric view. This is likely a reflection of the differences between the smaller, more unitary nations of the European Community, compared with the federal structure in Canada, the United

States, Australia, and other countries. For example, the holidays for Canada do not include Queen Victoria's Birthday (nominally May 24, but always celebrated on a Monday), nor important regional holidays such as the August 1 Civic holiday celebrated in most, but not all provinces, nor the Quebec festivities of St. Jean Baptiste on June 24. We might expect to see some important economic effects in Australia from a holiday (in Victoria State) such as Melbourne Cup Day. There should also be some interesting interactions between the Southern Hemisphere seasons and the common Christmas and Easter holidays.

More general metadata is not, as far as we know, well handled yet in any commonly available forecasting package. We definitely want to be able to deal with serious events that influence the phenomena we wish to forecast. Some of these are:

- Severe weather, such as tornadoes, hurricanes, ice storms, floods and droughts, and possibly some partially predictable cycles such as that underpinning the El Niño phenomenon;

- Political events, such as national elections, which are either predictable, such as the American Congressional and Presidential elections, or episodic, such as the parliamentary general elections in the UK, Canada, and similar systems;

- Major sports events, such as the Olympics (largely predictable timing), the World Cup for soccer (football, except in North America), or similar events that will perturb traditional patterns of activity;

- Other events that preoccupy a significant proportion of the community. We can think of the concerns with the Y2K computer 'bug' and its effects on demand for programming help, emergency supplies, and other products or services. In 1993, we were in Southern California when the verdict in the case of police officers charged with beating Rodney King was anticipated, and people were staying home to keep out of harm's way should there be a violent outburst in reaction to one outcome or another.

- Changes in data definitions, methods of collection, or similar systemic adjustments. The level shift that we observe in the Quarterly Traffic Fatalities in Canada data is an example, although we do not have documentation of the exact nature of the change. On the other hand, Dagum et al. (1988) identify several problems for seasonal adjustment in the early 1980s, namely, several years where Easter is 'early' and a change in Canadian retail opening hours that expanded trading hours in evenings and on weekends.

We believe that our examples above provide a range of situations that forecasters should include as metadata in preparing forecasts. Some are clearly predictable. Others, such as long-term weather patterns may have partial predictability, although timing of their impact may be difficult to forecast.

Missing data in seasonal series

When observations in time series are missing, most of the methods we have discussed in this book cannot be applied. What should the practical forecaster do? Unfortunately, a lot of the wisdom on the subject is in unpublished or internal technical reports. (See, for example, references in the TRAMO/SEATS manual, Gomez and Maravall, 1997.) A recent published discussion, although applied outside business, is Delicado and Justel (1999).

As a practical matter, we advise imputing values for single missing observations, then using the standard tools for forecasting with attention to detail in evaluating the results. The imputation can be done relatively simply by interpolating in the series, or graph, of observations for the appropriate season. The tricky part comes in the need to compute all the measures of fit *omitting* the time points for which data have been imputed. It is also a good idea to flag such points on all graphs. This is extra work, and we have found our own students very reluctant to make the effort, although failure to do so truly destroys the credibility of their forecasting efforts. We must not make decisions based on data that are really our own inventions.

It is relatively easy to modify a series of residuals to leave out the points where data were imputed. We simply create a *mask* series that has a 1 for every real data point and a 0 wherever we have imputed data. Call this series $M(i)$, $i=1, 2, ..., n$. If our residuals are in series $e(i)$, then the series $e'(i) = e(i) * M(i)$ will contain only the 'real' values. Moreover, there are

$$n' = n - \text{sum}(M(i))$$

such values. The e' series is, unfortunately, unsuitable for computing correlograms, that is, ACF and PACF. If one really must have such a measure, then we could collapse the e' series (i.e. shift the data down whenever there is a 'missing' data point) and compute the correlograms from the collapsed series, but there are obvious negative issues in so doing.

Choosing the trend and seasonal filters

Much of the discussion relating to governmental seasonal adjustment has related to the choice of the particular ways in which the unobserved trend and seasonal components of seasonal data are obtained by filtering the data. In this section, we present a couple of the many ideas that have been suggested in this domain. We make no claim that our discussion is comprehensive.

Moving medians and SABL – Seasonal Adjustment Bell Labs

Developed by W. S. Cleveland and others at Bell Laboratories (Levenbach and Cleary, 1981, Chapter 19; Cleveland *et al.* 1982) this approach uses moving medians rather than moving means to accomplish the smoothing in a time series decomposition. In practice, a fairly comprehensive study (Fischer, 1995) showed the trend/season models estimated by SABL to be rather unstable as new data are added over time. This is undesirable from the perspective of government statisticians, since the evolution of a particular time series is then seemingly erratic.

As well as moving medians, moving averages may be replaced by moving trimmed means or other measures of location (central tendency) which are robust against outliers. We believe that practitioners of forecasting could sensibly use such ideas in their own procedures for time series decomposition where outliers are likely in data. For example, if we were wanting to use decomposition to forecast demand for a seasonal product or service such as an ice-cream vending operation, we should anticipate that there will be days when the weather affects our sales up or down greatly. It then makes sense to use outlier-resistant smoothing methods. See Hoaglin *et al.* (1983).

Weighted moving averages

Even using traditional ratio to moving averages methods, there are plenty of choices. Moving averages (Chapter 13), need not be simple sums of the data, but may be weighted; that is,

$$\text{WMA}(n, t, y) = (\sum_{i=1}^{n} w_i\, y_{t+1-i})\,/\,n \qquad \text{where } \sum_{i=1}^{n} w_i = 1$$

that is, the weights are normalized. One way to do this (Chapter 13), is to compute moving averages of moving averages. Thus, applying an M period moving average to a K period one gives

$$\text{WMA\_special}(y_t) = \text{MA}(M, (\text{MA}(K, y_t))).$$

Such compound moving averages were very popular before the widespread availability of computers, because they could be evaluated on adding machines (the weights did not have to be entered). Now it makes more sense to use weighted forms directly.

Some other considerations

Calendar adjustments

Many of the special 'bells and whistles' in packages such as X-11 and X-12 are a result of attention to particular problems; for example, holiday timing on the calendar, trading days, and other matters. However, if we recall the tourist arrivals in Israel (Exercise E9.4) we see that good forecasts rely on adjustment of the time axis to the Jewish calendar. Anyone working with data that may be related to a non-Gregorian system of time measurement should think carefully about changing or transforming the data to the appropriate calendar system.

Despite being wholly predictable, calendars are not easy to merge into forecasting. Many workers have attempted to incorporate calendar knowledge into forecasting and related analyses. Cleveland and Devlin (1980, 1982) consider detection and modelling of calendar effects. Various authors contributed to an interesting study of possible seasonal patterns in homicides in Canada (edited by Gentleman and Whitmore, 1985). Unfortunately, at the time of writing, we have not been able to locate the data for this case study, which would provide an interesting homework problem.

Sometimes we do not know the length of the periodicity of some 'seasonal' phenomenon. Determining the frequency is of interest if we wish to forecast what is going on. This is easy if we observe continuously. Unfortunately, most measurements are made periodically. There may be interference between the frequency of measurement and the frequency of the underlying cycles. If we measure the height of water on a dock each day at the same time we may see cycles, but we are very likely to miss the tidal pattern overall. Burr (1975) considers this problem, but it seems his promise that 'computational aspects will be discussed in a future paper' was never realized.

Variation in the model components

The classical time-series decomposition (TSD) model of Chapter 14 has a fixed (linear) trend and fixed seasonal factors. On the other hand, the Winters' method allowed these components to vary. Even if we use fixed components, we may want to see if this approach is reasonable. Cleveland is well known for his work on statistical graphics (see the references in the Bibliography under his name), and some of the choices made in SABL are well worth noting, even though a consensus appears to be emerging that the overall method is less successful than modified versions of more-traditional approaches. In particular, the 'component by season' plot is extremely helpful in visualizing the changing pattern(s) of seasonality. We first saw this graph when reviewing Forecast Plus in the 1980s (Nash and Walker-Smith, 1989b). Unfortunately, although the concept is simple, drawing such graphs is not a trivial task.

Note that the component by season plot includes a forecast of the seasonal factor for the next main cycle (year). From Figure 18.1, we see that the first and fourth quarters have increasing fatalities relative to the 'average', but that the increase is much more marked in the first quarter. Similarly, the second and third quarters show a declining pattern, again relative to the 'average'.

Seasonal adjustment for small to medium organizations

For most businesses and government offices, particularly municipal or regional administrations, the heavy machinery of the previous section is not, in our opinion, a wise choice. There is a danger that users forget to think about the situation that is being forecast. The flexibility of the methods may hide the effect of changes in the underlying phenomenon or situation of interest, such as new legislation or re-drawing of boundaries. Users should stay close to their data and its origins. Therefore, we recommend, unless there are clear indications to the contrary, that most users should employ fairly simple methods such as classical time-series decomposition or Winters' seasonal smoothing method. These are easy to use and understand. If we apply both to a given data series and obtain similar results, we gain confidence in the reliability of predictions, while differences suggest we look carefully for possible changes in the trend or season pattern.

For organizations with large amounts of data and a relatively stable underlying situation, such as a regional power utility in an area with limited new construction (or where such developments can be separated out of the data), we can and should consider use of the more complex approaches to seasonal modelling. Moreover, in the case of a power utility, we would want to incorporate adjustments for weather, since extremes of heat and cold will increase demand. In such a case, we would want to model the temperature-adjusted demand, with the seasonal modelling program further adjusting the data for calendar and trading-day effects.

Exercises

E18.1. Develop a table of the number of Sundays in July in the period 1960–1999. (Hint: Many software packages for time planning offer a monthly calendar that can be adjusted. However some, as in the Microsoft Windows Date/Time facility, will not allow you to go back in time far enough.) Those who are interested in Management Information Systems may wish to think of how to program the task suggested here as well as similar ones useful in forecasting, such as the number of weekend plus public holiday days in a given date period.

E18.2. (Class discussion) Put yourself in the position of the manager of a small rural (or urban) municipality, a public utility, or similar institution. Define your scenario, then develop a strategy for forecasting demands for the services you must provide.

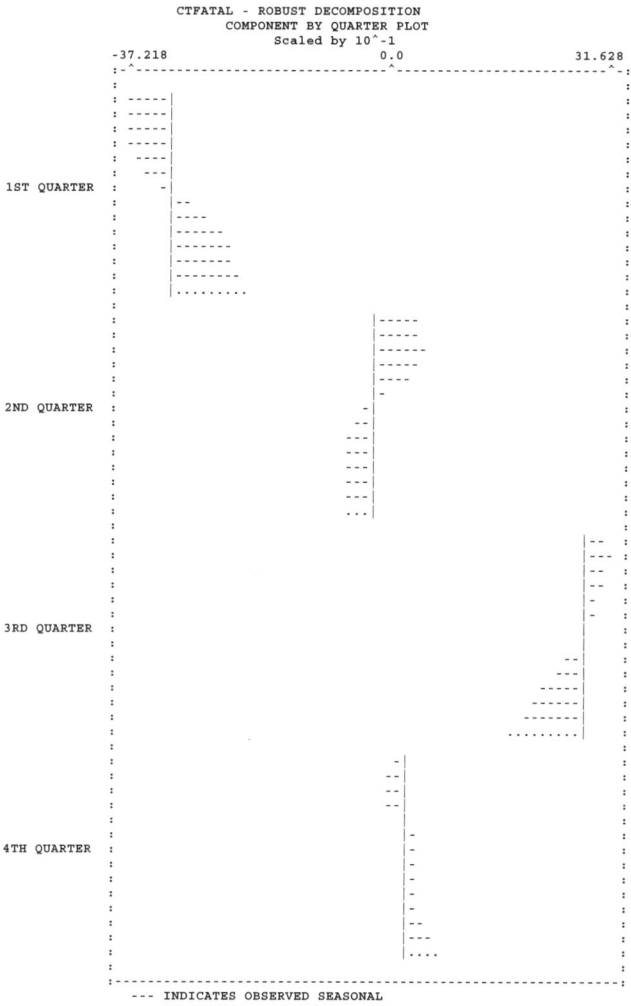

Figure 18.1. Component by season plot from SABL output of Forecast Plus. Quarterly Traffic Fatalities in Canada.

19

Mixed and extended models

Overview

There are always tools that do not quite fit neatly in the toolbox. This chapter is about the forecasting tools that are worth knowing about, but cannot be easily included in other chapters. Some of these are full-blown approaches to forecasting, while others are 'tricks of the trade'.

We deal with the latter first. Primarily we will be interested in approaches to dealing with 'wrinkles' that arise in standard methods; that is, in adapting standard methods to non-standard situations. Almost all applications of the ideas will be *ad hoc* and suitable for the particular situation only, but it is important to recognize that forecasters are continually making adjustments and improvements to their models and results.

The rest of the chapter is directed to describing some techniques that often arise in the forecasting literature. We do not intend to provide a working knowledge of these techniques, as we believe that they are too complicated for everyday use by managers. Nevertheless, managers may be consumers of the results of forecasting efforts by others. Our intent is to present a brief outline of the approaches and their main uses.

Sequential modelling – modelling residuals

Throughout the book so far, we have been concerned to discover ways to explain and predict the patterns in data. In the most obvious, but fortunately common cases, we use a trend and a seasonal pattern to do this. In Chapter 10, we introduced some 'simple' *ad hoc* methods that found an approximate trend or seasonal pattern, then removed this from the data in order to calculate the season or trend; that is, the other pattern component.

Such an approach, in its second step, is essentially modelling residuals. That is, we found an approximate model and used it to compute residuals. These residuals became the data for the next step of the modelling process. We call this *sequential modelling* (the term is our own, and not necessarily in general use).

Students quite like this approach, but may be surprised that the final model found by putting the sequentially estimated parts together is *not* the same as a model where all the parts are estimated simultaneously. The route we take is important, although we hope that the resulting models are not too different in their behaviour, even if the particular parameters are not the same. You can see this quite clearly by the different seasonal factor or seasonal shift models presented for the Quarterly Traffic Fatalities in Canada data in Chapter 9. The models are different, although they all do quite well in fitting the data.

We recommend estimating models that are to be used for forecasting from their full specification. That is, the sequential modelling we are discussing here is *not* used to find trial or main models used to generate validation or final forecasts. Instead, we use the sequential modelling as a way to discover the structure of the situation at hand.

Example: predicting population growth of a city

Let us consider the task of forecasting the population of a city by a municipal planner who wishes to make provision for roads, accommodation etc. We will use data for Calgary. The data (see Table 19.1) used here was provided by two of our students, Mojgan Alapour and Cheng Cheng Xi, but the calculations are our own. A very simple regression of population against year (1976–1997) gives the results.

```
The regression equation is
pop1 = -26807892 + 13819 yr2

Predictor        Coef        StDev         T         P
Constant     -26807892      1109405    -24.16     0.000
yr2             13818.6        558.5     24.74     0.000

S = 16619       R-Sq = 96.8%      R-Sq(adj) = 96.7%
```

Clearly, we have a very good fit of the data with this simple model. Note that we did not even bother to use 'year - 1975' as our independent variable, although this is usually a good idea to avoid numerical rounding errors with some software.

Can we improve on this model? We would like to do so, because the standard error (S) above is still nearly 17,000. If our population follows a Gaussian error model, we can expect to be 'wrong' in our forecasts by this amount – either plus or minus – about 1/3 of the time, and 17,000 people is quite a few to accommodate. By comparison, the Telus Convention Centre web page (www.calgary-convention.com/hotels.htm) dated 1 November 1999 states:

> Accommodating - YES! There are presently 3,700 hotel bedrooms in Calgary's downtown core and another 5,000 in surrounding areas; citywide room count is expected to reach 10,000 by the year 2000.

In an attempt to predict the population, we attempt to find simple, and hopefully predictable, variables that will provide explanations of the remaining variation in the population. We now need to learn more of the dynamics of the city of Calgary as a community. That is, what type of city is it, does it have a higher or lower proportion of young (and growing) families or is it a retirement community, why do people move to or away from the city? We shall not carry out the full analysis here. Moreover, we have not found a single term sufficient for our students to ever do more than a superficial analysis, given their other obligations. The choice made by Alapour and Xi was to use building permits as a measure of economic activity. They then recoded this information into a simply 'boom', 'normal' and 'bust' indicator, that is, 1, 0 or -1, using the histogram of the building permits. They decided that if less than 11,500 permits were issued in a year, it would indicate low or 'bust' economic conditions, while more than 19,000 permits would indicate a 'boom'. Other times would be 'normal'. There is plenty of opportunity for discussion and argument. Indeed, without serious criticism and defence, we do not feel an approach like this should be used. Moreover, we believe that it should be possible to get similar boom/bust measures from several indicators, because otherwise we are not finding a proxy for the subjective 'pull' that a city may exert on those who may migrate there.

Table 19.1. Population of Calgary and some ancillary data

year	pop	buildprm	boombust	year	pop	buildprm	boombust
1976	470,043	13,755	0	1987	647,285	14,038	0
1977	487,569	14,806	0	1988	657,118	8,786	-1
1978	505,637	16,693	0	1989	671,138	11,213	-1
1979	530,816	18,288	0	1990	692,885	10,375	-1
1980	560,618	19,584	1	1991	708,593	9,138	-1
1981	591,857	21,396	1	1992	717,133	10,946	-1
1982	623,133	14,375	0	1993	727,719	10,209	-1
1983	620,692	14,269	0	1994	738,184	9,633	-1
1984	619,814	12,080	0	1995	749,073	8,812	-1
1985	625,143	12,670	0	1996	767,059	11,245	-1
1986	640,645	11,745	0	1997	790,498	14,630	0

Using the 'boom-bust' variable we carry out a second regression model, finding:

```
The regression equation is
pop1 = -29051000 + 14950 yr2 + 16296 boombust

Predictor        Coef        StDev         T         P
Constant     -29051000     1404998     -20.68     0.000
yr2            14950.4       708.1      21.11     0.000
boombust         16296      7114        2.29     0.034

S = 15093       R-Sq = 97.5%      R-Sq(adj) = 97.3%
```

We have improved a little on our fit as measured by R_squared, but still have a standard error of 15,000. The boombust variable is, however, significant (although part of the output we have not included suggests that the 1997 observation has high influence). We would therefore not use this model. Suppose, however, we consider how migrants make the decision to move to Calgary. It is a city with demands for both specialized and general labour. We may presume that those with specialization move to a pre-arranged job, while at least a portion of the general workers move speculatively. (Both authors have direct family experience of the latter!) In either case, there will be a 'decision making' period for employers and potential employees, and this will involve perceptions of the economic conditions prevailing. Thus, we may expect some lag or delay in the influence of 'boom' or 'bust' conditions. We therefore try 1 and 2 year lags on the `boombust` variable and find the following results.

```
The regression equation is          The regression equation is
pop1 = -30375366 + 15619 yr2         pop1 = -28399493 + 14624 yr2
         + 26017 boomlag1                      + 19069 boomlag2

S = 11658                            S = 11916
R-Sq = 98.3%      R-Sq(adj) = 98.1%  R-Sq = 97.9%      R-Sq(adj) = 97.7%
```

We now have a much better fit (although we still cannot fit the 'error' in available hotel rooms!). Note that the coefficient of the `boombust` variable is bigger than this error. In other words, the variable appears to 'explain' approximately 20,000 more or less people. The lag of one year seems to be best, but two years is not far behind. An 18 month lag might be best, but our data series make this awkward to use. Sociologists may have some models of human decision-making about movement that could be helpful in deciding which lag to use.

We have not considered other indicators on the boom/bust, but should do so since we need a forecast of this indicator. The building permit approach provided a quantitative measure that was consistent across time and was, importantly, easily available. What we really need are measures that are similar to those used by people deciding to move or not to move to Calgary. These are likely to be media and related reports of economic and related activity. These are commonly published in the business sections of newspapers and in the business press, as well as in television 'profiles'. However, from the point of view of the forecaster, they require a lot of effort to gather and organize to provide a suitable time series, particularly retrospectively. Nevertheless, for the purposes of providing a '1, 0, -1' figure for forecasting one to two years in the future (or a little further if we have a model based on a lagged version of this indicator), we feel that the generation of such information is relatively straightforward.

For the practitioner who is charged with providing such forecasts on a regular basis, the gathering and indexing of the information needed for modelling and forecasting is an ongoing task. Moreover, familiarity with the particular situations that influence the population of a city will aid in refining the models and tools.

Econometric models and their uses

An old joke says that an economist is someone who guesses about the future wealth of a country or its citizens, while an econometrician is someone who uses a computer to do the same job. As with many jokes, it holds a germ of truth. This is that econometrics attempts to measure and model economic activity using theories and principles that have been established or postulated to apply. That is, we take so-called 'laws' of economics and use them as the basis for mathematical models. These models may then be used for forecasting. This, in capsule form, is why we will be interested in econometrics. However, for practical reasons detailed in this section, it is our opinion that these methods are of limited value to most people wanting to forecast, or rather, they do not offer sufficient benefit compared to their cost of operation.

Identification

Suppose we have theory or accumulated experience to suggest that at a given time point t there are at least n functional relationships that we can describe between n *endogenous* variables y_t and some number of *exogenous* or lagged endogenous variables that we collect as the vector x_t. Endogenous variables will be determined by our model(s) in simulation and forecasting; exogenous variables will come from sources outside the model(s). The relationships between the variables we will also collect as a vector of functions

$$G(y_t, x_t) = u_t$$

where u_t is the collected set of stochastic disturbances, 1 for each of the relationships $G()$. Usually the relationships are linear, but log-linear forms are also common; nonlinear relationships are generally not used as they drop us into difficult statistical theories. The *identification* problem comes about when we try to get the endogenous variables on their own at the left-hand side of the equation, as in the forecasting form of ARIMA models, namely

$$y_t = Q(x_t, y_t) + T(u_t)$$

where $Q()$ is a vector of functions such that the ith function is not dependent on the ith element of y_t and $T()$ is an appropriate transformation of the stochastic elements.

In theory, it is possible that a number of equations of this form will all work just as well, meaning a single model cannot be identified. If the models are expressed in terms of linear relationships, then there are a number of 'rules' that place restrictions on the structural parameters and allow the model to be identified. Clearly this sort of modelling will not be a 'quick and dirty' approach to forecasting.

Closer to what has been discussed in this book, an important choice is that of the exogenous and endogenous variables that are appropriate for modelling our own particular problem. A second, and equally important, choice will be that of the form of the relationships. These choices will hopefully lead to identifiable systems of equations (Kmenta, 1986, p. 661; Johnston, 1972, Chapter 12). Our limited experience in this field suggests that a good understanding of the real-world situation generally leads to model equations that are 'not too bad', while including every variable that might conceivably be important simply bogs the process down under too much choice.

Estimation

Having chosen an identifiable model, we now need to estimate the particular parameters such as regression coefficients that make it fit available data. Ordinary Least Squares (OLS) estimates, carried out one equation at a time, are still often used. However, this will tend to ignore the interactions between the different equations of the model. A variety of techniques have been devised to overcome this difficulty, as well as to try to take into account the variance–covariance structure and autocorrelation structure in the system (Kmenta, 1986, Chaper 13; Intrilligator, 1978, Chapter 11). A different perspective is provided by Hendry (1993).

Simulation and forecasting

Once a model has been estimated, we can supply values for the exogenous data and then solve for the endogenous values. This is generally called simulation. The computation of the y_t given a set of x_t (the u_t are generally taken to have 0 expectation) can be quite tricky and involves the solution of sets of simultaneous nonlinear equations if the model is anything but purely linear. Econometricians are often heard complaining that the model 'did not converge', implying that the solution to the equations was not obtained.

In the jargon of econometrics, forecasting is often taken to refer to the computation of a single element of y_t. Thus, forecasts may require a set of 'forecasts' for other endogenous variables along with our exogenous inputs.

Revision

New data are always becoming available. Thus, a large portion of an econometrician's effort is devoted to revising and updating the model and checking that it still 'works'.

When are econometric models useful?

For the practical forecaster who wants to obtain useful forecasts as an adjunct to other work and interests, econometric models are not likely to be a tool of choice. Typically, they take a large amount of human and computational effort to build, test, revise and use, along with prodigious volumes of data, much of which may have to be purchased. One of us once had the responsibility (but not the authority!) for the computational and data budget of a large econometric modelling activity. This effort involved about 20 personnel, most of them holding either a doctorate or a master's degree, and used nearly $1 million in computer time in the late 1970s. Fortunately, computer resources have become very much cheaper. Salaries, however, have become more costly.

The forecasts of individual variables from econometric models generally are not as accurate as those from simpler, univariate techniques such as those discussed earlier in this book (Makridakis *et al.*, 1998, p. 525). This should not surprise us. Econometric models are trying to provide a consistent picture of a whole economy or sector of an economy, so changes in one variable affect many others. We have a large, coordinated set of movements through time in a (hopefully) consistent and reasonable manner. This is where econometric models shine: we can test the effect of policy alternatives such as changes in interest rates set by central banks, taxation rates of different governments, changes in transportation or energy costs, or flows of migrant workers.

Few forecasters will have the luxury of the resources to build econometric models of the scale needed to carry out policy analysis. Nevertheless, one may hope to benefit from the work of others. In particular, the outputs of econometric models may provide useful forecasts of variables that are *inputs* to our own models; for example, in the population forecasting of Calgary discussed above. Thus it is useful to be able to read reports of econometric models, evaluate their relevance and utility to our own work, and to be able to extract information from such reports.

Sources of advice

Where can we find reports and commentary on econometric models? The major players are:

*   government departments whose mandate involves structural analysis, for example, finance departments and central banks, as well as agencies responsible for particular economic sectors, such as human resources (labour), agriculture, transportation and industry;

*   large financial institutions, mainly banks;

*   research institutes and think-tanks, such as The Conference Board (Canada and USA);

*   consulting firms specializing in economic analysis and advice.

Dynamic regression models

In Chapter 12, we introduced regression models for forecasting. In Chapters 15 and 16 we presented ARIMA models and their uses. It is fairly obvious that the ideas can be combined, and this idea has been variously proposed from different perspectives. In fact, it goes by several names. *Dynamic regression* seems to be the currently popular name, but the models have also been called *transfer function models* or (our preference, since it is descriptive of what is going on) *MARMA*, or *Multivariate AutoRegressive Moving Average* models. The 'I' of ARIMA is dropped. Chatfield (1996) also refers to *VAR* or *Vector AutoRegressive* models as well as *VARMA* for *Vector ARMA* models.

Clearly, if we wish to include previous prediction or modelling errors in our process for forecasting along with values of several variables, possibly from several time lags, then this is the type of model we will want. The trouble is finding suitable tools with which to estimate the models and then forecasting with them. Like ARIMA models, dynamic regression models may need some work to transform them to a form suitable for forecasting. Worse, we may have difficulty deciding which variables, errors and lags to include, because the statistical properties of our estimates may require very sophisticated mathematical analysis to unravel.

From the perspective of the practical forecaster, the major difficulty with dynamic regression models is the forecasting of the future 'disturbances' or errors. If these are 'small', we can approximate them to be zero. Indeed, we generally assume that the disturbances or errors have mean zero, and that they are random within some distributional shape, so that the expected value is zero. However, in many cases, as with a simple random walk model, which may be written

$$y_t = y_{t-1} + e_t$$

it is the error or disturbance that actually gives the 'interesting' shape to the outcome. Note that this model may be categorized as an ARIMA 0 1 0 model.

For the readers we intend as our main audience, that is, students or managers in business or administration, we would recommend building models that incorporate multiple variables and previous errors in a sequential modelling manner. That is, we can take the residuals from a regression modelling exercise and attempt to model these residuals with an ARIMA model, hopefully a very simple one. The alternative of trying regression on the residuals of an ARIMA model may also make apparent sense in some circumstances, but we feel that ARIMA sometimes overfits data so that the causal effects of the regression are largely masked by this overfitting.

Clearly, this approach is sub-optimal. We really want to bring in all the modelling components at once. One could think of minimizing the sum of squared residuals with a brute-force minimization approach (see Chapter 20). This may give a good fit, but rarely will we have good diagnostic statistics about the model. Is it unique? Are the parameters stable? Indeed we only recommend trying this in extreme circumstances.

This is not to imply that there is not a lot of interesting work going on in this area (see, for example, Harvey, 1989, or West and Harrison, 1999). The difficulty is in finding good explanations of the principles and practices, and especially good ways to present the results of such modelling to non-technical managers. Moreover, we believe the biggest objection is that there may be several equally 'good' models that are or appear to be totally different in structure and method of estimation.

A final concern is that the books cited give no examples of how actually to do the work using any software tools. We feel that it is highly unlikely our readers are going to want to develop or find appropriate tools themselves. In fact, for forecasters interested in attempting to use dynamic regression models, we would recommend looking at modern software for econometrics. Much of this is quite expensive to acquire and use, although some (semi-)freely available tools do exist, for example, Doornik (1996).

State space modelling

The ideas of dynamic regression and many other approaches to forecasting can be recast by taking a *state space* viewpoint. This posits that a system of interest has a (finite) number of states, each with properties we can measure, although we usually cannot measure the state variable itself.

A very simple example, is at the base of one of our early efforts in forecasting milk production per cow (Nash, 1977b; Nash and Teeter, 1975). A CAUTION: most state space models use a very different notation and perspective, and their authors may feel this example is not representative of the main stream of state space modelling. We modelled dairy production so that there were just two production regimes: 'old' and 'new'. The 'old' regime used breeds of cows, management practices and agricultural technology that were presumed to produce at a rate of 3950 pounds of milk per cow per year (we will use the units provided to us), while the 'new' regime could produce 15,000 pounds per cow per year. These figures were derived from long-standing statistics under a Record of Production program that monitored a quite large number of individual animals. The mean production per cow per year in a given population of cows where $x(t)$ represents the fraction that belong to the 'new' regime in year t is

$$p(t) = 3950 + x(t) * (15000 - 3950)$$

The data, apparently for the period 1950 to 1972 (23 periods) is (in pounds of milk per cow per year): 6111   6220   6141   6086   6007   6080   6074   6217   6500   6743   6594   6734   6759   6935   6993   6935   7218   7400   7553   7851   7854   7738   7930.

We seek a model for the evolution of the state(s) rather than of the measured outcome. Clearly, we need a mechanism for the state to evolve. In our case, we felt that dairy farmers were quite unlikely to change from 'new' to 'old' regimes, and set the probability of such a change in any year as zero. Furthermore, the working life of a dairy cow, we were informed, is about four years. In any one year, only about 25% of the cows will be replaced, assuming farmers do not send a producing animal to slaughter. This puts an upper bound on the probability of a change from 'old' to 'new' at about 0.25. We also thought there was a 'band wagon' effect towards the 'new' regime. Thus, we hypothesized that the probability of change from 'old' to 'new' in the single year t was of the form

$$b0 + b1 * x(t)$$

in year t, which yields a 'forecast' for year $(t + 1)$ of

$$x(t + 1) = x(t) + (1 - x(t)) (b0 + b1 * x(t))$$

that is, the proportion of cows under the 'new' regime is equal to those in that regime in the previous period (since none go back to the 'old' regime), plus the fraction of those who were in the 'old' regime who shift to the new. (Note that Nash, 1977b, has fixed $b0 = 0$. The notation in that paper is rather different from that used here.) The model value of the production per cow $p(t)$ is then found from the model value of $x(t)$ by the equation given above. We need to find $x(0)$, $b0$ and $b1$ to be able to compute our model and hence forecasts, and these are calculated recursively. While we could use special purpose nonlinear modelling packages (Chapter 20), we will use the optimization features of spreadsheet software (Excel or Quattro, also discussed in Chapter 20). Such an exercise gives $x(0) = 0.1804394$; $b0 = -0.013577$ and $b1 = 0.100138$ with a residual sum of squares of 399,705. This translates into a RMS error of approximately 132 pounds of milk per cow per year, or less than 2% of the average production per cow. Figure 19.1 presents this graphically.

Clearly, we can envisage much more complicated models, using many states with much less obvious linkages between the state vector, which is generally unobservable, and the measured data. Part of the confusion for non-specialist readers of such material is that often the same or equivalent equations are derived from different perspectives, so that the jargon and notation may vary, even when the model is essentially reducible to a familiar form. Indeed, the production per cow example above arose out of several fields of study, namely, logistic growth curves, Markov chain modelling, finite difference equations, and agricultural economics.

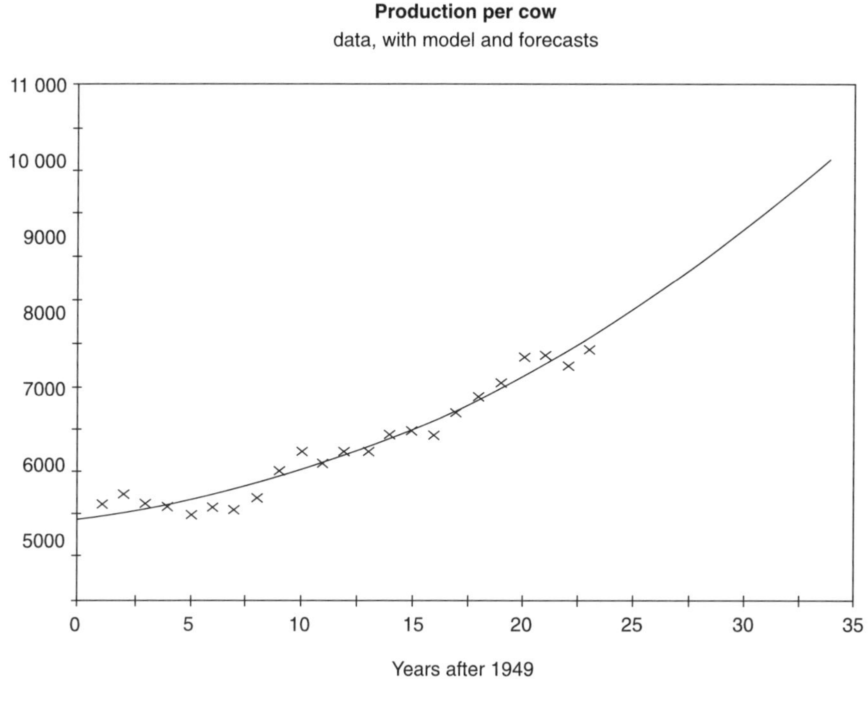

Production per cow
data, with model and forecasts

Years after 1949

Figure 19.1. Illustration of a simple model based on transitions between two states.

State space models are the basis of the Kalman filter (see Harvey, 1989, or Chatfield, 1996), which has been a popular tool in signal processing, although it has seen limited use in forecasting. Like dynamic regression modelling, the form(s) of the models are often difficult to explain in simple language, so that they are hard to explain to non-technical audiences in a way that allows for useful incorporation of subject matter experience and expertise.

Exercises

E19.1. (Library or Internet research) Find examples of econometric models that are in use and discuss their applicability to possible business forecasting issues.

E19.2. Discuss and develop a conceptual structure for a state space model of a system of interest. Examples:

 Evolution of book selling, from retail stores to the Internet.

 Shifts in mode of energy supply for the production of electricity.

 Patterns of post-secondary educational enrollment, e.g. mature vs. school leaving students.

Nonlinear regression modelling

Motivations: nonlinear forecasting models

Modelling data is finding equations or rules that let us predict our data. Typically we use equations that have pre-specified functional forms. These are not always appropriate to the situation at hand, and we now consider methods that let you specify the functional form of your model. Conventionally, such functional forms are analytic functions involving variables and parameters. However, it is also possible to use models involving ordinary differential equations (Nash and Quon, 1996), or even partial differential equations or integro-differential equations, although we have no experience with the latter two types of models for forecasting. We have, however, used functions evaluated by simulations or by computational rules rather than functional expressions. The methods used for carrying out such modelling are powerful, but it is very easy to misuse them. This chapter draws on several of the authors' previous works, including Nash (1979), Nash and Walker-Smith (1987) and Chapter 32 of the manual for Data Desk 5, to which one of us (JCN) contributed.

A note of warning: nonlinear regression is concerned with the estimation of models of whatever functional form. It developed as a subject in the 1960s. In the 1980s, the elucidation of fractals, chaotic systems, and what are sometimes called nonlinear time series can result in confusion, especially if the topic is simplified to 'nonlinear modelling'. For example, the book edited by Casadagli and Eubank (1992) entitled *Nonlinear Modelling and Forecasting* has almost nothing to do with the subject of this chapter, and is unlikely to be useful to business or administrative forecasting. It is, in fact, directed more to situations where there is a large amount of data, as in signal processing, with some overlap with the topics in Chapter 21.

What is a nonlinear model?

We start with data in a variable Y. This is believed to be explainable by using some functional relationship on variables $X1$, $X2$, and so on. In a forecasting situation, one of the variables will usually represent time. There may only be one X; in fact, this is the most common case for many forecasting problems. In Chapter 10 we have already encountered a linear form as a model that attempts to describe Y in terms of p predicting variables $X1$, $X2$, ..., Xp by estimating the $p+1$

parameters $b0$, $b1$, ..., bp using multiple linear regression. In such cases there is 1 parameter for every X variable plus a constant $b0$. In nonlinear models, we still have parameters b (we will collect them together as a vector and use a single symbol in bold face) and variables X (again a collection when we do not use an additional label), but the relationship can be any expression we can describe in a computational form. We will call the computed result variable y_hat. Once we can compute y_hat, we can form residuals as the difference between y and y_hat.

We estimate the parameters b by adjusting their values to make the size of the residuals as small as possible. Doing this efficiently and reliably is the core of nonlinear modelling. Students and other newcomers to data modelling often assume that there is a unique 'best' set of values for the parameters b. This is usually the case for linear models, although collinearity may lead to situations where only certain linear combinations of the parameters are unique. With nonlinear models, it is common that the objective function – the measure of lack of fit – has multiple (local) minima. A typical objective function is the residual sum of squares with respect to the parameters b. We really do not want to have to check all the possible minima, so must use external information – quite often the meaning of the parameters in the model in the context in which we are using it – to restrict the allowed values of the parameters. We therefore recommend that nonlinear modelling be carried out with *bounds*, i.e. constraints on the parameter values. Not all software available makes such constraints easy to impose.

An example

Consider trying to forecast the number of weeds per square metre in research on better methods of weed control (Nash, 1979, p. 121). The data are provided for 12 time periods (years or growing seasons; details were not provided to us). From the shape of the data (Figure 20.2), we suspect that an appropriate model for this problem is exponential growth. Graph log(weed count) against time (Figure 20.1) shows an 'almost' straight line, although there is some decrease in the slope at the end. Of course, the weeds have a finite amount of soil on which to grow. The count is for a fixed plot so it represents a density of weeds. They will eventually crowd each other and we expect an upper limit. This is best modelled by some form of S-curve or *sigmoid function*.

There are a number of ways to provide such a function, but we were told to use a *3-parameter logistic function*. One form of this is:

$$y_t = b1 / (1 + b2*\exp(-b3*t))$$

Let us jump to the solution, which unfortunately cannot be found from the tools provided in the distributed form of Minitab (versions 9 or 12, Student Edition for Windows). We computed estimates for the bs using our own software (Nash and Walker-Smith, 1987). t is the time variable. The three parameters $b1$, $b2$ and $b3$ decide the particular shape of the curve. Indeed, we can see quite simply by substitution of different values for the $b1$, $b2$ and $b3$ that as long as $b2$ and $b3$ are positive, then $b1$ defines the upper limit or *asymptote* for the modelling function.

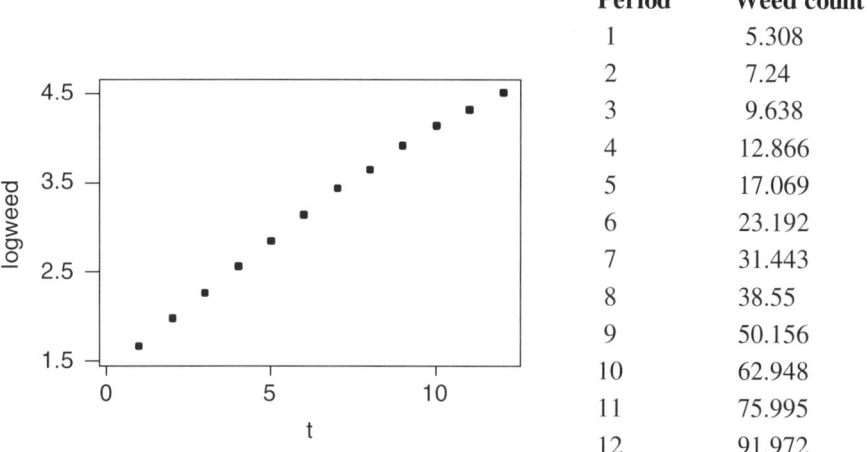

Period	Weed count
1	5.308
2	7.24
3	9.638
4	12.866
5	17.069
6	23.192
7	31.443
8	38.55
9	50.156
10	62.948
11	75.995
12	91.972

Figure 20.1. Log(weeds) versus time.

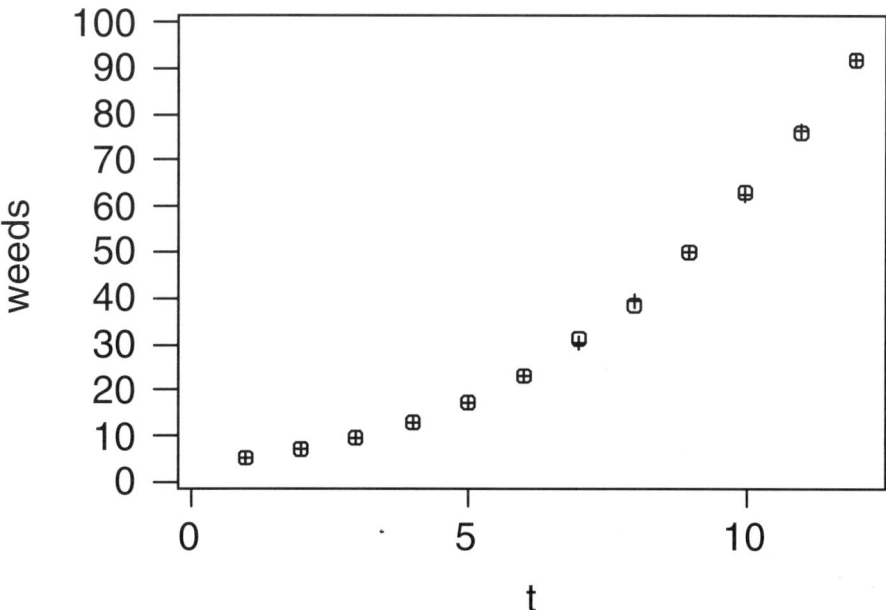

Figure 20.2. Weed density (o) and 3-parameter logistic model fits (+).

It is clear that we have a very good fit over the time span of the data. However, we could also fit an exponential model

$$y_t = b1 * \exp(b2 * t)$$

(These *b*s are *not* the same as those in the logistic model.) This model will also fit very well.

What happens if we advance time into the future? The graphs diverge rapidly. We should point out that to get an image that did not portray the logistic as simply a smudge along the time axis, we had to reduce our range for t to 1:20 from an original 1:50 choice. The exponential growth model is truly *not* a good choice for many real-world situations, since it has this explosive nature. The change is in the order of magnitude, which humans rarely appreciate fully. Even with much experience, we find our intuitions about its pattern of change with time can mislead us.

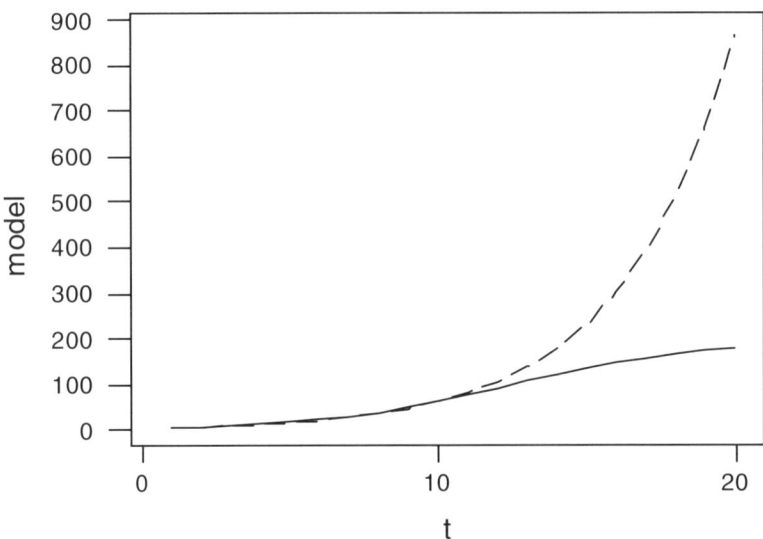

Figure 20.3. Extrapolation of logistic and exponential models.

Objective functions, constraints and parameters

We estimate nonlinear models by adjusting the parameters b to reduce the size of the deviations between the data and the modelling function. Making the residuals 'small' requires some way of measuring their size. The most common approach, as before, is to try to minimize the sum of the squared residual values – this is called nonlinear least squares fitting if one is interested in approximation, or nonlinear regression if one takes a statistical perspective.

Besides the least squares criterion – that is, the sum of squared residuals is used to measure their 'size' – we can use many other *loss functions* or *objective functions* that measure how badly our model fits the data. This can be important in forecasting. Some years ago we gave several short courses in forecasting to public-utility workers who had to plan new facilities for residential and business developments. One worker raised the very interesting question, loosely quoted as:

If we provide a forecast that turns out to be too small and there is not enough equipment in place when customers wish to be connected, we'll lose our jobs. If we forecast too high, and equipment ends up sitting idle, we don't get our annual increase. How should we do our forecasts?

Clearly, these forecasters want to bias their predictions toward the 'high' side. One way would be to prepare a forecast using traditional methods, using the least squares loss function, then 'add a little more' to provide a cushion. Another approach is to use a different loss function, which the framework of nonlinear modelling allows, even when the model itself is a traditional simple or multiple linear regression form. We want an *asymmetric* loss function that provides a higher penalty in one direction than another. Here we want to move towards 'small' residuals, but recall that, in word form, we define

Residual = data - model

Thus, a 'high' forecast has negative residuals, while a 'low' one – that gets the worker fired – has a positive residual. We therefore define our loss function to be the sum of squares of the residuals redefined as follows:

Residual contribution   = residual   if the residual is negative (high forecast)
                           = W*residual   if the residual is positive (low forecast)

where W is some small number, such as 1.5 or 2. This will make a negative residual 'penalize' the particular set of parameters used to try to fit the data by between 2.25 and 4 times as much as a positive residual of the same magnitude. The literature on asymmetric loss functions is quite small, but includes our own contribution (Nash and Walker-Smith, 1986).

Methods for nonlinear fitting

This is not the place to provide a treatise on nonlinear modelling (see Nash and Walker-Smith, 1987; Nash and Nash, 1996). However, readers should recognize the general nature of nonlinear modelling, which has no *closed-form* solution. Instead, we must use *iterative* methods that start with some guess of the parameter values, compute the objective function, and then use various ways to proceed 'downhill' or otherwise choose 'better' sets of parameters.

There are vast differences in the efficiencies of different methods for nonlinear modelling. Moreover, a method that is extremely quick to find solutions for one problem may be hopeless on the next. Our experience is that a well-implemented version of Marquardt's approach (Marquardt, 1963; Nash, 1977a; Nash, 1979) is very effective. It does, however, need the user to supply first derivatives of the modelling function with respect to the parameters. Business students struggle to hide anxiety attacks when words like 'derivative' or 'calculus' are used.

Therefore, we only recommend this method to students in conjunction with pre-worked models or else with software that automatically uses numerically approximated derivatives.

The Marquardt approach is only applicable to least squares problems, so where we want to use a more general loss function (the asymmetric one above, for example), we need a different algorithm. In Data Desk 5 (Data Description Inc.) we suggested what is called a *variable metric* method (Nash, 1979) due to Roger Fletcher. This also requires first derivatives, but this time of the objective function with respect to the parameters. Once again, we recommend that numerical approximations be used. Some modern software can compute analytic derivatives for us. It is unlikely, at the time of writing, that such software will be readily available to our readers.

There are also *direct search* methods such as those of Nelder and Mead or Hooke and Jeeves. Both are described in our books (Nash and Walker-Smith, 1987; Nash, 1979) with software that actually implements them. We like the Nelder–Mead method, which has proven surprisingly robust in a wide variety of settings and in the hands of many different workers. The direct search methods have the great advantage of needing no derivative information. In fact, they are heuristic, and we often explain how they work in classes to business students.

The most widely available optimization tools are those built into most modern spreadsheet software. The (edited) output in Figure 20.4 was easily generated with Quattro Pro 7. Microsoft Excel gives equivalent results. Our only caution: the optimization technology in spreadsheets is hidden from view, and its quality and behaviour can be unpredictable. The user must keep a careful eye on results to ensure that they make sense. Graph your results and watch the output. In particular, the default convergence settings and other controls of the optimizers in spreadsheets may need adjustment to get good results from particular forecasting problems. Generally, we use much smaller tolerances than the defaults. Also, there are examples where the methods give results that can be clearly shown to be non-optimal. Unless the results are ridiculous, we may not reject them. If the models generated by the optimizer are satisfactory in that they 'work' for our problem, we could still use these sub-optimal solutions to the estimation problem. However, if the optimality is important, we recommend using different software tools.

Our Quarterly Traffic Fatalities in Canada data provides another example. For this exercise, we will use the trial modelling period and a very simple multiplicative seasonal decomposition. We load the data and a time period counter into the spreadsheet, then set up our parameters: the trend intercept, the trend slope, and the four seasonal factors. We must put values in these cells to be able to compute the model components. This is equivalent to choosing starting values for the nonlinear optimization methods discussed above. We chose the values 800, -20, 0.8, 1, 1, and 1.25 for the six parameters. Then we created a column for the trend, which is just the intercept plus the slope times the time index. The seasonal factors must be set up four at a time, and we must point to the *location* of the parameters rather than their value. The same comment applies to the trend, of course. This requires that we use *absolute* cell references in our spreadsheet. (Typically, a cell address has a letter and a number for the column and row respectively. By using B7 in a formula, we force the use of cell B7's value, even if the formula is copied or moved. This form of absolute addressing applies in all the spreadsheet software we have used.)

```
hobbopt   Fit 3-parameter logistic to Hobbs using Quattro 7 optimizer
  b1      196.211253267425
  b2      49.09698750369      Sumsquares=    2.58728052663621
  b3      0.313561364864066

  t    weeds             fitted              resid
  1    5.308        5.31997046813965    -0.0119704681396504
  2    7.24         7.20729015157432    0.032709848425684
  3    9.638        9.73002991696561    -0.0920299169656076
  4    12.866       13.0747095553199    -0.208709555319905
  5    17.069       17.4614600654696    -0.392460065469646
  6    23.192       23.1341052809891    0.0578947190109176
  7    31.443       30.3368403929077    1.10615960709233
  8    38.558       39.2732644016389    -0.715264401638926
  9    50.156       50.0478594584152    0.108140541584767
 10    62.948       62.599368279035     0.34863172096496
 11    75.995       76.6479705713632    -0.652970571363156
 12    91.972       91.6858878655313    0.286112134468738
```

Figure 20.4. Fitting a 3-parameter logistic curve to the Hobbs weed data with the optimization capabilities of Quattro Pro 7.

Once we have the trend column and the seasonal factor column, we multiply them and form our fitted model column. Subtracting model from data gives us a residual column, from which it is easy to form the residual sum of squares. We then point to the cell containing the sum of squares as our objective function in the optimizer tool, and the cells of the parameters as the values to be adjusted. As indicated, we find it usual to have to set the convergence tolerances much tighter than the defaults (usually under *options* in the optimizer). Our edited results are shown Figure 20.5. Note that the R_squared value suggests that this approach has done quite well. We check our results by drawing the graph of the data and model in Figure 20.6.

The diligent reader will note a label 'prodq - 1' in Figure 20.5. The number to the right of this label is, in fact, the product of the four seasonal factors minus 1. This quantity should be zero if our seasonal factors are *normalized*. The optimizers in Excel and Quattro can accept constraints such as this. (We want a value near zero.)

Tips for choosing model parameters and bounds

When nonlinear modelling 'works' as in the Hobbs example, it is surprisingly straightforward. Indeed, we began work on this example at 16:40 one afternoon. The write-up, with formatting, spacing and graphic was in place by 17:20, a period of only 40 minutes. The modelling itself took less than 10 minutes. It was the transfer into our word processing files that took most of the time.

Unfortunately, nonlinear modelling often does *not* 'work' smoothly. There may be multiple optima, many of which are clearly unsuitable for our needs. The optimizer may stubbornly refuse to adjust parameters to regions where we believe their values should lie. We can waste a lot of

sfopt3a.xls		Optimize model parameters for the trial fatalities data				
intercept	1126.0981					
slope	-5.0769		271527.09 = sumsquares			
q1	0.6855		212.73109 = RMSE (adj.)			
q2	0.9974		0.9003681 =R^2			
q3	1.2753				meandata= 991.11538	
q4	1.0418				TSS= 2725303.3	
prodq-1	-0.0916241					

t	fqtdata	fits	resids	sfact	trend	data - mean
1	624.00	768.43	-144.43	0.69	1121.02	-367.12
2	1049.00	1113.06	-64.06	1.00	1115.94	57.88
3	1324.00	1416.66	-92.66	1.28	1110.87	332.88
4	1172.00	1152.04	19.96	1.04	1105.79	180.88
. . . .						
5	691.00	754.51	-63.51	0.69	1100.71	-300.12
50	752.00	870.00	-118.00	1.00	872.25	-239.12
51	1013.00	1105.89	-92.89	1.28	867.18	21.88
52	858.00	898.16	-40.16	1.04	862.10	-133.12

Figure 20.5. Application of spreadsheet optimization (Microsoft Excel) to fitting a multiplicative time series model to the Quarterly Traffic Fatalities in Canada data. Corel Quattro Pro gave almost identical output. The rightmost column is used to compute the total sum of squares so we can get the R_squared value.

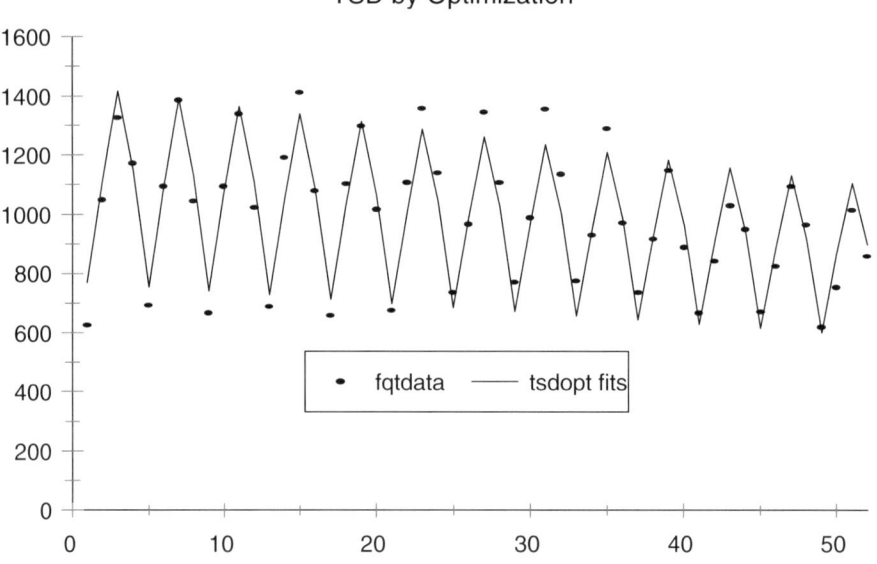

Figure 20.6. A seasonal decomposition for the Quarterly Traffic Fatalities in Canada data by spreadsheet optimization. (This graph was prepared with Corel Quattro Pro 8.)

time if we do not know what we are looking for. Therefore, it is almost always useful to have some idea of the parameter values. In the seasonal modelling example, we could even use our ruler forecast approach to get quite good starting values for the trend intercept and slope and a quick 'by-eye' judgement will give us rough seasonal factor values. As mentioned above, these should average or multiply to 1. (Traditionally, seasonal factors are normalized to average to 1, even though they multiply our trend line.) In our example above, we forced the seasonal factors to multiply to 1. Figure 20.7 shows the sets of parameters with and without this constraint.

Optimization model parameters for the trial Quarterly Traffic Fatalities in Canada data

	with normalization	no normalization constraint
intercept	1126.098	352.501
slope	-5.077	-1.589
q1	0.686	2.190
q2	0.997	3.186
q3	1.275	4.074
q4	1.042	3.328
	-0.09162 =	product of quarterly factors - 1
sumsquares	271527.09	271527.09

Figure 20.7. Constrained and unconstrained quarterly seasonal factors.

A very useful class of constraint is a *bound* on a parameter. We may want a parameter to be positive (especially if we must divide by it), so could require that it be greater than or equal to some small number such as 1.0E-5 (= 0.00001). The scale will depend on the problem. Upper bounds are useful too. For example, we may be sure that there is an upper bound on the number of weeds per square metre. Optimization methods benefit from the use of such bounds, but precise values are not needed. Mainly, the optimizer is restricted from taking 'wild' steps during its iterations. The restrictions of the bounds prevent excursions into areas of the parameter space that have no meaning in the context of the problem at hand, so we avoid inadmissible solutions or parameter values that make the function difficult to compute correctly.

In the development of *Data Desk 5*, our belief that bounds should be applied in all nonlinear estimations was mirrored in the *visual fitting* tools that *Data Desk* supplies. This feature is possibly unique to *Data Desk*, and allows the user to adjust parameters using specially designed controls while watching one or more interactive and dynamic graphical displays of the modelling situation. The restriction on the search domain generally gives much faster convergence of our iterative methods to a solution. Good starting values are also useful in saving time and effort.

WARNING: we can carry the use of bounds too far. If we put the upper and lower bound on a parameter very close together, the iterative methods may bounce back and forth between the bounds, a condition referred to as 'hemstitching'. This slows convergence.

Growth curve examples

There are many situations where we want to predict the growth of a market or similar quantity. We have already seen this in the production per cow example of Chapter 19 and in the weeds example of this chapter. In Chapter 6 we were also interested in such curves for studying market penetration. The market for technology products and services provides two published examples. Wang and Kettinger (1995) were interested in predicting the growth of cellular telephone communications. Rai *et al.* (1998) want to predict Internet diffusion by forecasting the number of Internet host computers. Both these papers unfortunately have some 'bugs' in their forecasting efforts. In the case of Wang and Kettinger, this is largely in editing graphs for publication, though some other concerns are presented in Nash (1996b). In the case of Rai *et al.*, we had some difficulties approximating their calculations with either *Stata*, our own software (Nash and Walker-Smith, 1987), or the Quattro numerical optimizer tool.

The data for the Rai example is taken directly from their paper (by cut and paste from the Adobe Acrobat version on the ACM Digital Library). This data, corrected as indicated for (apparent) typographic errors, is listed in Table 20.1. Looking at the data, apart from the apparent typos, we may be suspicious that the precision drops from five to two digits between periods 27 and 28, where it seems that the data collection started to use thousands of hosts.

Table 20.1. Data for the growth in Internet hosts August 1981 – July 1987. Published data was corrected for observations 30 and 43, that is, 1st quarter 1989 and 2nd quarter 1992, where there was a decimal point instead of a comma as digit separator i.e. 80.000 instead of 80,000 and 890.000 rather than 890,000

Period	Number of Hosts	Period	Number of Hosts	Period	Number of Hosts	Period	Number of Hosts
1	218	17	1961	33	159000	49	2056000
2	225	18	2221	34	197500	50	2217000
3	233	19	2926	35	236000	51	2757948
4	279	20	3853	36	274500	52	3212000
5	344	21	4780	37	313000	53	3864000
6	409	22	8641	38	376000	54	4852000
7	475	23	13968	39	455500	55	5747000
8	540	24	19295	40	535000	56	6642000
9	628	25	24622	41	617000	57	8057000
10	727	26	28863	42	727000	58	9472000
11	826	27	30932	43	890000	59	11176500
12	925	28	33000	44	992000	60	12881000
13	1024	29	56000	45	1136000	61	14513500
14	1258	30	80000	46	1313000	62	16146000
15	1493	31	105000	47	1486000	63	17843000
16	1727	32	130000	48	1776000	64	19540000

Rai *et al.* decided to set aside some data for validation purposes. Beneath the table in their paper is a note:

The period before January 1994 was used to estimate model parameters, and the subsequent periods were used to assess the predictive ability of the models.

Slightly later, in the body of the paper, we find:

We used nonlinear regression with data from August 1981–January 1994 to estimate parameters each of the three models. In addition to statistical fit, predictive validity should govern model selection, as it reflects a true test of model performance. Mead [6] observed almost 15 years ago that ability of a growth curve to forecast future development is the crucial requirement, and thus it is desirable if this ability can be evaluated on some available data.

We have left in the comment about *Meade* (correct spelling), since this does not accord with the references to the paper, nor to any paper known to us, providing further evidence of weak editing and proof-reading of this paper, possibly outside the authors' control.

The comment with the table implies that 49 observations are used to estimate the model, while the body-text suggests that 50 points are used. It turns out the authors actually used 49 points. They attempted to fit three types of models, logistic, Gompertz and exponential, but the functional form is not listed in their paper. We inferred the form from the values of parameters given in a table of parameter estimates, later confirmed by electronic mail from one of the authors. The models are:

- Logistic: $y_t = M / (1 + \exp(B - c*t))$

- Gompertz: $y_t = M * \exp(-\exp(B - c*t))$

- Exponential: $y_t = A * \exp(B * t)$

As with so many forecasting matters, the devil is in the detail.

Exercises

E20.1. Use spreadsheet optimization to estimate the models for the Rai *et al.* data (see the PFM Web site to avoid having to enter it).

E20.2. Estimate a logistic curve for the milk production per cow data of Chapter 19.

E20.3. Try to set up the state space modelling of milk production per cow, as in Chapter 19, using a spreadsheet optimizer.

Artificial neural networks

New and improved?

One of the difficult aspects of any technology – and forecasting methods are a technology – is that someone is always tinkering. Selon Volaire, *Le meilleur est l'enemi du bon*. For the user of forecasting methods, there is always the nagging doubt that the 'new' method might do better. Taking our own advice, we caution that 'new' methods must be judged on their merits:

- Are they easy to use and apply?

- Can we understand how and why they work?

- Do they work well, and in what situations?

- Do the people promoting these methods have their own agenda? That is, are they academics who need to publish papers for career advancement, or marketing managers of software companies? Or are they users of forecasting trying to get the job done?

Obviously, there will be improvements in tools and understanding over time. In this chapter we take a look at some of the directions. However, we are not convinced that any of the techniques we present in this chapter should be considered 'established' as good forecasting tools, although they seem to work well in the hands of some practitioners.

Artificial neural networks (ANNs) have become fashionable in the last decade as tools for attacking a variety of prediction and classification problems, ranging from computer recognition of images of human faces to the forecasting of seasonal time series.

Underlying ideas

The essential idea that forms the basis of ANNs is that of mimicking the processing of animal brain functions. That is, there are inputs of information, processing and outputs in the form of results. All the work is done in *nodes*, also called *neurons*. The input nodes generally receive a single number input which is then passed on (*fed forward*), usually to every node in the next *layer* of the 'network'. The next layer may be the last, or output, layer, or it may be a *hidden* layer. The more hidden layers, the more complex the underlying forecasting model.

Generally, each node of a given layer sends its outputs to all nodes of the next. These outputs therefore become the inputs of the next layer, but we must sum them together. There is no reason to give each of these inputs equal weight, and indeed we form a weighted sum. Then each node processes its input value and transmits the result to the nodes of the next layer, unless it is an output node, in which case we have one of our results.

Clearly there are lots – in our opinion too many – choices.

- What weights should we use for summing the inputs?

- Should we even use a sum for the inputs, or would a product or more complicated function be appropriate?

- What should the processing nodes do with the input? This *transfer function* is commonly a sigmoid function like the logistic, but other functions are possible.

- How many hidden layers should we use?

- Is there any reason to connect all nodes of one layer to all nodes of the next?

- When is it appropriate to include 'extra' hidden nodes, often called *bias* nodes? These are commonly used (DeLurgio, 1998, p. 684) as a way to tune the transfer function.

These choices mean that ANN methods can often do an extremely good job of modelling various time series. The difficulty is in getting the choices right, and doing so without a lot of effort. The weights, at least in ANN systems specified with simple weighted sums of inputs and sigmoidal transfer functions, are chosen by using a simple *back-propagation* algorithm that adjusts the weights based on the current errors in the outputs.

We would like to give a concise presentation of ANN methods that will allow our readers to make use of them, but we feel they are not yet a practical method for managers. There are too many choices. Nevertheless, other authors (DeLurgio, 1998) clearly feel that ANN methods are an important development for forecasting, as well as for other areas of study (Lippmann, 1987).

Published examples

A number of authors have applied neural nets to forecasting applications. Some of these have been forecasts of a binary variable, such as the state (good/bad) of a company's financial health (Lacher *et al.*, 1995; Lenard *et al.*, 1995). This a classification rather than a forecasting problem.

Although not directly a forecasting problem, the ranking of sports teams clearly has a role in predicting their performance. There is even an American Statistical Association Section on Statistics in Sports. Given the many potential mechanisms for the interaction between teams, neural nets are an obvious candidate for such ranking, and Wilson (1995) has looked at (US) college football team rankings.

Direct forecasting applications of neural networks include:

- Quarterly earnings in the utility industry (Lee and Chen, 1990);

- Forecasts of futures trading volume – a sort of meta-forecast (Kaastra and Boyd, 1995), or the volatility of the stock market (Donaldson and Kamstra, 1996);

- Exchange rates (Kuan and Liu, 1995).

More generally, neural nets have been applied as 'helpers' to other methods. Sohl and Venkatachalam (1995) use them to select appropriate forecasting models from different candidates. This is essentially a classification problem. Gupta and Lam (1996) used neural network approaches to estimate missing values in data sets. In practice, we have observed a danger in 'filling in' missing data, as discussed in Chapter 18.

Assessment of utility of neural networks

Where will neural nets fit in the toolkit of the practical forecaster? We have them as a 'to be watched' item. Some workers believe that they are the tool of choice. Kuo and Reitsch (1995) clearly take this view. Tal and Nazareth (1995) report that, in 1994, the Canadian Imperial Bank of Commerce 'replaced its index-based Leading Indicators with a neural network-based system'.

On the other hand, the development of neural networks has come from the computer science community. Along with so-called 'fuzzy logic', neural networks re-introduce ideas that statisticians and probabilists have worked with for many decades, and the 'new' results usually fail to refer to the preceding body of work. Because the details are complicated, it often takes a lot of work by clever researchers to make clear the similarities and differences. Brian Ripley at the University of Oxford has done considerable work in this area (e.g. Ripley, 1993, and works on his own web site). Some of the concerns arise from the particular issues raised in the application of ANN methods. We want to simplify the process of estimation of the NN models, called *training* in the jargon of the subject – see Lactermacher and Fuller (1995). As in other methods for forecasting, the generation of the forecast is one matter, while producing an interval forecast is another. De Veaux *et al.* (1998) propose one way to do this, but the approach is clearly not for the novice and we judge it unlikely to appear in readily-available general-use software soon. Closer to the present interest, the study by Faraway and Chatfield (1998) compared traditional and neural network approaches to forecasting the series referred to as the 'airline data' (see the PFM Web site). Their conclusions reflect our own concerns.

- The flexibility and complexity of ANN modelling gives too much scope for disastrous mistakes by the novice user.

*   It is very easy (by adding more hidden layers) to improve 'fit' of the models, but just as we can improve fit by including more variables and parameters in a regression model, we may worsen our performance during the validation period.

*   ANN models are difficult to explain and interpret. Giving meaning to the nodes and their weights is not at all like using the sizes of seasonal factors to explain higher or lower sales at different times of the year.

The work of a student, Xuecong Luo, in comparing ARIMA and ANN methods as applied to forecasting Australian electricity utilization gave very similar conclusions to that of Faraway and Chatfield, which should not be surprising since the two time series have some similarities. Adya and Collopy (1998), while finding the published reports gave mixed results, felt that there were hopeful signs, particularly in some carefully prepared comparisons. Tkacz (2000) reported that ANN methods showed no improvement for short term forecasts, but held some promise for multi-period ahead predictions. His conclusion is worth repeating:

> neural networks will not necessarily be superior in every situation, and ... our results support this proposition.

Exercises

E21.1. (Library or Internet search) Find recent examples of the use of ANNs for forecasting or prediction. From the reports can you work out how to perform and assess the forecasts.

E21.2. (Library or Internet search) Build a table of features claimed by software for Neural Network prediction and modelling. Note that you will find many offerings.

Building the forecast report

Grading our own work

As teachers, consultants and practitioners, we are frequently surprised at the rarity of self-criticism by our students, clients and colleagues. We are *not* talking here of *self-deprecation*. There is no intent to suggest that anyone should diminish either their abilities or their work. The purpose of examining and grading our work is to assess its value and to improve it.

The one-page critique

In teaching forecasting and similar courses where project assignments are used as learning and assessment vehicles, student critiques of each others' work have proven useful in allowing the professor to see alternative viewpoints to student work. They have also been popular with students as a tool for learning and evaluation. They help students understand how their work will be marked, and we recommend looking at assignments or other forecasting reports in the 'critique' framework as a practical way to focus your own work. Few of our readers will make their career in forecasting, but many will have to assess, quickly and reliably, forecasts that others have prepared. The following are guidelines for carrying out critiques.

- *Format*: one typewritten page. The work reviewed and critic should be identified.

- The *text* should summarize:
    The MAJOR STRENGTHS of the work,
    The MAJOR WEAKNESSES of the work, and
    Your MAIN REACTION to it.

A marking scheme

A different view of a forecasting report can be provided by 'marking' it. Over the years we have found the following marking scheme to work extremely well. While the critique above emphasizes overall reaction to the content and message of the report, the marking scheme forces one to balance the presentation and technical elements with the need to convince the reader of the soundness of the work.

Our marking scheme, by tradition, has a 20 point total. Since we usually award half-points, we are really marking out of 40. Such is human perception.

- *Presentation.* We give a maximum of 5 points for presentation, that is, general neatness of the work, absence of grammatical, spelling or typographical errors, and the appropriate use of diagrams, tables and references. Appropriate length is also important, especially in course work where we impose limits to save the professor's sanity. Overall, we prefer well-presented handwriting to laser-printed rubbish.

- *Technical content.* A maximum of 10 points is available for technical content. Is the material correct and comprehensive within the guidelines of the assignment? In course work, the student may, if reasonable, have explained restrictions on the topic. Is the technical discussion understandable?

- *Overall success.* Up to 5 points are available. How well does the whole work succeed in convincing you of the validity of the author's thesis? Is the work interesting and fresh, even if it is a report of other work, or a pile of correct but rehashed ideas?

We often ask students to append a 'mark', in the 5/10/5 form just indicated, to their critiques of other students' work. The purpose of this is to promote a better understanding of the marking scheme and improve the reports. However, a distribution of points very different from the professor's judgement suggests that the student critic has a very different appreciation of a report than the professor. The 'marks' students give are generally much less generous than those of the professor.

We mark the student reports and critiques together. The critiques are marked out of 5 points (but we award half points, so it is actually 10!). A good critique is neither too harsh nor too generous. The goal is fair criticism.

While the marking scheme and one-page critique are specifically tools for student assessment, they have a wider application when we appraise our own work before submitting it to the view of others. Both the one-page critique and the marking scheme help to avoid egregious mistakes while promoting good practice.

Words, numbers and pictures

The marking scheme also provides some guidance for the preparation of reports in an efficient fashion. The suggested elements of good presentation and the goal of convincing the reader imply that we should structure our work so the reader can easily understand it and our conclusions.

How should we go about preparing the forecast report? We often advise students to 'write' the report first! This is, of course, a suggestion to cast on paper or in a word processor an outline of what we want to achieve, possibly with a statement of the situation, the elements of the conclusions we hope to be able to reach, and place-holders for the main elements of our work such as:

- Executive summary

- Introduction and statement of situation

- Data sources and, where appropriate, a form of the data or time plot, with time horizons

- Candidate methods

- Trial models and their analysis and validation

- Main models and forecasts

- Conclusions and assessment

- References and supporting material

Upon this skeleton we add as much flesh as we need, and preferably no more. It is tempting to spend a lot of time searching for the 'perfect' source of information. This can generate a great deal of 'data' but very little useful information. It can also be discouraging to have to discard hard-won material. If the forecast is being constructed by a team, it helps to choose one of the team members to be the 'editor' who informs the others of the needs of the report and keeps the effort focused.

Generalities and specifics

We conclude with a few ideas, both general and specific, that we feel are relevant to anyone wishing to attempt forecasting. First, we echo the sentiments of Sir Maurice Kendall (1973):

> In practical time-series analysis there are few golden rules, and indeed few general rules which can be applied without detailed thought about the nature of the series and the purpose of studying it.

So our first generality is to be suspicious of generalities, but always to look for the meaning and origin of data. Out of this concern to know where data come from and why, we focus on the specifics of:

- Units of measurement;

- Appropriate recording of the time point or time period to which data applies; and

- Careful recording of special events and other meta-data.

When dealing with seasonal data series for which we have a reasonable body of data, we feel that it is almost always worthwhile trying either Winters' method and classical time-series decomposition. Even if we later find other models or methods that are superior, we gain a baseline for judgement. Only rarely have we found more complicated models superior for forecasting. This is because they may require us to forecast other variables; for example, if we have introduced explanatory variables into a regression model. Or we may be using an ARIMA model, which could be arithmetically simpler for the computer, but awkward to explain to non-technical audiences.

For non-seasonal data, we like to use the Holt 2-parameter double exponential smoothing to provide a baseline method for forecasting if the data appear to have a simple trend. For patterns that do not seem to be simple trends, we may have to use nonlinear or ARIMA models depending on the context.

With limited data, we are always going to have great difficulty preparing forecasts that are in any way accurate. The forecaster must then spend most of his/her effort in developing 'data' that allow at least some estimate of the future situation to be prepared. We urge readers to do this by seeking as well-founded a model as possible. Without a model, it is difficult to structure thinking, and wishes and guesses are translated into 'forecasts'.

Exercises

E22.1. Apply the one-page critique to a recent article that presents a forecast or forecasting exercise.

E22.2. Prepare a one-page critique of one of your own reports or that of a colleague.

Appendix

Tips, tricks and scripts

1. Importing text data into spreadsheet and statistical software   in Chapter 2

Data in the word processing file fragment below is awkward to cut and paste into a spreadsheet.

```
note fq98byq.mtb                          1987    674   1107   1357   1138
note Quarterly Traffic Fatalities in      1988    734    967   1345   1106
Canada                                    1989    768    988   1355   1135
note J C Nash jcnash@uottawa.ca           1990    774    929   1288    970
note Q1 is jan-mar, etc.                  1991    734    916   1147    888
read c1-c5                                1992    666    841   1030    949
1975   1043   1457   1999   1564          1993    670    823   1093    964
1976    840   1377   1679   1411          1994    618    752   1013    858
1977    786   1340   1735   1392          1995    668    834   1021    815
1978    821   1353   1822   1433          1996    697    643    926    813
1979    985   1460   1836   1582          1997    630    683      *      *
1980    969   1371   1849   1272          end data
1981    961   1369   1732   1321          name c1 'yr'
1982    624   1049   1324   1172          name c2 'q1'
1983    691   1095   1386   1044          name c3 'q2'
1984    666   1095   1338   1021          name c4 'q3'
1985    686   1190   1411   1078          name c5 'q4'
1986    657   1102   1296   1016
```

The problems are (1) the column break, and (2) the spaces between the numbers are not recognized by some spreadsheet software so each 'row' of data is treated simply as a line of text. We found the quickest solution (though not necessarily the most elegant) was to copy the data into a plain text editor and save the material as a file, e.g. qd.txt. We used the text editor to remove the column break. So data could be read into a spreadsheet (we used Quattro Pro 7) we repeatedly replaced two spaces by a single space until there were only single spaces in the file. Then we converted all spaces to tab characters. Quattro then opened the file with the numbers each in a single cell.

2. Linear interpolation and inverse linear interpolation   in Chapter 4

Interpolation is used to find approximate values of a function (e.g. a model or forecast) in a table for arguments between tabulated arguments. Inverse interpolation is used to find a value of an argument for a given value of the function. That is, when points $(x1, f(x1))$ and $(x2, f(x2))$ are given, interpolation finds $f(x)$ where $x1 < x < x2$; inverse interpolation finds a value x when $f(x)$ is given. There are many methods (Fröberg, 1965). Here we will look at linear interpolation only.

Example: linear inverse interpolation. What is the value of xc giving cumulative probability of 0.9 for a standard Gaussian variate? We want to find the value of a standard Gaussian variate X (one having mean 0 and standard deviation 1) so that the probability such a variate falls between 0 and xc is 0.4 (or a probability between $-\infty$ and xc of 0.9). Call the probability function (between 0 and x) $f(x)$, and from standard tables (e.g. Aczel, 1996) find $f(1.28) = 0.3997$ and $f(1.29) = 0.4015$. We want to find the value of z so that $f(z) = 0.4000$;

In Figure A.1, the curved line is $f(x)$. The values $x1=1.28$, $x2=1.29$, $y1=0.3997$ and $y2=0.4015$. We really want to get the true xc, and if we have a good graph like the figure below, we could simply read off the value. We use interpolation because we cannot draw the curve. In its place, we use a function that we can draw and use in calculations. The simplest of these

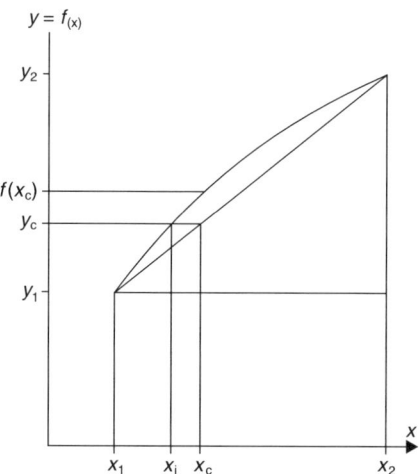

Figure A.1. Linear interpolation between tabulated numbers.

'curves' is a straight line, leading to *linear interpolation*.

The line we draw sets up the triangle of the points given by $(x1, y1)$, $(x2, y2)$ and $(x2, y1)$. The *similar triangle* with vertices $(x1, y1)$, $(xi, y1)$, and (xi, yc) allows us to carry out the following calculations.

$$y2 = f(x2); \quad y1 = f(x1)$$

$$\frac{(yc - y1)}{(y2 - y1)} = \frac{(xi - x1)}{(x2 - x1)}$$

So

```
0.4000 - 0.3997        xi - 1.28
---------------   =   -----------
0.4015 - 0.3997       1.29 - 1.28
```

Or   xi = 1.28 + 0.01 * .0003 /.0018 = 1.2817

We wanted $xc = f^{-1}(yc)$, the inverse function at yc. Instead we got xi, which is 'not too bad'.

Using a similar 'interpolation', we can adjust observations to jurisdictional boundaries. For example, a business in Ottawa will have customers from both Ontario and Quebec (different provincial jurisdictions) and may want to know the size of the customer base from published populations or number of households of municipalities. However, the business may not feel its customer base covers the entirety of all the municipalities. By setting a proportion of each municipality that the business judges to be within its service region, we can compute numbers of households that we then aggregate over the municipalities in both provinces.

3. Time point conversions   in Chapter 4

Suppose that the first observation in a series with periodicity L is drawn from year $y1$ and season number $s1$. We must have the three pieces of information $(L, y1, s1)$ in order to be able to convert between the period number k and the year and season given by y and s. We now compute y and s from k as follows:

$$y(k) = y1 + INT((k + s1 - 2) / L)$$

where the INT() function *truncates* the magnitude of its argument to the nearest whole number. The manipulation is a little tricky because we start actually *in* y1, even though the first period number will be 1 and $s1$ cannot be smaller than 1. This is the reason we need to subtract 2. We divide by L to count the number of years to move forward. Having counted forward from the starting year, we can figure out the season using the formula

$$s(k) = k - L * INT((k + s1 - 2)/L) + s1 - 1$$

This is designed to drop the years out of the period number, then adjust for the starting season.

Unfortunately, the definition of INT() is not the same in all computing systems. Our definition is that $INT(x)$ is the integer whose *magnitude* is less than or equal to the absolute value of x. Thus $INT(1.5) = 1$, while $INT(-1.5) = -1$. This is the definition that appears to be used by Lotus 1-2-3 and Quattro, as well as by the Watcom Fortran compiler. However, we note that Microsoft Excel (as well as an older MS-BASIC interpreter) displays -2 for $INT(-1.5)$. This is more like the FLOOR( ) function of MATLAB or Minitab. Where it exists, you may use a function to ROUND( ). In Lotus and Quattro, we need to specify the number of decimals to ROUND to. To get integers, this should be zero, and we use $ROUND(x-0.5, 0)$ to get the integer part of x. This is *not* the same as the INT( ) function for negative x. In Quattro Pro and MS-Excel, we have been able to use the TRUNC() function rather than INT(). Mostly we will not be troubled by negative numbers, but it pays to be aware that there may be such differences. When they bite us, they are never 'minor' differences.

To reverse the computations above, we compute k from y and s using:

$$k(y, s) = L*(y - y1) + s - s1 + 1$$

As an exercise (timechg.wk1 on the PFM Web site), set up a column of 'period indices', e.g. 1, 2, ..., 100. Convert these to quarterly seasons starting with 1959-4 (4th quarter of 1959 is period 1) as well as to monthly seasons starting with 1972-6 (June 1972 is period 1). Then check your work in each case by converting back to period numbers.

4. Merging seasonal data into a single series   See Table 2.1

We show how to do this by the example of merging data stored as separate seasons for the Quarterly Traffic Fatalities in Canada data. We generally want data in a single time series in time-period order. The Quarterly Traffic Fatalities in Canada data are available via a Minitab executable script (fq98byq.mtb on the PFM Web site) that sets up *separate* series for each of the four quarterly seasons 1 = Jan–Mar, 2 = Apr–Jun, 3 = Jul–Sep, 4 = Oct–Dec. The script below (fq98tots.mtb) does the job. Note that the command `echo` displays the script lines in the Session Window of Minitab as the script is run. This will be copied to any output file if the `outfile` command has been activated. Alternatively, you can scroll back through the Minitab Session Window to see the operation of the script, and especially any messages from Minitab.

```
echo
note fq98tots.mtb
note    Create single time series for Quarterly Traffic Fatalities in Canada)
note This version begins with the data in separate series for each quarter.
note
note Now read in the data -- we include the whole of a file
note   to input the data
note --------- start of included file: fq98byq.mtb -------------
note FQ98BYQ.MTB
note Quarterly Canadian road fatalities
note Q1 is jan-mar, etc.
read c1-c5
1975   1043   1457   1999   1564
1976    840   1377   1679   1411
...
1997    630    683      *      *
end data
name c1 'yr'
name c2 'q1'
name c3 'q2'
name c4 'q3'
name c5 'q4'
note --------- end of included file -------------
note Our data is in 4 columns, 1 for each quarter
note We want a series with all the data in sequence
note 1975Q1, 1975Q2 ... 1997Q2
note
note One way to do this is to write the data, then read it back.
write 'temp' c2-c5
note Above command outputs ONLY the data, 4 elements per row, 1 row/year
note We MUST get rid of unnecessary columns
erase c1-c5
set 'temp' c1
note And then we read it back into a single column (C1)
name c1 'qftl'
note And name the column -- always a good idea.
note
set c2
4(1:23)
end
name c2 'yearno'
set c3
23(1:4)
end
name c3 'qtrno'
note Now could save this as a new file
write 'allftl.txt' c2 c3 c1
note -------- end of manipulation ----------------
```

Here we merge the columns by writing the data to a file and reading it back in a suitable order. This use of files is a helpful trick often overlooked by novice data analysts. Another way is to use a 'sort' capability to carry out the reordering. See script fq982ts2.mtb on the PFM Web site. Here is an excerpt with the key operations.

```
...
note Our data is in 4 columns, 1 for each quarter
```

```
note We want a series with all the data in sequence
note 1975Q1, 1975Q2 ... 1997Q2
note Observations in column 1 will go to positions 1 5 9 13 ...
note in the new series.
note Those in column 2 will go to positions 2 6 10 etc.
note We create series with the right index values
let k12=4*n(c2)
set c12
1:k12/4
end
note Now set up corresponding series for the other quarters. Here we
note do NOT have to use SET, since we can just add 1, 2, or 3 to the
note index in c12.
let c13=1+c12
let c14=2+c12
let c15=3+c12
note Now we want to put the data together. We use the 'stack' command,
note which has a very particular syntax. You could use the mouse and
note the window commands, but sometimes it is helpful to be able to use
note a script. This is especially useful for monthly series.
note Our command is
note    stack (c2 c12) (c3 c13) (c4 c14) (c5 c15), put in (c10 c11)
note which results in c10 being c2  on c3   on c4   on c5
note and              c11 being c12 on c13 on c14 on c15
stack (c2 c12) (c3 c13) (c4 c14) (c5 c15), put in (c10 c11)
note You can check the results by printing them out. We deliberately
note chose the column numbers c12-c15 to correspond to c2-c5 to avoid
note typing errors.
note Now sort the data on the index in c11, carrying along the data in c10.
sort c11 c10, put in c21 c20
note Now could save this as a new file
write 'allftl2.txt' c21 c20
```

5. *Separating a time series into seasonal series*   see Table 2.1

The operation just performed can be reversed. In Minitab, the `unstack` command can be used to do this. First we establish a column of indices (or subscripts in Minitab's jargon), and use this to indicate which data elements go in which columns. See fq98ts2q.mtb on the PFM Web site. The key parts of this are as follows.

```
name c1 'ftldtat'
set c2
25(1 2 3 4)
end data
name c2 'qtrnum'
note c2 is our set of quarters.
note It is longer than c1, so we adjust length.
let k1=n(c1)+1
let k2=n(c2)
delete k1:k2 c2
note
note Now separate the data into quarters.
unstack c1 c11-c14;
subs c2.
name c11 'ftl-q1'
name c12 'ftl-q2'
name c13 'ftl-q3'
name c14 'ftl-q4'
note -- end of manipulation -
```

6. *Trading day adjustments*   in Chapters 9, 18

Trading day adjustments are mostly a scaling of data for the time that business activity is taking place. We divide our data by some measure of the number of days or hours of 'activity'. Forecasts have, of course, to reverse this process. We presume we know how to do this since it is under the control or at least the understanding of management. We suggest the following examples.

- Trading days are useful for such activities as financial markets, which typically (at least until Internet trading started) run on a fixed timetable for the days of operation, and do not function on weekends and public holidays. While we have looked at the possibility of automating some calculations in this area (Nash and Nash, 1986), it is likely simpler in most cases to take a calendar or calendars and simply cross out the non-working days.

- For retail stores, it is likely more sensible to apply an 'hours of operation' scale. Here a possible approach is to prepare a calendar file in a spreadsheet, making sure weekdays, holidays and special events are noted, and add up the hours of operation for each of the 'periods' corresponding to our forecasting data.

- In the tourism industry we may wish to use the 'number of weekends' if we are given monthly or quarterly data. Since 'long weekends' intervene, we suggest that the variable to construct is the 'number of weekend days in the reporting period'. This allows a weekend that has Saturday as month-end to contribute that day in one reporting period. It also counts three days for the typical long weekend. Other fields of application include communications, power and other utilities affected by people being 'off' work, and even such issues as road fatalities, which may increase as people travel more for leisure and recreation. Fatalities, in fact, are often associated with night-time hours on weekends because of a fairly strong association with alcohol usage.

7. The ruler forecast   in Chapter 10

Using a time plot, we simply take a ruler and, by eye, position it on the graph to follow the general trend in the data points. Then we draw a line to extrapolate to future time points and read the forecast value(s) directly from the graph. This is crude but effective. It serves as a preliminary estimate, giving us the scale and approximate value we should expect from other methods.

The ruler can also be used to provide crude 'error' bounds on our forecasts. By placing the ruler so that it just touches the uppermost data points and drawing an extrapolation line, then repeating this action with the lowermost ones, you get an envelope for a linear extrapolation. This will be unsatisfactory for a highly curved pattern, but it serves as a guide to variability in the data.

Seasonal pattern can be forecast using a trace of the curve on a clear acetate sheet of the type used for overhead projectors. The position of the centre or trend line should also be marked. The acetate is then shifted forward in time, keeping it aligned with the centre or trend, until the parts of the tracing for 'old' time periods more or less match up with a later cycle. The time points on the trace that are in the 'future' then provide a fair forecast of the future pattern.

8. *Long or large scale (multi-page) time plots*   in Chapter 10

Most software aims to display graphs on a single page. This may be unsuitable for visual analysis of 'long' time series, that is, series with a large number of observations, especially when we need to prepare ruler forecasts or make annotations. Therefore, we need multi-page graphs and/or graphs printed on large or long paper.

One method we have found useful is to prepare a character plot into a large file and then print this file. The first task is to prepare the file. We can turn on character graphics in Minitab (although there is some threat that this feature will be made obsolete) via the command `gstd` (which Minitab calls 'standard' graphics, as opposed to 'professional' graphics invoked with the `gpro` command). Issuing the command `tsplot` still tries to put the time axis across our page, breaking the graph into pieces as needed so that it can be displayed on an 80 character wide line display. These breaks make it difficult for our eyes to see patterns or deviations from patterns. Also, we have the issue of the time index being in a form that is not tied to calendar periods. On the other hand, the character graphs are 'dumped' into the (plain text) log file (or `outfile` in Minitab), so we have to make no special effort to save them. We do, however, have to remember to print them with a monospaced font to render them useful.

We can change the Minitab character graph page size with the `height`, `width`, and `ow` commands. (`ow` stands for Output Width.) This may allow the time plot to be displayed in one panel, but there are limits to the number of characters per line (the line width as measured in characters). Unfortunately, Minitab uses `width` for some character graphs, such as those from `plot`, and `ow` for others (including those from `tsplot`). `height` is limited to 400 lines.

We can use these 'tricks' to `plot` graphs with time as the *vertical* axis. This graphs our quantity of interest *across* the 'page' and the time, *in decreasing order*, down the page. We have, on occasions, created a new variable $t' = (T - t + 1)$, where T is the number of time points in our series, and graphed t' versus y_t, where y_t is our quantity of interest. Clearly, the t' axis is not helpful to us directly, but the graph will have the appropriate direction when viewed 'sideways'.

We can print graphs prepared this way as text files directly to 'old' continuous-paper printers. For page printers, we need to import the graph file into a word processor, choose a monospaced font, and set margins and page breaks so that we can remove (or fold back) top margins of pages and tape the pages together.

Printing sections of a 'large' graph can be achieved with high-resolution graphical software. The principal issues we need to address are:

- all sections of the graph must be printed on the same scale and with appropriate origins for the x and y axes, so that the 'pieces' can be put together correctly;

- some aid to registration of the sections of the graph are helpful. That is, some small lines or axis ticks that can be lined up when we paste the pages together are essential.

Only in Matlab do we know how to do this automatically, and acknowledge the help of Chuck Packard of Mathworks for helping us find this capability.

9. Multiple plots
in Chapters 9, 17

Multiple plots place several graphs on the same scale. Here is the command script that Minitab used (as seen from the OUTFILE) to generate Figure 9.2.

```
GPro.
Plot 'q1'*'year' 'q2'*'year' 'q3'*'year' 'q4'*'year';
   Connect;
   Title "Canadian traffic fatalities by quarter over time";
   Footnote "q1 solid, q2 dash, q3 dot, q4 dot-dash";
     Center;
   Overlay;
   Minimum 2 0;
   Axis 2;
   Label 'fatalities'.
```

This forces overlay of the plots. Minitab offers another method to prepare multiple graphs, i.e. by grouping, which can provide for multiple plots and annotated legends. In our opinion, the facility for drawing such graphs is so important that it ought to be easier to use. Some of the Minitab graphs associated with the *ad hoc* forecasts in Chapter 10 used graphs prepared via the grouping mechanism. As a template for readers to use, here is the script to prepare Figure 10.10.

```
echo
note fig10-10.mtb -- Put all seasonal forecasts together
read c1-c8
    1        91     1025.73    1069.78    1008.98    1046.00    972.39    1027.69
    2        92      832.46     840.29     821.30     816.52    794.88     803.49
    3        93      567.84     522.02     580.62     517.61    543.87     478.36
    4        94      773.31     770.29     775.50     765.90    730.55     731.92
    5        95     1000.81    1048.63     984.34    1024.91    941.36    1003.41
    6        96      812.11     819.15     801.11     795.44    769.30     779.21
    7        97      553.87     500.88     566.26     496.52    526.23     454.07
    8        98      754.17     749.15     756.20     744.82    706.66     707.64
    9        99      975.89    1027.49     959.69    1003.83    910.32     979.12
   10       100      791.76     798.01     780.93     774.36    743.73     754.92
end data
name c1 'tafter'
name c2 'period'
name c3 'Ammodel'
name c4 'Bmmodel'
name c5 'Cmmodel'
name c6 'Dmmodel'
name c7 'Emmodel'
name c8 'Fmmodel'
stack c3 c4 c5 c6 c7 c8 c11;
subs c12.
stack c1 c1 c1 c1 c1 c1 c13
name c11 'forecasts'
name c13 'tperaftr'
code (1) "A" (2) "B" (3) "C" (4) "D" (5) "E" (6) "F" c12 c14
GPro.
Plot c11*c13;
Connect   C14;
Title "Forecasts by ad hoc Seasonal Methods";
ScFrame;
ScAnnotation;
Axis 1;
Label "period after 1997-2";
Axis 2;
Label "Fatalities".
GStd.
```

Multiple plots are easy to prepare with spreadsheet software, but fine control of the scale, axis origins, plotting symbols and line types can be awkward, with inconvenient software defaults.

10. Level adjustment for sudden change   in Chapter 9

Sometimes data take a sudden jump, for example, when data recording mechanisms are altered or jurisdictional boundaries change. The Fatalities data has such a drop. When we are interested in the seasonal pattern of the fatalities, it may be useful to make an adjustment for the change in level. One possibility of several choices we could use is to average observations 1 to 28 and average observations 29 to 56. We label these avg(1:28) and avg(29:56). Then subtract [avg(1:28) - avg(29:56)] from the first 28 observations. We show how to do this in script qftladj.mtb below. We omit the data entry part of the script.

```
Note qftladj.mtb
note   Adjust level for first 7 years of Quarterly Traffic Fatalities in Canada
note c1 is original data
note c2 is ex-post estimation data
note c3 is the ex-post validation data
note c4 is the period number for the data
note c5 is the period number for the ex-post estimation data
note c6 is the period number for the ex-post actual data
note ----------- end of setup. We will use this part many times ------
note
note Now we adjust the first 7 years of data.
copy c1 c11
copy c1 c12
delete rows 29:90 c11
delete rows 57:90 c12
delete rows 1:28 c12
let k11=mean(C11)
let k12=mean(c12)
print k11 k12
let k9=k11-k12
set c13
28(k9) 62(0)
end
let c9=c1-c13
name c9 'adjqftl'
write 'adjqftl.txt' c4 c9
note We really need to adjust the data in column c2 as well, since it
note will be used for ex-post modelling.
```

11. Compressed scale plots   for Chapter 9

We have been concerned above that the time scale of graphs may be too coarse, but sometimes it may be too expanded to see long-term trends in data. Then it is useful to use tools that prepare scatterplots rather than time plots, so we can then control the axis scaling to provide the compression or extension desired. We strongly recommend viewing time series data at several scales while exploring it. In Minitab as well as Microsoft Excel and Corel Quattro, scale is provided by specifying the minimum and maximum axis values for each of the x and y axes.

12. Trend line calculations for seasonal data   for Chapters 9, 14

It is temptingly easy to compute trend lines in statistical packages such as Minitab. We simply run a *regression* of the data against time. The graph in Figure 10.2 showed why this may be dangerous with seasonal data. This is especially so if we are planning to use ratios of data to the corresponding trend line value as raw seasonal factors in forecasting. Below, we present the Minitab output that was used to create the example in Figure 10.2, which has an increasing seasonal pattern. Other patterns that have one 'end' of the pattern predominately higher or lower than the other will give similar difficulties. After the example, we show a better way to proceed.

```
MTB > set c1
DATA> 1:12
DATA> end
MTB > name c1 'time'
MTB > set c2
DATA> 3 (1 2 3 4)
DATA> end
MTB > name c2 'spattern'
MTB > let c3=3+0.1*c1+c2
MTB > name c3 'seasdata'
```

Here we should perform a time plot of the data (omitted). The regression for the trend line is

```
MTB > regr c3 1 c1;
SUBC> coeff c4.
The regression equation is
seasdata = 4.82 + 0.205 time
```

Predictor	Coef	Stdev	t-ratio	p
Constant	4.8182	0.7131	6.76	0.000
time	0.20490	0.09690	2.11	0.061
s = 1.159	R-sq = 30.9%		R-sq(adj) = 24.0%	

The coefficients of the trend line have been saved in column C4 so we can compute the model into column C5. (We could also do this in other ways in Minitab, but prefer to show how it can be done as an explicit calculation.)

```
MTB > print c4
C4      4.81818   0.20490
MTB > let c5=c4(1)+c1*c4(2)
MTB > name c5 'tline1'
```

Now let us try to avoid the trend 'inflation'. Note that the data we created above in column C2 that counts 1, 2, 3, 4, 1 ,2, 3, 4, etc. is both a seasonal pattern AND an index for the four seasons we have here. We will now use this to separate our data into separate seasons.

```
MTB > unstack c3 c11-c14;
SUBC> subs c2.
```

We also compute the average data for each 'year'.

```
MTB > rmean c11-c14 c15
MTB > name c15 'yravg'
```

To correspond to the average yearly data, we need an average time index for each year, of which we have just three here. Then we compute the 'mid-year point' in time (expressed in terms of the period index). In each year, the average quarterly index is $2.5 = 0.25*(1+2+3+4)$, which we then offset from the start of each year to get the variable `midyrpt`.

```
MTB > set c6
DATA> 1:3
DATA> end
MTB > name c6 'yrindex'
MTB > let c16=2.5+(c6-1)*4
MTB > name c16 'midyrpt'
MTB > print c15 c16
  ROW   yravg   midyrpt
    1    5.75      2.5
    2    6.15      6.5
    3    6.55     10.5
```

Note that the middle of year 1 for quarterly data has time point 2.5. For monthly data, the middle would be at period 6.5. For daily data over a 7-day cycle, the middle is day 4. Thus, for odd-length cycles, we can get mid-cycle points that correspond to actual periods, but for even-length we will get artificial points.

Now let us compute the trend line over this *annualized data*.

```
MTB > regr c15 on 1 c16;
SUBC> coeffs c17.
The regression equation is
yravg = 5.50 + 0.100 midyrpt
Predictor          Coef        Stdev       t-ratio          p
Constant        5.50000      0.00000          *            *
midyrpt        0.100000     0.000000          *            *
s = 0                    R-sq = 100.0%    R-sq(adj) = 100.0%
```

We can use this new model to compute predicted values for the trend at the original time points.

```
MTB > let c7=c17(1)+c1*c17(2)
MTB > name c7 'tline2'
```

We can compare the two different trend lines by their coefficients or by the data in variables tline1 and tline2, which have been graphed in Figure 10.2.

```
MTB > name c4 'coeff1'
MTB > name c17 'coeff2'
MTB > print c4 c17


  ROW     coeff1   coeff2
    1    4.81818      5.5
    2    0.20490      0.1
```

Here are the Minitab commands to prepare the graph in Figure 10.2..

```
MTB > Plot 'seasdata'*'time' 'tline1'*'time' 'tline2'*'time';
SUBC>    Connect;
SUBC>    Overlay.
MTB > Plot 'seasdata'*'time' 'tline1'*'time' 'tline2'*'time';
SUBC>    Connect;
SUBC>    Footnote "Solid line is data";
SUBC>    Footnote "Dashed line is simple trend";
SUBC>    Footnote "Dotted line is trend of yearly average vs. centered time";
SUBC>    Overlay.
```

13. Equivalence of additive seasonal models using different base seasons   in Chapter 12
Let us now show the equivalence of additive models using different base seasons. To emphasize the issue of placement of the seasons, our quarterly data in this case starts with Winter. We have edited the output for brevity. For information, here is the data series: qdata.dat

```
96.342 112.485 129.082 108.731   96.278 118.736 141.813 110.955 103.336 124.943
147.537 118.398 109.022 127.916 142.556 122.236 113.820 134.705 153.062 132.728
116.211 142.948 159.041 139.318 120.616 146.533 165.546 138.027 123.561 150.006
166.425 140.196 129.124 148.475 175.645 151.503 133.191 158.188 176.052 149.603
140.327 166.361 186.584 157.713 147.411 174.157 191.145 166.465 156.726 171.253
192.393 169.156 155.808 177.873 197.581 178.429 166.480 183.120 205.288 172.112
169.298 189.006 207.436 180.843 170.465 190.447 214.142 189.714 175.487 196.625
213.825 191.573
```

```
MTB > name c1 'qdata'
MTB > set c2
DATA> 1:72
DATA> end
MTB > name c2 't'
MTB > ow 100              (This allows us to display the entire time plot in one panel.)
MTB > gstd                (This command starts character graphics.)
MTB > tsplot c1           (We use a small font to get the whole diagram on the page.)
```

```
qdata   -
        -                                                                 7   1
        -                                                       9   3
 200.0+                                                   5                   0
        -                                            7   1        2   6 8   2
        -                                 3                   8       4
        -                        5   9        6          4 6   0           9
        -                7   1             2       8 0 2     7   1     5
 160.0+              3                 8       4       9
        -             9            0   4 6   0                  3
        -   7   1   5       2   6       2         1   5
        -             8 0     4     8         7
        -   3       0   4             3
 120.0+       6       2   6     1   5   9
        -  2 4     8     3   7
        -           9
        -  1     5
          +---------+---------+---------+---------+---------+---------+---------+
          0        10        20        30        40        50        60        70        80
```

```
MTB > note   Seasonality of 4          MTB > name c5 'fall'
MTB > note Use season 1 as base        MTB > note season 1 is Winter
MTB > set c3                           MTB > regr c1 4 c2-c5;
DATA> 18(0 1 0 0)                      SUBC> residuals c11.
DATA> end                              MTB > set c6
MTB > name c3 'spring'                 DATA> 18(1 0 0 0)
MTB > set c4                           DATA> end
DATA> 18(0 0 1 0)                      MTB > note try new model
DATA> end                              MTB > name c6 'winter'
MTB > name c4 'summer'                 MTB > regress c1 4 c2 c3 c5 c6;
MTB > set c5                           SUBC> resids c12.
DATA> 18(0 0 0 1)
DATA> end
```

Let us display the two models one above the other to show their similarities:

```
qdata = 91.5 + 1.23 t + 20.4 spring + 38.7 summer + 12.6 fall    <-- based on Winter
qdata = 130  + 1.23 t - 18.3 spring - 26.1 fall    - 38.7 winter <--   based on Summer
```

We have not printed out the 'fit' information. For both models we find

$$s = 2.891 \qquad R\text{-sq} = 99.1\% \qquad R\text{-sq(adj)} = 99.1\%$$

We can write down the coefficients for the model based on Summer and show how they relate to those based on Winter :

	Winter base	Summer base
$b0$: Constant	91.5036	130.242 = 91.5036 + 38.7379
		= $b0$(Winter) + $b3$(Winter, Summer)
$b1$: t (slope)	1.23244	1.23244 (same)
winter	(base)	-38.7379 = -$b3$(Winter, Summer)
spring	20.4495	-18.2884 = 20.4495 - 38.7379
		= $b2$(Winter, Spring) - $b3$(Winter, Summer)
summer	38.7379	(base)
fall	12.647	-26.0909 = 12.6470 - 38.7379
		= $b2$(Winter, Fall) - $b3$(Winter, Summer)

14. Drawing spread-level graphs   in Chapter 9

The spread-level graph is intended to compare the dispersion in different groups (e.g. seasons) of data. We follow the discussion in Chambers *et al.* (1983). Often we first take the logarithm of data. Then we form the square root of the absolute deviation of each observation from its relevant group median, that is, if $X(i,j)$ is the ith observation in the jth group, and $M(j)$ is the median of this group, we form $D(i,j) = \mathrm{sqrt}(\mathrm{abs}(X(i,j) - M(j)))$. We then graph $D(i,j)$ versus the $M(j)$ and draw a regression line of this data over the points (e.g. a fitted line plot). The macro sprlevgr.mac for Minitab is on the PFM Web site.

15. Computing measures of fit   in Chapters 3, 8, 16

Different forecasting methods have developed difference conventions for reporting how well their models fit the data. Suppose y_t for $t = 1, 2, ..., n$ is our data, for which $\mathrm{model}(t)$ is the corresponding model. The model is assumed to have k parameters. The residuals are given by

$$e_t = y_t - \mathrm{model}(t)$$

The *Total Sum of Squares* or TSS $= \displaystyle\sum_{t=1}^{n} (y_t - \bar{y})^2 = (n - 1) * \mathrm{SD}^2$

where SD is the sample standard deviation of the data series.

The *Residual Sum of Squares* or RSS $= \displaystyle\sum_{t=1}^{n} e_t^2$

The *Coefficient of Determination* is R_squared $= (1 - \mathrm{RSS} / \mathrm{TSS})$

The *Mean Squared Deviation* is MSD $= \mathrm{RSS} / n$

The *Mean Squared Error* is MSE $= \mathrm{RSS} / (n - k)$

Note that the MSE and MSD definitions given are the ones we use, but we suspect that other workers may not distinguish them precisely. Sometimes RSS is called SSE for *Sum of Squared Errors*. In the case of ARIMA modelling and some other forecasting methods, we have only n' observations because lags, moving averages or differencing reduce the available model points.

The RSS will be computed using only n' points, and MSD and MSE will be computed using n'. We can therefore compute the R_squared in various ways, of which some are

$$\text{R\_squared} = (1 - (\text{MSD} * n')/(n * \text{SD}^2)) = (1 - (\text{SSE} * (n / n') / \text{TSS})$$

Another, not quite equivalent approach is to use

$$\text{R\_squared\_alternate} = (1 - \text{MSE/SD}^2)$$

The *Mean Absolute Deviation* or $\text{MAD} = \dfrac{1}{n} \sum_{t=1}^{n} |e_t|$

The *Mean Absolute Percentage Error* or $\text{MAPE} = \dfrac{1}{n} \sum_{t=1}^{n} 100 * |e_t / y_t|$

In both the last cases, the sum is performed over a reduced set of residuals (i.e., n' observations) if necessary. We can also find the *worst case* situations, but need to note whether we are using the magnitude of the residuals or not. That is, we could note

$$\text{Maximum residual} = \max_{t=1...n} (e_t)$$

$$\text{Minimum residual} = \min_{t=1...n} (e_t)$$

$$\text{Maximum absolute residual} = \max_{t=1...n} (|e_t|)$$

The *best case* is the *Minimum absolute residual* $= \min_{t=1...n} (|e_t|)$

We can also report maximum and minimum percentage or absolute percentage errors, simply by dividing by the appropriate data values y_t.

Bibliography

ABRAHAM, B. AND LEDOLTER, J. (1983) *Statistical Methods for Forecasting*. New York: Wiley.

ACZEL, A. D. (1996) *Complete Business Statistics*. 3rd ed. Chicago: Irwin.

ADYA, M. AND COLLOPY, F. ED (1998) How effective are neural networks at forecasting and prediction? A review and evaluation, *Journal of Forecasting* vol.17, 481–495

ARMOLAVICIUS, R. J. *et al.* (1988) Technological forecasting for the telecommunications industry. Paper submitted to ITC-12, Torino, Italy, June, 1988.

ATKINSON, A. C. (1985) Plots, *Transformations and Regression: an Introduction to Graphical Methods of Diagnostic Regression Analysis*, Oxford: Oxford Science Publications.

BAILAR, B. A. AND LANPHIER, M. C. (1978) *Development of Survey Methods to Assess Survey Practices*. Washington: American Statistical Association.

BARNETT, V. (1974) *Elements of Sampling Theory*. London: The English Universities Press.

BASS, F. M. (1980) The relationship between diffusion rates, experience curves, and demand elasticities for consumer durable technological innovations. *Journal of Business*, vol. 53, no. 3, pt. 2, S51–S75.

BELL, W. R. AND HILLMER, S. C. (1983) Modeling time series with calendar variation. *Journal of the American Statistical Association*, vol. 78, no. 383, pp. 526–534.

BLACKMAN, A. W. JR. (1971) The rate of innovation in the commercial aircraft jet engine market. *Technological Forecasting and Social Change*, vol. 2, 269–276.

BLOOM, M. F. (1995) The next generation: a world forecast for the year 2020. *The Journal of Business Forecasting*, Summer, pp. 10–14.

BOWERMAN, B. L. AND O'CONNELL, R. T. (1979) *Time Series and Forecasting*. North Scituate, Mass: Duxbury Press.

BOX, G. E. P. AND PIERCE, D. A. (1970) Distribution of the residual autocorrelations in autoregressive-integrated moving-average time series models. *Journal of the American Statistical Association*, vol. 65, pp. 1509–1526.

BRANCH, M. C. (1998) Future view: why we simulate. *The Futurist*, vol. 32, no. 3, p. 52, April.

BROWN, L. (1998) Food scarcity: an environmental wake-up call. *The Futurist*, vol. 32, no. 1, pp. 34–38, Jan–Feb.

BUNN, D. W. (1988) Combining forecasts. *European Journal of Operational Research*, vol. 33, pp. 223–229.

BURR, S. A. (1975) On detecting a periodic event by means of periodic observations I. *Mathematics of Computation*, vol. 29, no. 129, pp. 57–65 January.

CASADAGLI, M. AND EUBANK, S. (1992) Nonlinear modeling and forecasting. *Proceedings Volume XII*, Santa Fe Institute Studies in the Sciences of Complexity, Redwood City CA: Addison-Wesley.

CHAMBERS, J. M., CLEVELAND, W. S., KLEINER, B., AND TUKEY, P. A. (1983) *Graphical Methods for Data Analysis*, Belmont, CA: Wadsworth.

CHATFIELD, C. (1995) Editorial: Positive or negative? *International Journal of Forecasting*, vol. 11, pp. 501-502.

CHATFIELD, C. (1996) *Analysis of Time Series: an Introduction*. 5th ed. London: Chapman & Hall.

CHATFIELD, C. (1997) Forecasting in the1990s. *The Statistician*, vol. 46, no. 4, pp. 461–473.

CHATFIELD, C., KOEHLER, A., ORD, K., AND SNYDER, R. (1999) A new look at exponential smoothing, Statistics Research Report 99.02, Bath, UK: University of Bath, Department of Mathematical Sciences .

CHICAGO MANUAL OF STYLE (1993) 14th ed. Chicago: University of Chicago Press.

CLARKE, G. M. AND COOKE, D. (1998) *A Basic Course in Statistics*, 4th ed. London: Arnold.

CLEVELAND W. S. (1985) *Elements of Graphing Data*. Monterey, CA: Wadsworth.

CLEVELAND, W. S. (1993) *Visualizing Data*. Murray Hill, NJ: AT&T Bell Laboratories.

CLEVELAND, W. S. AND DEVLIN, S. J. (1980) Calendar effects in monthly time series: detection by spectrum analysis and graphical methods. *Journal of the American Statistical Association*, vol. 75, pp. 487–496.

CLEVELAND, W. S. AND DEVLIN, S. J. (1982) Calendar effects in monthly time series: modeling and adjustment. *Journal of the American Statistical Association*, vol. 77, pp. 520–528.

CLEVELAND, W. S., DEVLIN, S. J., AND TERPENNING, I. J. (1982) The SABL seasonal and calendar adjustment procedures. In O. D. Anderson (ed.) *Time Series Analysis: Theory and Practice I*. Amsterdam: North Holland.

COMMUNICATIONS OF THE ASSOCIATION FOR COMPUTING MACHINERY, vol. 39, no. 11, November 1996 issue, devoted to Data Mining and Knowledge Discovery in Databases.

DAGUM, E-B. (1983) The X-11-ARIMA seasonal adjustment method. Report 3-0002-511, Ottawa: Statistics Canada.

DAGUM, E-B., HUOT, G. AND MORRY, M. (1988) Seasonal adjustment in the eighties: some problems and solutions. *Canadian Journal of Statistics*, vol. 16 Supplement, pp. 109–126.

DAVIDSON, J. D. AND RESS-MOGG, LORD WILLIAM (1991) *The Great Reckoning: How the World Will Change in the Depression of the 1990s*, New York: Summit Books.

DELICADO, P. AND JUSTEL, A. (1999) Forecasting with missing data: application to coastal wave heights, *Journal of Forecasting*, vol. 18, pp. 285–298.

DE VEAUX, R. D. *et al.* (1998) Prediction intervals for neural networks via nonlinear regression. *Technometrics*, vol. 40, no. 4, pp. 273–282.

DELURGIO, S. A. (1998) *Forecasting Principles and Applications*. Boston: Irwin/McGraw-Hill.

DIEBOLD, F. X. AND LOPEZ, J. (1996) Forecast evaluation and combination. In G. S. Maddala and C. R. Rao (eds) *Statistical Methods in Finance*, vol. 14 of *Handbook of Statistics*, pp. 241–268, Amsterdam: North Holland.

DIEBOLD, F. X. (1997) *Elements of Forecasting in Business, Government and Finance*. Cincinatti, OH: South-Western College Publishing.

DONALDSON, R. G. AND KAMSTRA, M. (1996) Forecast combining with neural networks. *Journal of Forecasting*, vol. 15, 49–61.

DOORNIK, J.A. (1996) *Object-Oriented Matrix Programming using Ox*. London: International Thomson Business Press. See also http://www.nuff.ox.ac.uk/Users/Doornik/.

DOORNIK, J.A. AND OOMS, M. (1996) A package for estimating, forecasting and simulating ARFIMA models. http://www.nuff.ox.ac.uk/Users/Doornik/.

DRAPER, N. R. AND SMITH, H. (1981) *Applied Regression Analysis*, 2nd ed., New York: Wiley.

DRUCKER, P. (1998) The future that has already happened. *The Futurist*, vol. 32, no. 8, pp. 16–18, November (excerpt from Stone, N., 1998).

DUONG, Q.P. (1989) The combination of forecasts: a ranking and subset selection approach. *Mathematical Computing and Modelling*, vol. 12, no. 9, pp. 1131–1143.

DUVIVIER, J. F. (1971) Technological advances and program risks. *Technological Forecasting and Social Change*, vol. 2, pp. 277–287.

EASINGWOOD, C., MAHAJAN, V. AND MULLER, E. (1981) A nonsymmetic responding logistic model for forecasting technological substitution. *Technological Forecasting and Social Change*, vol. 20, pp. 199–213.

EUROSTAT (2000) Seasonal Adjustment Interface DEMETRA for Tramo/Seats and X-12-ARIMA: *User Manual Version 1.4* (Release Version: February 2000), Luxembourg: Eurostat.

FARAWAY, J. AND CHATFIELD, C. (1998) Time series forecasting with neural networks: a comparative study using the airline data. *Applied Statistics*, vol. 47, Part 2, pp. 231–250.

FAIRGRIEVE, J. AND BRANNEN, K. (1998) Idaresa - a tool for construction, description and use of harmonised datasets from national surveys. *COMPSTAT'98*, R. Payne and P. Green (eds.) Heidelberg: Physica-Verlag, pp. 299–304.

FISCHER, B. (1995) Decomposition of time series: comparing different methods in theory and practice, Version 2.1, Eurostat Report, March/April 1995.

FOOT, D. K. WITH STOFFMAN, D. (1998) *Boom, Bust, and Echo 2000: Profiting from the Demographic Shift in the New Millennium*. Toronto: McFarlane, Walter & Ross.

FRÖBERG, C-E. (1965) *Introduction to Numerical Analysis*. Reading MA: Addison-Wesley.

GARDNER, E. S. (1985) Exponential smoothing: state of the art. *Journal of Forecasting*, vol. 4, pp. 1–28.

GENTLEMAN, J. F. AND WHITMORE, G. F. (Section eds) (1985) Case study in data analysis no. 4: Temporal patterns in twenty years of Canadian homicides. *Canadian Journal of Statistics*, vol. 13, no. 4, pp. 261–291.

GIBALDI, J. (1995) *Handbook for Writers of Research Papers*. Modern Language Association (MLA) of America.

GOLDSTEIN, R. (1993) Editor's notes. *The American Statistician*, vol. 47, no. 1, pp. 46–47. (The majority of these particular notes were provided by Paul Velleman.)

GOMEZ, V. AND MARAVALL, A. (1997) Programs TRAMO (Time Series Regression with ARIMA Noise, Missing Observations, and Outliers) and SEATS (Signal Extraction in ARIMA Time Series) Instructions for the User (Beta Version: November 1997) Madrid: Ministerio de Economía y Hacienda & Banco de España.

GUPTA, A. AND LAM, M. S. (1996) Estimating missing values using neural networks. *Journal of the Operational Research Society*, vol. 47, pp. 229–238.

HANKE, J. E. AND REITSCH, A. G. (1992) *Business Forecasting*, 4th ed., Needham Heights, MA:Allyn & Bacon.

HARRISON, P. J. (1965) Short-term sales forecasting. *Applied Statistics*, vol. 14, pp. 102–139.

HARRISON, P. J. (1967) Exponential smoothing and short-term sales forecasting. *Management Science*, vol. 13, no. 2, pp. 821–842.

HARRISON, P. J. AND STEVENS, C. F. (1971) A Bayesian approach to short-term forecasting. *Operational Research Quarterly*, vol. 22, no. 4, pp. 341–362.

HARVEY, A.C. (1989) *Forecasting, Structural Time Series Models and the Kalman Filter*, Cambridge: Cambridge University Press.

HENDRY, D. (1993) *Econometrics: Alchemy or Science?* Oxford: Blackwell.

HOAGLIN, D. C., MOSTELLER, F. AND TUKEY, J. W. (eds) (1983) *Understanding Robust and Exploratory Data Analysis*. New York: Wiley.

HOFF, J. C. (1983) *A Practical Guide to Box–Jenkins Forecasting*. Belmont, CA: Lifetime Learning Publications (Wadsworth).

HOGARTH, R. M. AND MAKRIDAKIS, S. (1981) Forecasting and planning: an evaluation. *Management Science*, vol. 27, no. 2, February, pp. 115–138.

HUGHES, B. B. (1985) *World Futures: a Critical Analysis of Alternatives*. Baltimore: The Johns Hopkins University Press.

INTRILLIGATOR, M. D. (1978) *Econometric Models, Techniques and Applications*. Englewood Cliffs, NJ: Prentice-Hall.

JOHNSTON, J. (1972) *Econometric Methods*, 2nd ed. New York: McGraw-Hill.

KAASTRA, I., AND BOYD, M. S. (1995) Forecasting futures trading volume using neural networks. *Journal of Future Markets*, vol. 15, no. 8, pp. 953–970.

KASTNER, J. (Director) (1998) *Ask a Silly Question*. Toronto: Canadian Broadcasting Corporation (video recording of one episode of the Witness documentary program).

KENDALL, M. G. (1973) *Time-Series*. London: Griffin.

KENDALL, M. G. AND STUART, A. (1968) *The Advanced Theory of Statistics, Volume 3: Design and Analysis, and Time-Series*. 2nd ed. London: Griffin.

KMENTA, J. (1986) *Elements of Econometrics*, 2nd ed. New York: Macmillan.

KUAN, C-M., AND LIU, T. (1995) Forecasting exchange rates using feedforward and recurring neural networks. *Journal of Applied Economics*, vol. 10, pp. 347–364.

KUO, C. AND REITSCH, A. (1995) Neural networks vs. conventional methods of forecasting. *Journal of Business Forecasting*, Winter, pp. 17–22.

LACHER, R.C. *et al.* (1995) A neural network for classifying the financial health of a firm. *European Journal of Operational Research*, vol. 85, pp. 53–65.

Lee, C-W. J. and Chen, C. (1990) Structural changes and the forecasting of quarterly accounting earnings in the utility industry. *Journal of Accounting and Economics*, vol. 13, pp. 93–122.

LENARD, M. J. *et al.* (1995) The application of neural networks and a qualitative response model to the auditor's going concern uncertainty decision. *Decision Sciences*, vol. 26, no. 2, Spring, pp. 209–227.

LEVENBACH, H. AND CLEARY, J. P. (1981) *The Beginning Forecaster: the Forecasting Process through Data Analysis.* Belmont, CA: Lifetime Learning Publications (Wadsworth).

LEONARD, D. (2000) One CEO's Nightmare: Bhopal ghosts (still) haunt Union Carbide. *Fortune*, vol. 141, no. 7, April 3, pp. 44, 46.

LIPPMANN, R. P. (1987) An introduction to computing with neural nets. *IEEE ASSP Magazine*, April, pp. 4–22.

LJUNG, G. M. AND BOX, G. E. P. (1978) On a measure of lack of fit in time series models. *Biometrika*, vol. 65, pp. 297–303.

MAHAJAN, V. AND PETERSON, R. A. (1978) Innovation diffusion in a dynamic potential adopter population. *Management Science*, vol. 24, no. 15, November, pp. 1589–1597.

MAKRIDAKIS, S., WHEELWRIGHT, S., AND MCGEE, V. (1983) *Forecasting: Methods and Applications.* New York: Wiley.

MAKRIDAKIS, S., WHEELWRIGHT, S., AND HYNDMAN, R. J. (1998) Forecasting: Methods and Applications, 3rd ed. New York: Wiley.

MARQUARDT, D. W. (1963) An algorithm for least-squares estimation of nonlinear parameters. *Journal of the Society for Industrial and Applied Mathematics*, vol. 11, pp. 431–441.

MCCULLOUGH, B. D. AND WILSON, B. (1999) On the accuracy of statistical procedures in Micrsoft Excel 97. *Computational Statistics and Data Analysis*, vol. 31, pp. 27–37.

MEADE, N. (1984) The use of growth curves in forecasting market development – a review and appraisal. *Journal of Forecasting*, vol. 3, pp. 429–451.

MEADE, N. (1985) Forecasting using growth curves - an adaptive approach. *Journal of the Operational Research Society*, vol. 36, no. 12, pp. 1103–1115.

MERINO, D. N. (1990) Development of a technological S-curve for tire cord textiles. *Technological Forecasting and Social Change*, vol. 37, pp. 275–291.

MINKIN, B. H. (1995) *Future in Sight: 100 of the Most Important Trends, Implications and Predictions for the New Millennium.* New York: MacMillan.

NASH, J. C. (1976a) *Some Methods for Solving Single Equation Linear Least Squares Regression Problems.* Ottawa: Agriculture Canada.

NASH, J. C. (1976b) *An Annotated Bibliography on Nonlinear Least Squares Computations with Test Problems*, Ottawa, ON: Nash Information Services Inc., May.

NASH, J. C. (1977a) Minimizing a nonlinear sum of squares function on a small computer. *Journal of the Institute of Mathematics and its Applications*, vol. 19, pp. 231–237.

NASH, J. C. (1977b) A discrete alternative to the logistic growth function, *Applied Statistics*, vol. 26, no. 1, pp. 9–14.

NASH J. C. (1979) *Compact Numerical Methods for Computers: Linear Algebra and Function Minimisation.* Bristol: Adam Hilger, 2nd ed. 1990. Japanese ed. Tokyo: Omsha, 1996.

NASH, J. C. (1989) A technological forecast of the role of personal computers in statistical survey methods. *Proceedings of Symposium 88: The Impact of High Technology on Survey Taking* (Ottawa, 24–25 Oct. 1988), J. Kovar & E. Doucet (eds), Ottawa: Statistics Canada, pp. 297–307.

NASH, J. C. (1994) Tools for including statistical graphics in application programs. *The American Statistician*, vol. 48, no. 1, February, pp. 52–57.

NASH, J. C. (1996a) Nonlinear estimation. Chapter 32 of the manual for *Data Desk 5*, Ithaca, NY: Data Description Inc, March, pp. 32/1–32/14.

NASH, J. C (1996b), Letter 'Iridium Project'. *Communications of the ACM*, vol. 39, no. 2, February, pp. 13–14.

NASH, J. C. AND TEETER, N. J. (1975) Building models: an example from the Canadian dairy industry. *Canadian Farm Economics*, vol. 10, no. 2, April, pp. 17–24.

NASH, J. C. AND NASH, M.M. (1986) Building a database of secular and religious holidays for world-wide use, *Canadian Journal of Information Science*, vol.11, no.3/4, pp.38–47.

NASH, J. C. AND WALKER-SMITH, M. (1986) Using compact and portable function minimization codes in forecasting applications, *INFOR*, vol. 24, no. 2, May, pp.158–168.

NASH J. C. AND WALKER-SMITH, M. (1987) *Nonlinear Parameter Estimation: an Integrated System in Basic*. New York: Marcel Dekker. This work is now combined with Nash and Walker-Smith (1989a) and re-published by Nash Information Services Inc. See http://www.nashinfo.com

NASH, J. C. AND WALKER-SMITH, M. (1989a) Nonlinear Parameter Estimation: Examples and Software Extensions, Ottawa, ON: Nash Information Services Inc.

NASH, J. C. AND WALKER-SMITH, M. (1989b) Forecasting (an introduction to a series of reviews of software for economic forecasting) and reviews of FORECAST PLUS and PRO*CAST. *PC Magazine*, vol. 8, no. 5, 14 March, various pages from 225–240.

NASH, J. C. AND QUON, T. K. (1996) Software for modelling kinetic phenomena. *The American Statistician*, vol. 50, no. 4, November, pp. 368–378.

NASH, J. C. AND NASH, M. M. (1995) Managing risks. *Chance*, vol. 8, no. 4, Fall, pp. 25–31.

NASH, J. C. AND NASH, M. M. (1996) Commercial Internet publishing – the practicalities. In *Electronic Publishing: Its Impact on Publishing, Education and Reading*, (C. T. Meadow, M. Weaver, F. Hébert, eds), Canadian Association for Information Science. *Proceedings of the 24th Annual Conference*, Faculty of Information Studies, University of Toronto, pp. 62–72.

NASH, J. C. AND NASH, M. M. (1997) *Managing Technological Risk*. Ottawa, ON: Nash Information Services Inc.

PEGELS, C. C. (1969) Exponential forecasting: some new variations. *Management Science*, vol. 12, no. 5, pp. 311–315.

PENNER, R. AND WATTS, D. G. (1991) Mining Information. *The American Statistician*, vol. 45, no. 1, pp. 4–9.

PIERCE, D. A. (1980) A survey of recent developments in seasonal adjustment. *The American Statistician*, vol. 34, no. 3, August, pp. 125–134.

PUBLIC WORKS AND GOVERNMENT SERVICES (1997) *Canadian Style*. Ottawa: Public Works and Government Services.

RAI, A. *et al.* (1998) How to anticipate the internet's global diffusion. *Communications of the ACM*, vol. 41, no. 10, October, pp. 97–106.

RIPLEY, B. (1993) Statistical aspects of neural networks. In *Chaos and Networks – Statistical and Probabilistic Aspects,* (O. Barndorff-Nielsen, J. Jensen and W. Kendall, eds). London: Chapman & Hall, pp. 40–123.

ROUSSEEUW, P. J., RUTS, I. AND TUKEY, J. W. (1999) The bagplot: a bivariate boxplot. *The American Statistician*, vol. 53, no. 4, December, pp. 382–387.

SANDE, I. G. (1982) Coping with reality. *The American Statistician*, vol. 36, no. 3, part 1, pp. 145–152.

SATIN, A. AND SHASTRY, W. (1983) Survey Sampling: a Non-mathematical Guide. Ottawa: Statistics Canada.

SELLERS, R. (1998) Nine global trends in religion. *The Futurist*, vol. 32, no. 1, Jan–Feb, pp. 20–25.

SOHL, J. E. AND VENKATACHALAM, A. R. (1995) A neural network approach to forecasting model selection. *Information & Management*, vol. 29, pp. 297–303.

STONE, N. (1998) *Peter Drucker on the Profession of Management*. Cambridge MA: Harvard Business School Press.

TAL, B. AND NAZARETH, L. (1995) Artificial intelligence and economic forecasting. *Canadian Business Economics*, Spring, pp. 69–74.

THOMAS, M. K. (1996) *Forecasts for Flying: Meteorology in Canada, 1918–1939*. Toronto: ECW Press.

TKACZ, G. (2000) Neural network forecasting of Canadian GDP growth, Technical report, Department of Monetary and Financial Analysis, Ottawa, ON: Bank of Canada, February.

TRYFOS, P. (1996) *Sampling Methods for Applied Research: Text and Cases*. New York: Wiley.

TRYFOS, P. (1998) *Methods for Business Analysis and Forecasting: Text and Cases*. New York: Wiley.

TRYFOS, P. AND BLACKMORE, R. (1985) Forecasting records. *Journal of the American Statistical Association*, vol. 80, no. 389, March, pp. 46–50.

TUFTE, E. R. (1983) *Visual Display of Quantitative Information*. Cheshire, CT: Graphics Press.

TUKEY, J. W. (1977) *Exploratory Data Analysis*. Reading, MA: Addison-Wesley.

UNWIN, A. R. (1998). Analysing real time series? *CTI Maths & Stats Newsletter*, vol. 9, no. 4, pp. 8–10.

UNWIN, A. R., AND WILLS, G. (1988). Eyeballing Time Series. In *Proceedings of the 1988 ASA Statistical Computing Section*. Alexandria, VA: American Statistical Association, pp. 263–268.

UNWIN, A. AND WILLS, G. (1999) Exploring Time Series Graphically. *Statistical Computing and Graphics Newsletter*, vol. 9, no. 2, pp. 13–15.

WANG, M. AND KETTINGER, W. J. (1995) Technical Opinion: Projecting the growth of cellular communications. *Communications of the ACM*, vol. 38, no. 10, October, pp. 119–122. See also Nash (1996b)

WEST, M. AND HARRISON, J. (1999) *Bayesian Forecasting and Dynamic Models*, 2nd ed. / 2nd Corrected Printing, New York: Springer.

WILKINSON, L. (1999) Dot plots. *The American Statistician*, vol. 53, no. 3, August, pp. 276–281.

WILSON, J. H. AND KEATING, B. (1998) *Business Forecasting*. Boston, MA: Irwin McGraw-Hill.

WILSON, R. L. (1995) Ranking college football teams: an neural network approach. *Interfaces*, Vol. 25, no. 4, July-August, pg. 44-59.

WINTERS, P. R. (1960) Forecasting sales by exponentially weighted moving averages. *Management Science*, vol. 6, pp. 324–342.

WRIGHT, D. J. (1985) Extensions of forecasting methods for irregularly spaced data. In *Time Series Analysis: Theory and Practice 7*, Anderson, O. D. (ed). Amsterdam: North-Holland/Elsevier, pp. 169–181.

WRIGHT, D. J. (1986) Forecasting data published at irregular time intervals using an extension of Holt's methods. *Management Science*, vol. 32, no. 4, April, pp. 499–510.

Index

ACF see Autocorrelation

ad hoc methods 111, 115, 120, 130, 148, 227, 231, 242, 279

Aggregation 9, 20, 46, 53, 68, 88, 235, 273

ARIMA 7, 18, 31, 45, 78ff, 133, 193ff, 208ff, 219ff, 246ff, 267

ASCII/ANSI 13, 21

Asymmetric loss function 257

Asymptote 25, 63, 254

Autocorrelation 6, 29ff, 76ff, 123, 128, 143, 202, 211ff, 227, 247

Average, medial 185

Bibliography 4, 35ff

Bitmap 138

Bounds 70, 113, 230, 254, 259, 277

Box-Pierce statistic 81ff, 128ff, 213

Boxplot 6, 18, 66, 92ff, 101, 126

Brown's 1 parameter double exponential smoothing method 166ff

Business cycle 25, 183

Character plot 113, 123, 278

Chi-squared test 3, 6, 80

Closed-form 257

Coefficient of variation 127

Collinearity 143, 154ff, 254

Command Syntax 133

Component, irregular 186

Confidence interval 3, 19, 31, 79, 93, 128, 141, 179

Constant term 76, 142, 196, 227, 254, 281

Correlation, Serial see Autocorrelation

Cross-impact analysis 53ff

Data examples, see PFM Web site

Data mining 13

Data subsetting 93, 104ff, 115, 129, 200, 213, 231

Decomposition, time series 7, 32, 45, 183ff, 210, 227, 229, 231, 233, 258, 271

Decomposition, singular value, 155

Delphi methods 52ff

Derivatives 26, 257

Descriptive statistics 3, 19, 104, 19, 148, 205, 223

Deseasonalize 117, 129, 180, 185, 216

Deterministic processes 5

Deviations 16, 18, 26, 29, 75, 117, 120, 152, 176, 188, 195, 212, 219, 227, 231, 256, 278, 284

Differences 64, 79, 99ff, 117, 119, 145, 154, 193ff, 208ff, 220, 240

Differential equations 5, 23, 253

Disaggregation see Aggregation

Dispersion 101ff, 191, 284

Distribution (probability distribution) 3, 31, 66, 75, 78ff, 91ff, 122, 143, 148, 201, 213, 227

Dotplot 92

Durbin–Watson Statistic 6, 84ff, 143, 148, 151

Error means 224

Error sum of squares 224

Errors 6, 16, 26, 78, 84, 88, 115, 120, 143, 153, 156, 176, 193, 219, 249, 265

Estimation 15, 61ff, 76, 118, 141, 152, 175, 198, 212, 235, 247, 256ff

Expert opinion 51, 70, 222

Exponential growth or decay 100, 211, 254

Exponential smoothing 7, 40, 162ff, 194, 216, 231, 271

Exponentiate 210

Extrapolation 5, 15, 23, 55, 62, 64, 90, 111, 124, 141, 178, 256, 277

First differences see Differences

Fit see Model fit

Freeware 134

Function, objective 254, 256

Futurists 49

Gaming 56

Gaussian or normal distribution 3, 18, 31, 75, 92ff, 122, 141, 148, 162, 201, 209, 213, 219, 227, 243, 272

Gompertz' function 62, 263

Groupware 52

Growth curve 62, 100, 251, 262

Heteroskedasticity 128, 143, 148, 209

Heuristic method 258

Histogram 6, 18, 31, 91ff, 128, 134, 148, 213, 219, 244

Holt's 2 parameter ES method 168, 231, 271

Homoskedasticity see heteroskedasticity

Horizon, forecasting 6, 10, 15, 53, 59, 61, 87, 104, 182, 218, 270

Hypothesis test 3, 6, 79ff, 140, 148, 219, 227

Imputation 7, 45ff, 99, 109, 237

Intercept 113ff, 120, 143, 161, 177, 187, 222, 229, 258

Internet chat 52

Interpolation 46, 98, 237, 272

Kalman filter 174, 252

Kernel density plot 92

Lag 31, 77ff, 194ff, 227, 245, 249

Leading indicators 61ff, 266

Level 10, 25, 61, 88, 101ff, 111, 144, 158, 162, 168, 188, 208

Level shift 158, 162, 175, 188, 230, 236

Linearity 143

Ljung-Box test, see (Modified) Box-Pierce test

Location 101, 104, 238

Logarithm 17, 24, 100, 153, 174, 208, 210, 284

Logistic growth curve 62, 251

MAD, or Mean Absolute Deviation 29, 174ff, 285 MAPE 174ff, 285

Marquardt's method 257

Maximum likelihood 153

Mean squared error, or MSE 17ff, 29, 122, 174ff, 203, 214

Mean or average 3, 24, 30, 32, 75, 95, 101, 104, 116, 124, 128, 185

Median 81, 93, 104, 115, 148, 176, 205, 223, 284

Metadata 47, 88, 136, 218, 234

Minimization, multivariate 141, 152, 176, 198, 249

Model 4, 15, 23, 56, 62, 75, 115, 127, 141, 193, 218, 233, 242, 245, 248, 250, 253

Model, additive 24, 33, 116, 144, 154, 156, 166, 174, 183, 209, 233, 282

Model, causal 5, 141, 156

Model, multiplicative 24, 33, 101, 107, 170, 183, 209, 233, 258

Model, nonlinear 26, 62, 76, 118, 123, 143, 153, 176, 198, 231, 246, 251, 253

Model, unstable 238

Modelling,, econometric 245

Modelling,, sequential 242

Moving average (simple, compound, weighted) 6, 158ff, 182, 183ff, 238

Moving median 237

MSD 176, 284

MSE, see Mean Squared Error

Multiplicative model see Model, multiplicative

Nonlinear least squares fitting see Model, nonlinear

Normal probability plot 31, 95, 122, 128, 209, 227

Objective function 256

Online bibliographic services 36

Optimization 32, 134, 153, 165, 176, 251, 258

Oscillations 200

Outliers 46, 88, 94, 113, 128, 175, 233, 238

PACF see Autocorrelation

Panels (of experts) see expert opinion and Delphi method

Pattern matching 5, 11

Pegels' classification 23, 90, 101, 107, 183, 234

Percentage error 18, 32, 120, 127, 176, 285

Periodicity 48, 127, 239, 274

Perturbation 27, 62

Pixels 138

Predictors 84, 142, 148, 155, 243, 281

Principal components 155

Principle of continuity 4, 15, 100, 119

Proxy 65, 244

Pseudo random numbers 33, 57, 81, 162, 182, 216

R_Squared 17, 75, 118, 129, 148, 154, 219, 234, 259, 284

Raster graphics 138

Reformat, file tools 137

Regressand 142

Regression, least squares 141, 24, 256

Regressors 142

Risk anticipation 690

RMSE 29

Ruler forecast 90, 111ff, 190, 261, 277

SABL 237

Sample instrument 60

Sample survey 53, 60, 66

Sampling 60, 78, 81

Sampling frame 56, 60

Scatterplot 93ff, 280

Seasonal adjustment 233

Seasonal factor 117ff, 124, 127, 170, 180, 185, 215, 229, 239, 258, 280

Seasonal shift 62, 114, 116ff, 127, 174, 180, 229, 243

Seasonality 24, 90, 101, 107, 149, 162, 170, 184, 194ff, 209, 234

Seed (random number) 162

Shareware 137

Sigmoid (function) 63, 112, 210, 254, 265

Simple Random Sampling 60

Simulation 23, 55ff, 64, 156, 246, 253

Singular Value Decomposition 155

Smoothing, exponential see Exponential smoothing

Standard Deviation (SD, STDEV) 3, 29, 77, 81, 94, 104, 120, 127, 215, 224, 229, 272, 284

Stationarity 7, 24, 79, 91, 158ff, 193ff, 208ff

Strategic management 68

Subsetting 93, 101, 104ff, 115, 129, 200, 213, 231

Summary statistics (numerical) 19, 70, 87, 106, 126, 151, 224

Systems analysis 55

Text editor 102, 136, 272

Time series, stationary see Stationarity??

Time series decomposition (TSD) see Decomposition, time series

Time point conversions 47, 274

Time (series) plot 90, 111, 123, 148, 172, 178, 209, 220, 270, 277ff, 280

Trend equations 113, 177

Trend line 19, 64, 111ff, 128, 144, 177, 186, 261, 277, 280

Trial modelling period 16

Validation period 18

Value, critical 31, 84, 148

Variability 10, 88, 101ff, 126, 141, 153, 179, 277

Variable, dummy 144ff, 183, 231

Variable, endogenous 143

Variable, exogenous 143

Variable, qualitative 49

Variance 3, 77, 104, 143, 153, 175, 209, 247

Vector 23, 29, 54, 138, 246, 248, 251

Weighted average 30, 98, 159, 222, 238

Winters' method 32, 170ff, 183, 219, 223, 227, 229, 231, 239, 271

Word processor 3, 34, 43, 103, 134, 136, 221, 259, 270, 272, 278